SPRINGER

LAB MANUAL

Springer-Verlag Berlin Heidelberg GmbH

Friedhelm Hildebrandt · Peter Igarashi (Eds.)

Techniques in Molecular Medicine

With 53 Figures, 3 in Color

Springer

DR. FRIEDHELM HILDEBRANDT
Universitäts-Kinderklinik
Mathildenstrasse 1
D-79106 Freiburg

DR. PETER IGARASHI, M.D
Section of Nephrology
Dep. of Internal Medicine
Yale University
2073 Laboratory of Medicine and Pediatrics
333 Cedar Street
New Haven, Connecticut 06620-8029
USA

ISBN 978-3-642-47808-6 ISBN 978-3-642-59811-1 (eBook)
DOI 10.1007/978-3-642-59811-1

Library of Congress Cataloging-in-Publication Data
Techniques in molecular medicine / Friedhelm Hildebrandt ; Peter Igarashi (eds.).
p. cm. - (Springer lab manual)
Includes bibliographical references and index.

1. Molecular biology Laboratory manuals.
2. Molecular diagnosis Laboratory manuals.
3. Medical genetics Laboratory manuals.
I. Hildebrandt, Friedhelm, 1957- . II. Igarashi, Peter. III. Series.
[DNLM: 1. Molecular Biology-methods. QH 506 T2548 1999]
QH506.T43 1999
572.8-dc21
DNLM/DLC
for Library of Congress

Cover design: design & production GmbH, D-69121 Heidelberg
Typesetting: Mitterweger Werksatz GmbH, D-68723 Plankstadt
SPIN 10124563 27/3136 5 4 3 2 1 0 - Printed on acid free paper

Preface

Standard molecular biology techniques are used widely in biomedical research and have had a major impact on the understanding of physiologic processes and disease mechanisms. They also play an increasing role in molecular genetic diagnosis and new therapeutic approaches.

This laboratory manual is designed for researchers who need reliable protocols for their laboratory work. It provides a step by step approach to standard methods of molecular biology. A wide range of topics is covered by protocols for immediate application in the laboratory. These include purification, isolation, labeling, hybridization, sequencing and enzymatic modification of DNA or RNA, polymerase chain reaction, cloning, libraries, and protein techniques.

The manual is based on experience with a practical course in molecular biology techniques, which was taught by the authors. Since most methods in molecular biology recapitulate in a test tube mechanisms that occur naturally in the cell, the first three chapters describe and illustrate basic mechanisms of molecular biology. Understanding more about these basic mechanisms will enable the researcher to apply techniques more specifically, based on the knowledge of their potential and limits. We hope that users find this manual a helpful guide on their road to exciting insights.

Freiburg,
Spring, 1999

Friedhelm Hildebrandt

Contents

Part I Molecular Biology Techniques

Chapter 1
Structure and Function of the Human Genome
FRIEDHELM HILDEBRANDT 3

Chapter 2
Enzymatic Modification of DNA and RNA
KAI-OLAF NETZER 24

Subprotocol 1: Restriction Endonucleases 24
Subprotocol 2: Exonucleases, Other Nucleases and Ligases 33
Subprotocol 3: Phosphatases – CIP Reaction 36
Subprotocol 4: Kinases 37
Subprotocol 5: DNA-Dependent DNA Polymerases –
E. Coli DNA Polymerase 1 and Klenow Enzyme 40
Subprotocol 6: T4 and T7 DNA Polymerase 42
Subprotocol 7: RNA-Dependent DNA Polymerases 44
Subprotocol 8: DNA – Dependent RNA Polymerases 47
Subprotocol 9: Terminal Deoxynucleotidyl Transferase (TdT) 49

Chapter 3
Purification and Nucleic Acids From Eukaryotic Cells
HANS GERD NOTHWANG AND FRIEDHELM HILDEBRANDT 50

Subprotocol 1: Extraction of Nucleic Acids 50
Subprotocol 2: Isolation of Genomic DNA From Mammalian Cells . . 56
Subprotocol 3: Isolation of Cytoplasmic RNA 60
Subprotocol 4: Rapid Preparation of Total RNA by Acid
Guanidinium Thiocyanate/Phenol/Chloroform/
Extraction 64
Subprotocol 5: Preparation of Poly(A)$^{+}$ RNA 67

Chapter 4
Electrophoresis of DNA and RNA Fragments
FENG QIAN AND GREGORY G. GERMINO 71

Subprotocol 1: The Recovery of DNA Fragments 75
Subprotocol 2: Electrophoresis of RNA 77

Chapter 5
Single-Strand Conformation Polymorphism (SSCP) Analysis
FENG QIAN AND GREGORY G. GERMINO 83

Chapter 6
Denaturing Gradient Gel Electrophoresis
KAI-OLAF NETZER 86

Chapter 7
Introduction of Pulsed-Field Gel Electrophoresis (PFGE)
FENG QIAN AND GREGORY G. GERMINO 96

Chapter 8
Pulsed-Field Gel Electrophoresis: Protocols
GUDRUN A. RAPPOLD, KARIN RIED, ALBRECHT KLINK,
ERCOLE RAO AND BIRGIT WEISS 103

Subprotocol 1: Preparation and Processing of High Molecular
DNA in Agarose Blocks 104
Subprotocol 2: Preparation of Size Markers 106
Subprotocol 3: Restriction Enzyme Digestion of DNA
in Agarose Blocks 108
Subprotocol 4: Gel Electrophoresis 110

Chapter 9
Radioactive Labeling of DNA and RNA Probes
PETER IGARASHI 114

Subprotocol 1: End-Labeling Using T4 Polynucleotide Kinase 114
Subprotocol 2: End-Labeling Using the Exchange Kinase Reaction . . 116
Subprotocol 3: Labeling 3' Ends with Terminal Transferase
(Forward Reaction) 117
Subprotocol 4: Random-Prime Labeling 118
Subprotocol 5: Riboprobes 121

Chapter 10
Hybridization Methods (Southern and Northern Blotting)
KAI-OLAF NETZER ... 126

Subprotocol 1: Preparation of Dot Blots for DNA ... 129
Subprotocol 2: Preparation of Southern Blots for Genomic DNA ... 131
Subprotocol 3: Hybridization of DNA Blots ... 135
Subprotocol 4: Preparation of RNA Dot Blots ... 139
Subprotocol 5: Preparation of Northern Blots ... 140
Subprotocol 6: Hybridization of RNA Blots ... 143
Subprotocol 7: Membrane Stripping ... 147

Chapter 11
Specimen Preparation Techniques for Cytogenetic Analysis by Fluorescence in situ Hybridization
STEFAN JOOS, MARTIN BENTZ, ANTON H. N. HOPMAN AND PETER LICHTER ... 148

Subprotocol 1: Cultivation of Blood Lymphocytes ... 149
Subprotocol 2: Preparation of Metaphase Chromosomes and Interphase Nuclei from Blood Lymphocytes ... 151
Subprotocol 3: Pepsin Digestion of Blood Lymphocytes ... 153
Subprotocol 4: Preparation of Fibroblasts for FISH (Adherently Growing Cells) ... 154
Subprotocol 5: Treatment of Blood Smears ... 155
Subprotocol 6: Bone Marrow Smears ... 156
Subprotocol 7: Clinical Material Obtained from Solid Tumors: Pretreatment of Slides with Poly-L-Lysine ... 157
Subprotocol 8: Preparation of Single Cell Suspensions from Solid Tumor Material ... 158
Subprotocol 9: Preparation of Frozen Sections ... 160
Subprotocol 10: Paraffin-Embedded Tissue Sections ... 161

Chapter 12
Chromosome Analysis by Fluorescence in situ Hybridization
MARTIN BENTZ, STEFAN JOOS AND PETER LICHTER ... 166

Subprotocol 1: Specimen Preparation ... 170
Subprotocol 2: Probe Labelling by Nick Translation ... 171
Subprotocol 3: Dot-Blot Assay to Test Labelling ... 174
Subprotocol 4: Probe Mixture and Denaturation for Fluorescence in situ Hybridization ... 175

Subprotocol 5: Denaturation of the DNA on the Slides 176
Subprotocol 6: Detection 178
Subprotocol 7: Chromosomal in situ Suppression (CISS) Hybridization 179
Subprotocol 8: Signal Amplification 181
Subprotocol 9: Multicolor Fluorescence in situ Hybridization 182

Chapter 13
Sequencing
KLAUS DEICHMANN 186

Subprotocol 1: Standard protocol for T7 DNA polymerase (Sequenase™) 194
Subprotocol 2: Standard Protocol for Sequencing with Taq Polymerase 197
Subprotocol 3: Denaturing Polyacrylamide Sequencing Gels 200

Chapter 14
Polymerase Chain Reaction
FRIEDHELM HILDEBRANDT AND IVA SINGH-SAWHNEY 207

Chapter 15
Cloning Vectors
ANNETTE DEICHMANN AND KLAUS DEICHMANN 226

Subprotocol 1: Transformation of Bacteria with Recombinant Plasmid Using the Ca^{++}-Precipitation Method (Mandel 1970) 234
Subprotocol 2: Transformation of Bacteria with Recombinant Plasmid by Electroporation (Dower 1988) 238
Subprotocol 3: Transformation of Bacteria with Recombinant Bacteriophage DNA 241
Subprotocol 4: "Mini-Prep" Method for Plasmid Isolation 245
Subprotocol 5: Chromatographic Method for Plasmid Preparation . . 247
Subprotocol 6: Large-Scale Preparation of Bacteriophage DNA (Replicative Form (RF)) 249

Part II Strategies for Gene Isolation and Characterization

Chapter 16
Cloning Strategies: An Overvierw
FRIEDHELM HILDEBRANDT 257

Chapter 17
Genomic Libraries
LUITZ F. ONUCHIC AND GREGORY G. GERMINO 261

Chapter 18
cDNA Libraries
LUITZ F. ONUCHIC AND GREGORY G. GERMINO 273

Chapter 19
cDNA Library Screening
JOHANN PETER HOSSLE 288

Subprotocol 1: Plating of a λgt11 cDNA Library 296
Subprotocol 2: Hybridization Screening - cDNA Sequences as Probes 298
Subprotocol 3: Immuno-Screening 300
Subprotocol 4: Purification of Phage Particles, DNA Isolation and Subcloning 306

Chapter 20
Expression Cloning with *Xenopus* Oocytes
DANIEL MARKOVICH, ANDREAS WERNER AND HEINI MURER 310

Chapter 21
Transgenic Animals
ECKART SCHOTT, MARTIN PAUL AND DETLEV GANTEN 319

Chapter 22
Expression in Eukaryotic Cells: Gene Transfer Using Retroviral Vectors
MARTIN MARX 337

Subprotocol 1: Synthesis of the Insert 339
Subprotocol 2: Purification 340
Subprotocol 3: Ligation of the Insert into the Vector 340
Subprotocol 4: Transfection of *E. coli* HB101 and Plasmid DNA Isolation 341
Subprotocol 5: Assay for the Presence of the DNA Insert 343

Subprotocol 6: Transfection of ψ2 Packaging Cells 344
Subprotocol 7: Infection of ψAM Packaging Cells 348
Subprotocol 8: Infection of Target Cells 348

Chapter 23
Positional Cloning and Linkage Analysis
FRIEDHELM HILDEBRANDT AND HEYMUT OMRAN 352

Part III Protein Methods

Chapter 24
Protein Techniques
ROBERT F. REILLY 367

Subprotocol 1: In Vitro Translation and Immunoprecipitation 367
Subprotocol 2: Western Blotting 375

Subject Index 379

Part I

Molecular Biology Techniques

Chapter 1

Structure and Function of the Human Genome

FRIEDHELM HILDEBRANDT

Structure of Nucleic Acids

The term "molecular biology" refers to the biology of **nucleic acids**. Methods of molecular biology make **in vitro** use for experimental purposes of naturally occurring mechanisms such as transcription or transformation. Most techniques of molecular biology are based on two basic principles:

- The use of nucleic acid-modifying enzymes
- Hybridization (base pairing) between two complementary strands of nucleic acids.

DNA-modifying enzymes are used to cut DNA at specific sites (restriction enzyme digestion) to recombine DNA fragments (ligation), or to amplify and sequence DNA fragments with the help of polymerases. **Hybridization methods** allow for specific selection of nucleic acid fragments of interest through base pairing with a defined probe. The probe is labeled radioactively or non-radioactively for detection of the cognate hybridization fragment.

The size of the haploid human genome (i. e. its complexity) is approximately 3×10^9 nucleotides. Figure 1 shows an electron micrograph of a metaphase chromosome, which has been treated with detergents leaving only a remnant of the protein scaffold with DNA spilling out from the chromosomal backbone. This structure gives an impression of the dimensions that we are dealing with when studying the sequences of individual nucleotide sequences with respect to the structure of a complete human chromosome, which measures approximately 10^8 nucleotides. In the following, the structure of the human genome is discussed, starting with its smallest struc-

Friedhelm Hildebrandt, Universitäts-Kinderklinik, Mathildenstrasse 1, Freiburg, 79106, Germany (*phone* +49-761-270-4301; *fax* +49-761-270-4533; *e-mail* hildebra@kkl200.-ukl.uni-freiburg.de)

tural elements, the nucleotides. This is followed in order of increasing complexity, by descriptions of the double stranded structure of DNA, nucleosomes, the structure of chromosomes and the full set of chromosomes. Finally, some functional aspects of the human genome are outlined.

Nucleotides

Nucleotides are composed of a nitrogenous base, a sugar, and a phosphate.

- Nitrogenous bases (Figure 2A): purines: Adenine (A), Guanine (G); pyrimidines: Cytosine (C), Thymine (T) (RNA: Uracil (U)); uncommon bases: dA, m^5C
- Sugar: 2'-deoxyribose (DNA), ribose (RNA)
- Mono-, di-, triphosphate residue

A schematic of the composition of nucleotides is given in Figure 2B: nitrogenous base + sugar = **nucleoside** (e.g. adenine +ribose = adenosine) nucleoside + phosphate = **nucleotide** (e.g. adenosine + mono-, di-, tri-phosphate = adenosine-mono-, di-, tri-phosphate (AMP, ADP, ATP)

Nucleotides are the primary building blocks of the nucleic acids DNA and RNA. For the nomenclature of nucleotides see Table 1.

Table 1. Nomenclature of nucleic acids

Base	Nucleoside	Nucleotide	Abbreviation	
			RNA	DNA
Adenine	adenosine	adenylic acid	AMP	dAMP
Guanine	guanosine	guanylic acid	GMP	dGMP
Cytosine	cytidine	cytidylic acid	CMP	dCMP
Thymine	thymidine	thymidylic acid		dTMP
Uracil	uridine	uridylic acid	UMP	

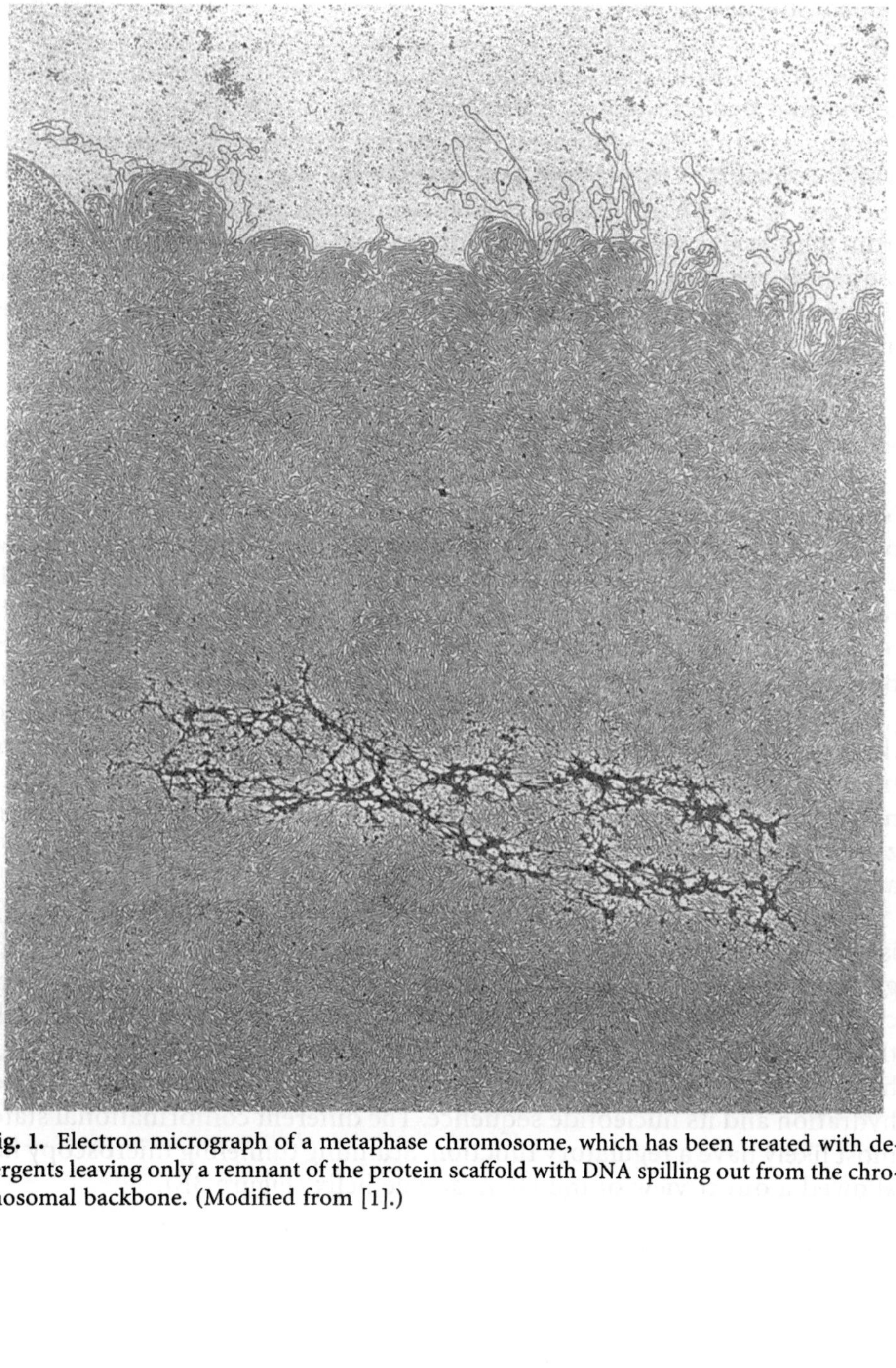

Fig. 1. Electron micrograph of a metaphase chromosome, which has been treated with detergents leaving only a remnant of the protein scaffold with DNA spilling out from the chromosomal backbone. (Modified from [1].)

Primary structure of DNA

Figure 2D: A triphosphate residue, bound to the 5'-OH of the nucleotide adenosine-5'-triphosphate, interacts with the 3'-OH of the nucleic acid and engages in a **phosphodiester bond** after cleavage of a pyrophosphate, which contributes the energy required for the reaction. The result of the reaction is shown in Figure 2C. Nucleic acids are synthesized starting from the 5' end proceeding towards their 3' end. As a convention, nucleotide sequences are written from their 5' end on the left to the 3' end to the right (Figure 2C). In double-stranded DNA this rule applies for the strand written in the upper row.

Secondary structure of DNA

Naturally occurring DNA is double stranded. J. Watson and F. Crick in 1953 provided the model for DNA of a **right handed double helix** containing two anti-parallel strands, meaning that the strands run 5' to 3' in opposite directions (Figure 3A). Sugars connected with phosphodiester bonds form the back bone for the structure, which resembles 2 intertwined spiral staircases. Base pairing occurs between complementary bases. C and G form 3 hydrogen bonds, whereas A and T only 2 hydrogen bonds (see Figure 3A). Since the glycosidic bonds between the sugars and the complementary bases are not exactly opposite each other, the double helix exhibits a **major and** a **minor groove** (Figure 3B). DNA binding proteins, such as restriction enzymes and transcription factors, recognize specific bases mainly via interaction with the major groove (Figure 3C).

The most common conformation of the double helix in aqueous solution is so called **B-DNA** (Figure 3D). There are 10 base pairs per turn. Since the glycosidic bonds of the bases can rotate freely, the double helix can undergo conformational changes. In its **A-Form** the major groove is shortened. In the rarely occurring **Z-Form** the double helix has switched to a left-handed helix (Figure 3D). The conformation of DNA is dependent on its state of hydration and its nucleotide sequence. The different conformational states most likely have a regulatory function. Scanning tunneling microscopy has allowed a direct view of the DNA double helix (Figure 3E).

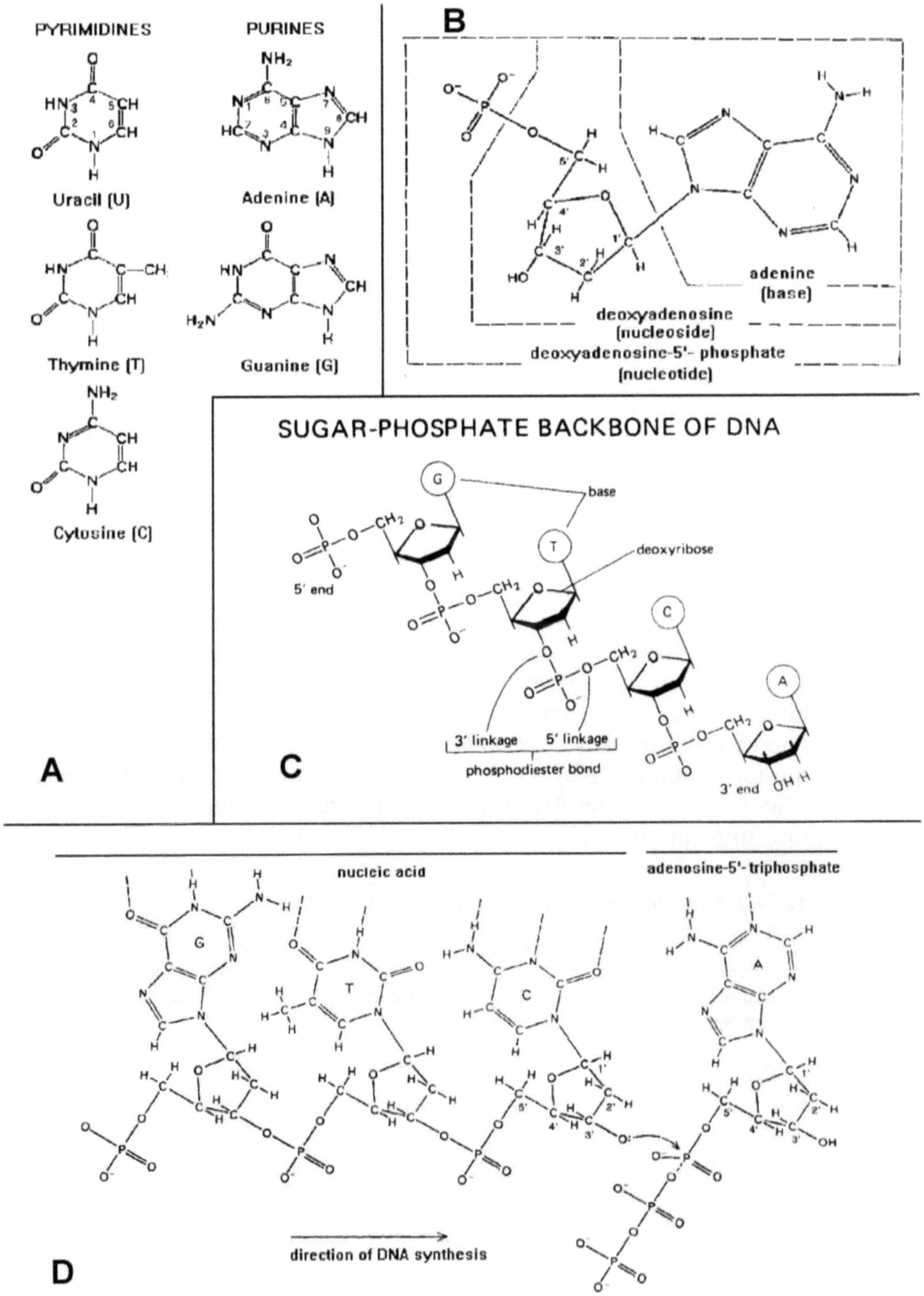

Fig. 2. Primary structure of DNA. For details see text. (Modified from [1,2 and 4].)

Tertiary structure of DNA

If a piece of double stranded DNA is unwound, the tertiary structure of a **superhelix** will result. (An overwound telephone cord may serve as an example of a superhelix from common experience.) In Figure 4A negative **supercoiling** is demonstrated in the example of a bacterial plasmid, a circular piece of double stranded DNA, which in its **relaxed form** forms a circle. If the plasmid is cut, the resulting linear piece of DNA is unwound and the two ends are then rejoined covalently, the energetically favoured conformation that results is a **right-handed superhelix** or a **left-handed spiral** will result. Most naturally occurring DNA adopts this negatively-supercoiled state. Now, if only one strand is cut, the negative supercoiling is dissipated and a relaxed circle of DNA re-forms. Electron micrographs of these structures are shown in Figure 4B.

RNA structure

The most important structural differences between DNA and RNA are emphasized in Table 2. RNA (and single stranded DNA) can spontaneously form an energetically more stable **secondary structure** via base paring within the same single stranded nucleic acid molecule. Figure 4C shows an example of "hairpin formation", which can occur in nucleotide sequences exhibiting dyad symmetry. Dyad symmetry of a sequence is present if a sequence is followed by its reverse complement (e. g.: AACGG-CCGTT). Hairpin formation can play a role as a transcription terminator in prokaryotes (Figure 4C). **Transfer RNA** undergoes extensive secondary structure formation to form an L-shaped molecule (Figure 4D).

Table 2. Some features distinguishing DNA from RNA

	DNA	RNA
Most abundant natural occurrence	double-stranded	single-stranded
Sugar	2'-deoxyribose	ribose
Nucleotides	A, C, G, and T	A, C, G, and U (also uncommon bases)
Chemical stability	relatively stable at room temperature	relatively unstable at room temperature (hydrolysed by RNases, alkali)

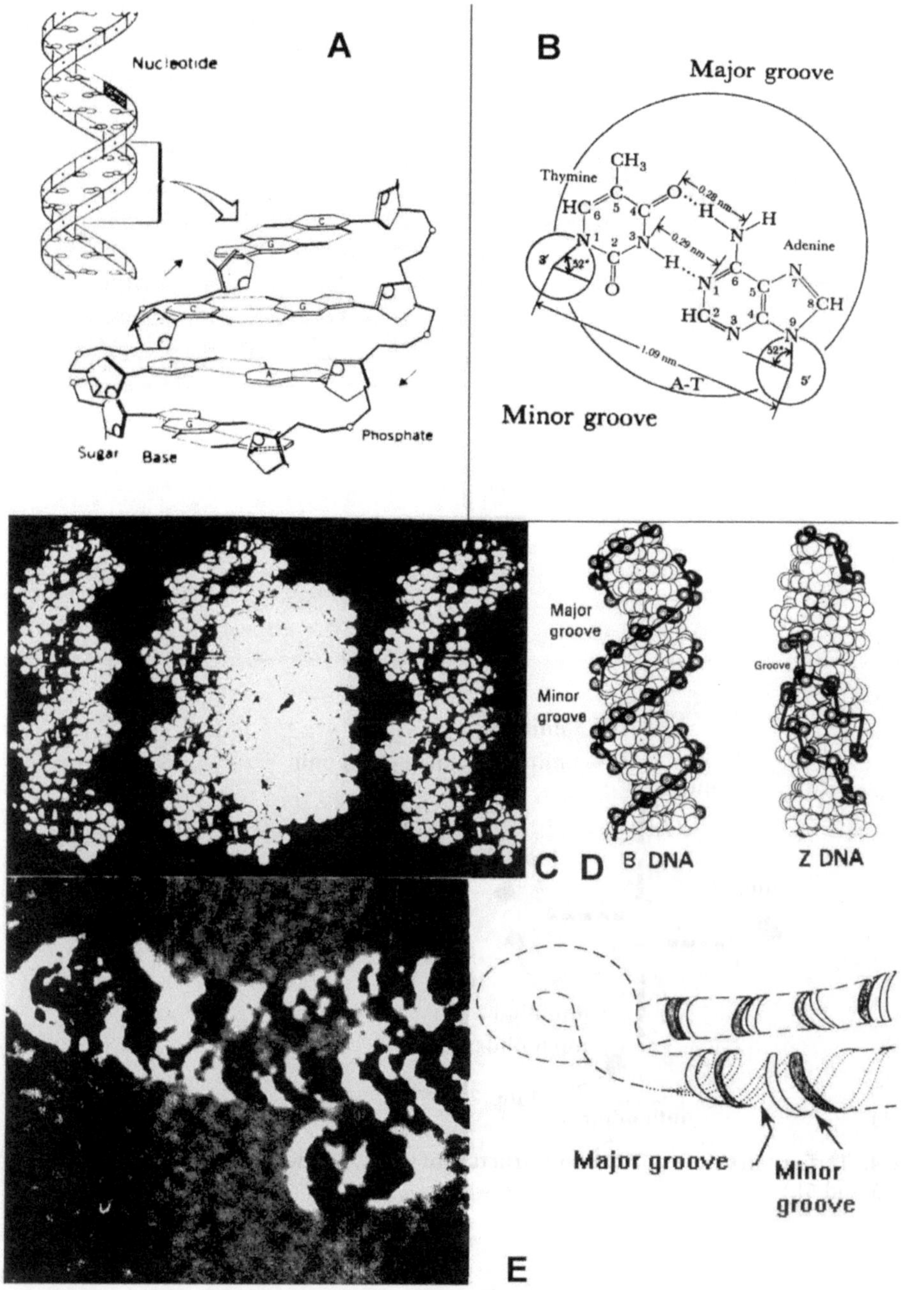

Fig. 3. Secondary structure of DNA. For details see text. (Modified from [2,3,6 and 7].)

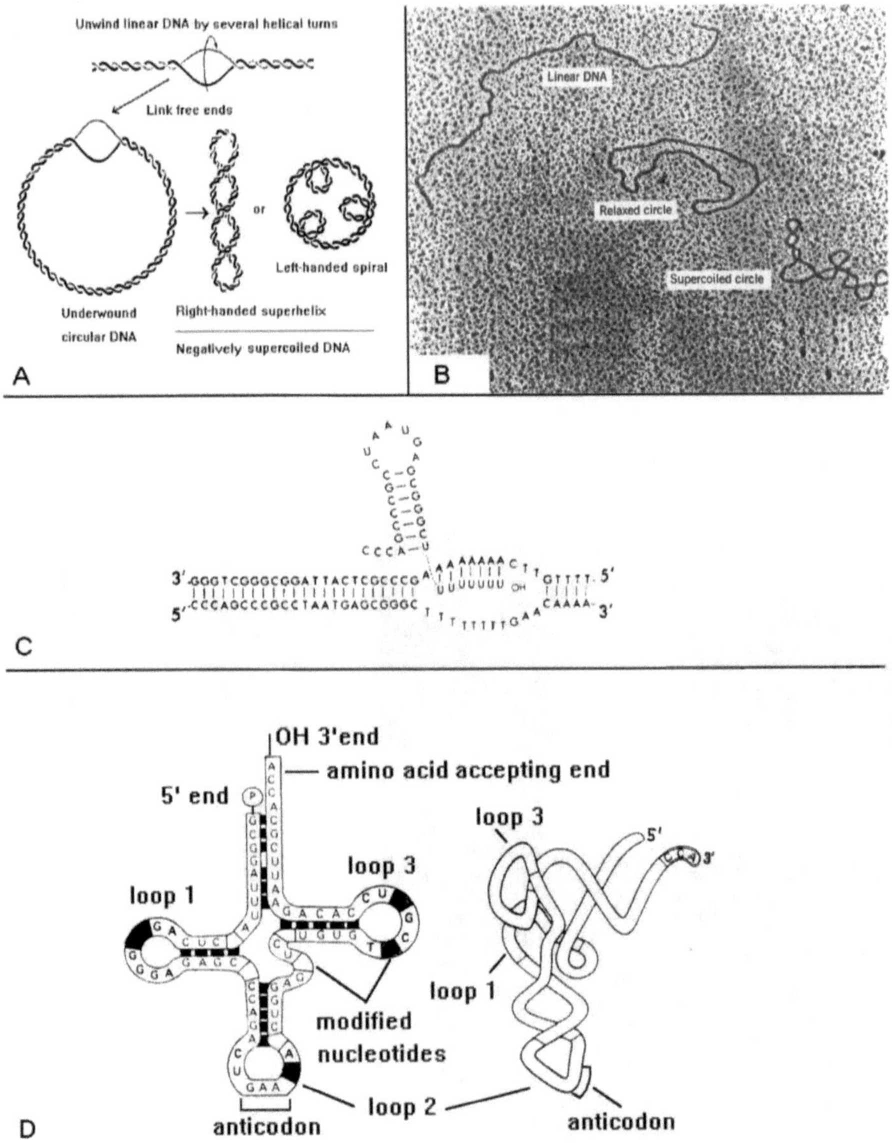

Fig. 4. Tertiary structure of DNA and structure of tRNA. For details see text. (Modified from [1,2,3 and 5].)

Organization of Chromatin

Structure of chromatin

DNA in chromosomes is associated with proteins called histones. Two molecules of histones H2A, H2B, H3, and H4 (Figure 5B) form a multimeric protein complex called a the **nucleosome** (Figure 5A). Double-stranded DNA is wound around the **nucleosome** in 2 turns over a length of 146 base pairs (Figure 5A). By electron microscopy consecutive nucleosomes appear as so-called "beads-on-a-string" or "10 nm-Filament" (Figure 5B,C). Under high salt concentrations 10nm filaments condense to form the "30 nm fiber" (Figure 5D,E). Histone H1 organizes the chromatin to form a **solenoid,** which is an assembly of 6 nucleosomes arranged in a bouquet-like structure (Figure 5E,F).

Chromosomes

During mitosis, higher degrees of chromatin condensation occur under the influence of non-histone proteins and lead to the formation of the **metaphase chromosome** (Figure 6A). Cytogenetic techniques permit the staining of specific chromosome bands which represent either AT-rich (dark) or CG-rich (light) regions of the chromosome. In Figure 6B the left half of each human chromosome shows staining during metaphase, whereas the right half shows a higher resolution as it is seen in prometaphase. This banding pattern, although species-specific, is highly conserved throughout evolution.

Figure 1 (see above) shows an electron micrograph of a chromosome that has been treated with detergents. A remnant of the protein scaffold can be appreciated, while long loops of DNA spill out from the scaffold.

Dimensions of the human genome

- The diploid set of human chromosomes is **46, XY** (i.e. 22 homologous autosomes: 22 from the father, 22 from the mother, plus 2 sex chromosomes: XY or XX)
- The length of the haploid human genome is 3×10^9 base pairs (bp)

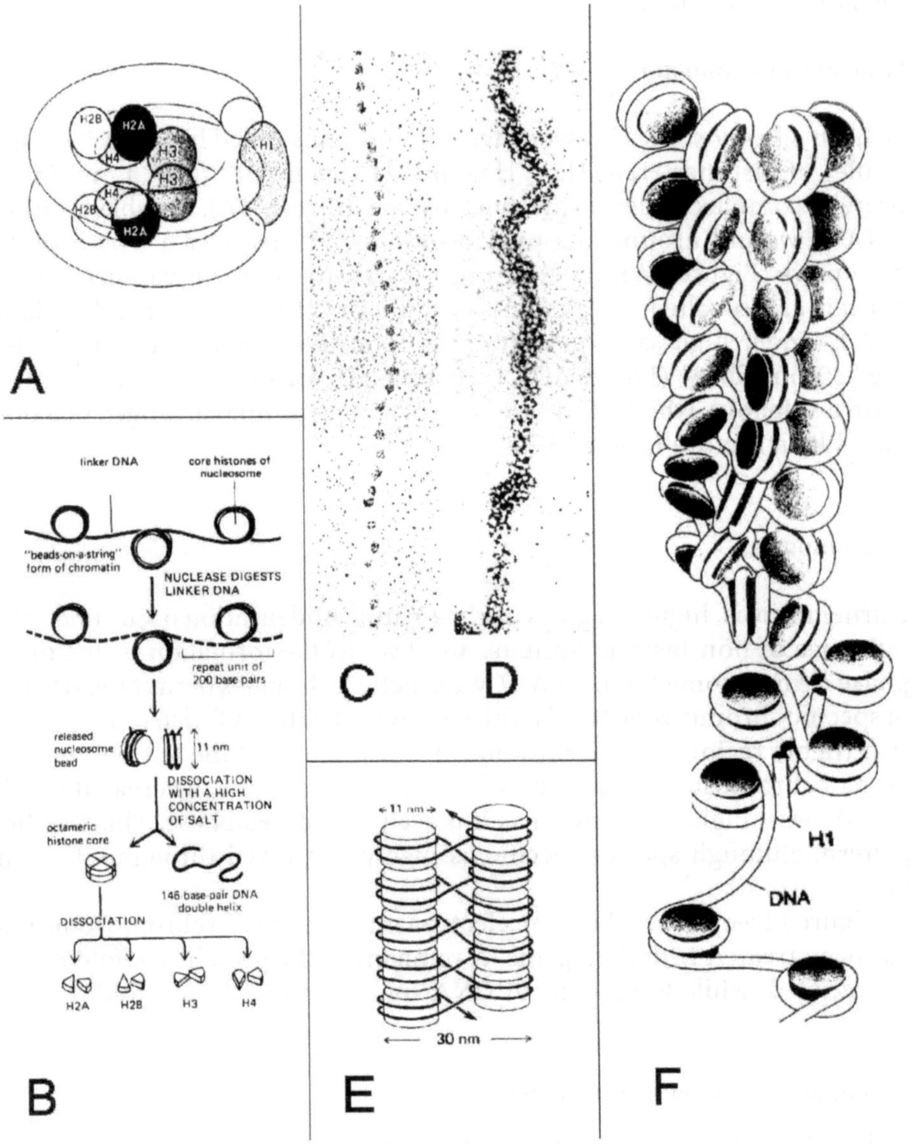

Fig. 5. Structure of chromatin. For details see text. (Modified from [1,3 and 4].)

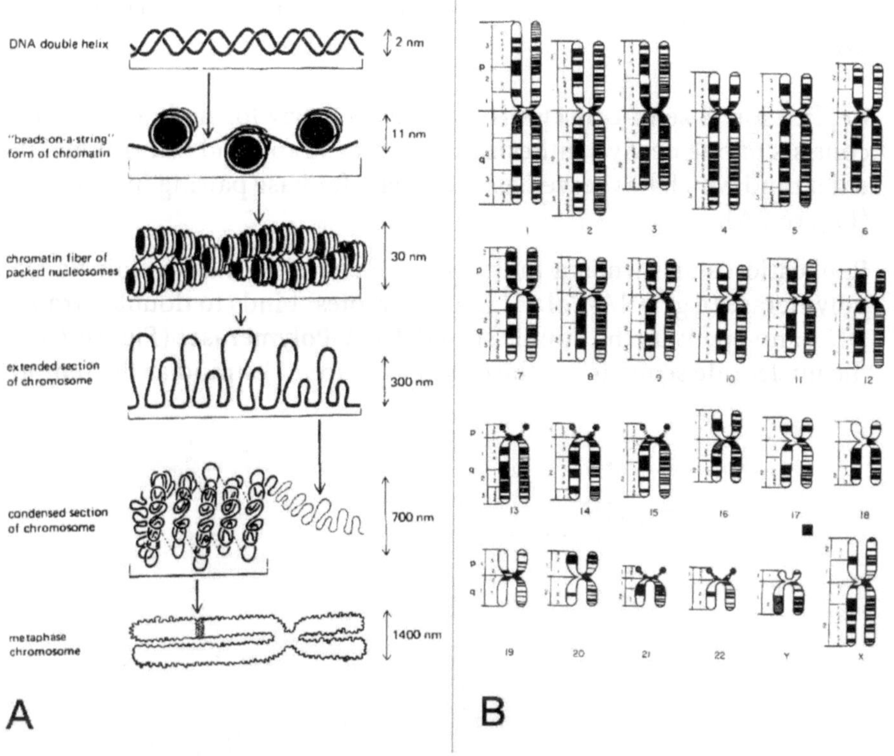

Fig. 6. Structure of chromosomes. For details see text. (Modified from [1].)

- There are about 80,000 expressed genes present in the human genome.
- scale for the dimensions of genomic DNA structures in base pairs is shown in Figure 7. For structural entities of different sizes the appropriate methodology for the study of its molecular structure is indicated.

Functions of the Human Genome

Replication

Cell division during **mitosis** and gametogenesis during **meiosis** require prior **replication** (or "reduplication") of the haploid set of chromosomes. The so-called bivalent chromosomes are built from the set of 46, XY. The bivalents each consist of **2 sister chromatids** which are joined at the centromer. Replication occurs in the S-phase of the cell cycle.

The following principles apply to replication of genomic DNA (Figure 8A):

- Replication is **semi-conservative,** i.e. each newly formed double strand consists of one newly synthesized daughter strand and one original parent strand which has served as a template for base pairing in replication (Figure 8A).
- Replication starts when a multi-enzyme complex, that contains a DNA polymerase (e.g. DNA-Pol. III in prokaryotes) binds to double-stranded DNA and opens up the double strand. **DNA-Polymerases** (Figure 8B) use the nucleotide sequence of the original ("parent") strand as the template

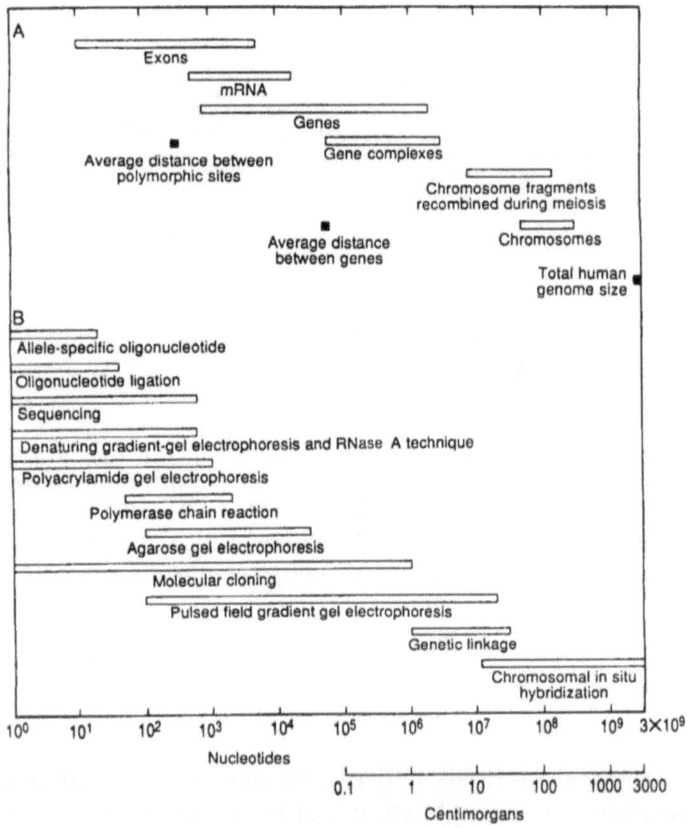

Fig. 7. Diagnostic techniques and the human genome. (A) The ranges of sizes of informational units of the human genome in nucleotides or centiMorgans. (B) The size ranges over which various diagnostic techniques are useful. (Modified from [7].)

for synthesis of a complementary strand. The daughter strand is elongated by sequential addition of dNTPs that are complementary to the parent strand. Phosphodiester bonds are found between the 5' phosphate of the dNTP and a free 3' hydroxyl of the nascent daughter strand. A polymerase always requires the presence of an oligonucleotide primer to supply the free 3'-hydroxyl group and start polymerization. Polymerases can work only in one direction synthesizing first the **5' end and then proceeding towards the 3' end** of the newly formed strand. Since the 2 template strands are in anti-parallel orientation, replication leads to a **"leading strand"** on one template strand and a **"lagging strand"** on the other template strand (Figure 8C). Since replication proceeds in the same direction on both strands, but polymerization of a daughter strand for the lagging strand proceeds from 5' to 3' as in the leading strand, the lagging strand has to be "stitched backwards". Short **RNA** primers serve as primers for the polymerase that binds to the lagging strand (Figure 8C). Starting from these primers so-called Okazaki fragments are polymerized. To finally connect the individual Okazaki fragments, the RNA primers are replaced with DNA by DNA polymerases exhibiting an exonuclease activity. Finally, a ligase joins the Okazaki fragments.

- Replication leads to a moving **replication fork** (Figure 8C) as demonstrable in electron microscopy (Figure 8D).
- The **error rate** of replication is about one in 10^8 - 10^{12} nucleotides. This is a relatively low error rate, owing to the semi-conservative mechanism and the presence of **DNA repair enzymes**.
- The cell cycle is shown in Figure 9A for a yeast as a representative eukaryotic cell. Progression through the cell cycle is **regulated** by so called **"cell division cycle genes"**. For example, Figure 9B shows how cyclin concentrations increase until the cdc2 kinase component of maturation promoting factor (MPF) is activated at metaphase, the midpoint of mitosis. The cyclin is then degraded and MPF activity drops as the cell divides. Then the cycle repeats.

Transcription and translation

Replication of DNA is a prerequisite for cell division and permits growth to occur in an organism. **Transcription** of genomic DNA into **messenger RNA** (mRNA), with the consecutive translation of mRNA into protein, represents

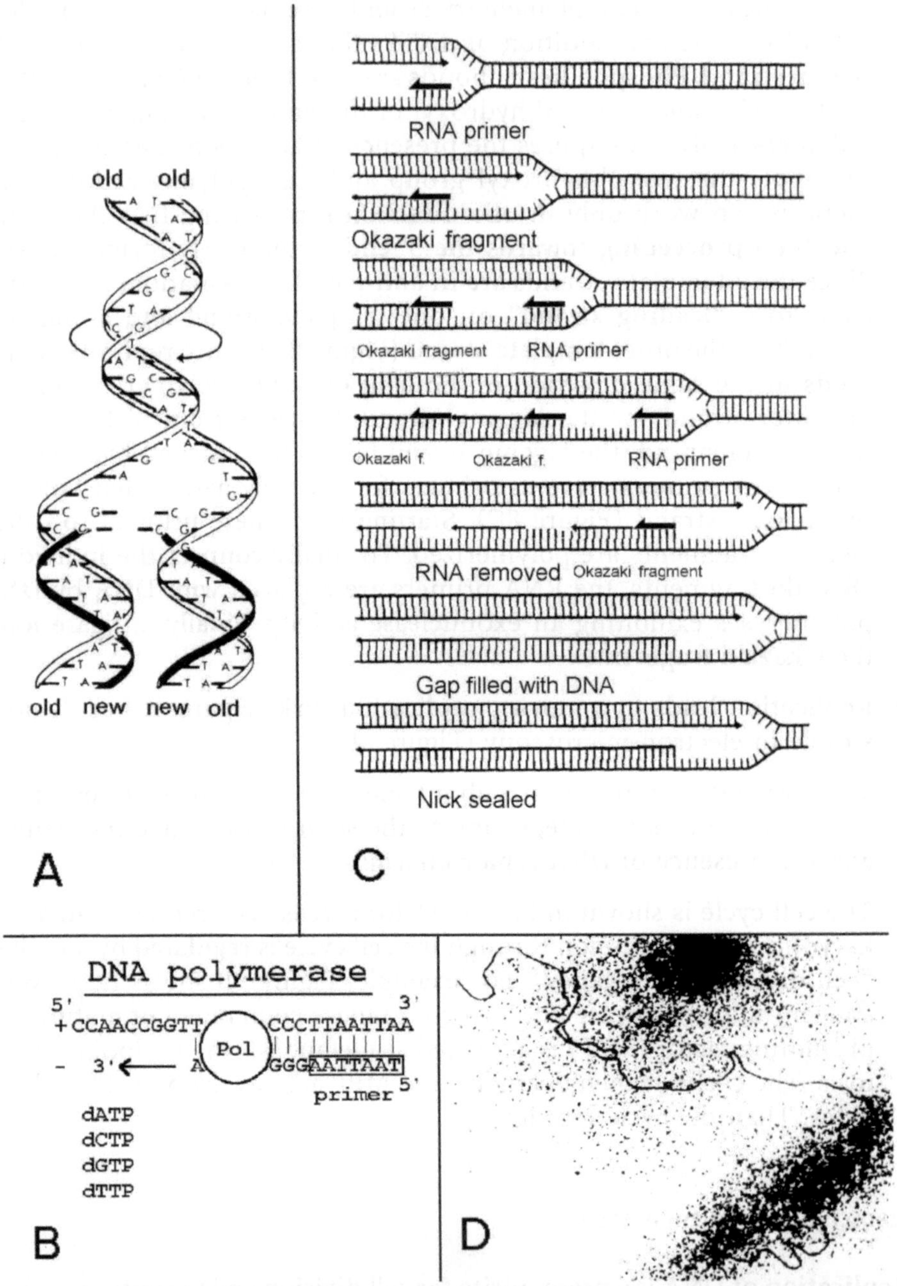

Fig. 8. Mechanisms of DNA replication. For details see text. (Modified from [3 and 5].)

the connection between genetic information and cell metabolism. The flow of information from genomic DNA to protein sequence has been itemized by Francis Crick (1956) in the **central dogma of molecular biology** (Figure 9C; see also Figure 10 and Figure 11B).

Transcription

Figure 10 shows Transcription starting at a **promoter**, which is a specific DNA sequence that is recognized by a multienzyme complex containing an **RNA polymerase** and **transcription factors.** The RNA polymerase then catalyses **transcription of DNA into mRNA.** As a convention, the strand of the genomic DNA that is identical in sequence to the mRNA is called the **(+) strand or sense strand.** The complementary strand is called the **(-) strand or antisense strand.**

Transcription initially leads to an immature **heteronuclear RNA (hnRNA).** hnRNA contains **exons** which encode amino acid sequence and are present in mature mRNA, and **introns** (or intervening sequences), which do not encode protein sequence and are absent from mature mRNA. Introns are **spliced** out by **spliceosomes** at splice-donor and splice-acceptor recognition sequences, and adjacent exons are joined to each other. Before hnRNA leaves the nucleus and enters the cytoplasm, a **poly(A) tail** is added at the 3' end by a poly(A) polymerase. The 5' end is protected by a 7-methyl-guanine, constituting the so-called **5'-cap**. Both mechanisms lead to higher stability of the mature mRNA. The mature mRNA is then **translated at ribosomes** into a polypeptide using the **genetic code** (see Figure 13). The ribosome in eukaryotes "docks" at the endoplasmic reticulum. In some instances posttranslational modification of the protein follows, such as cleavage of a signal peptide or glycosylation. The complete process from transcription to the final **gene product** (i.e. protein) is called **gene expression of the gene.**

In prokaryotes there is **co-linearity** between gene, mRNA and polypeptide. This co-linearity is interrupted in eukaryotes because of the intervening introns. Furthermore, the rule of co-linearity can be broken by the presence of alternative splicing and posttranslational modification.

Exon-intron boundaries can be visualized by electron microscopy using so called "R-loop" conditions (see above, Figure 9D).

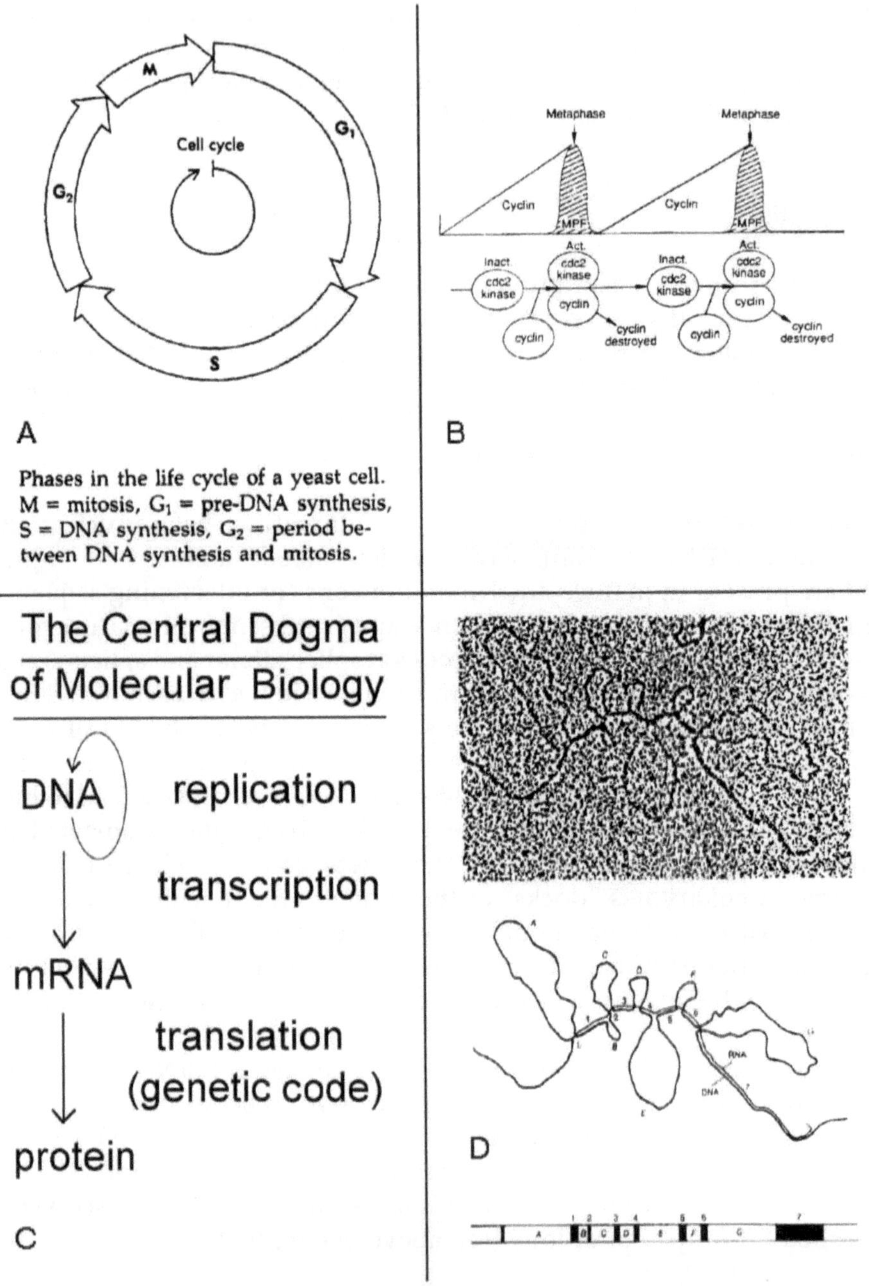

Fig. 9. Transcriptional mechanisms. For details see text. (Modified from [4 and 5].)

Regulation of transcription

Transcription is highly regulated by **transcription factors** (Figure 11A), which are proteins that recognize specific regulatory DNA sequences. Those **regulatory sequences**, called enhancers or silencers in most instances lie "upstream" from the transcription start site in the so called **5' flanking region.** The presence or absence of such protein factors in a cell leads to **tissue-specific** gene expression and allows the cell to function in a tissue specific-manner. In addition, these factors endow the cell with the ability to alter gene expression in response to external signals. The glucocorticoid receptor may serve as an example: After binding to the steroid hormone, the cytoplasmic receptor is translocated to the nucleus and binds to a glucocorticoid responsive element (GRE). The GRE functions as an enhancer for neighboring genes.

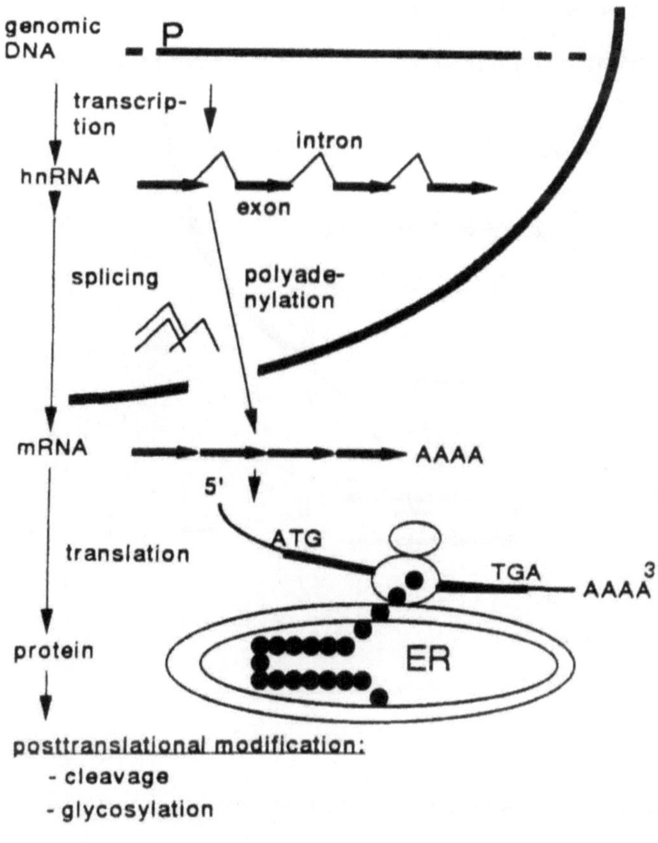

Fig. 10. Aspects of colinearity between gene and gene products. For details see text. P = promoter; arrows denote exons; hooks denote introns.

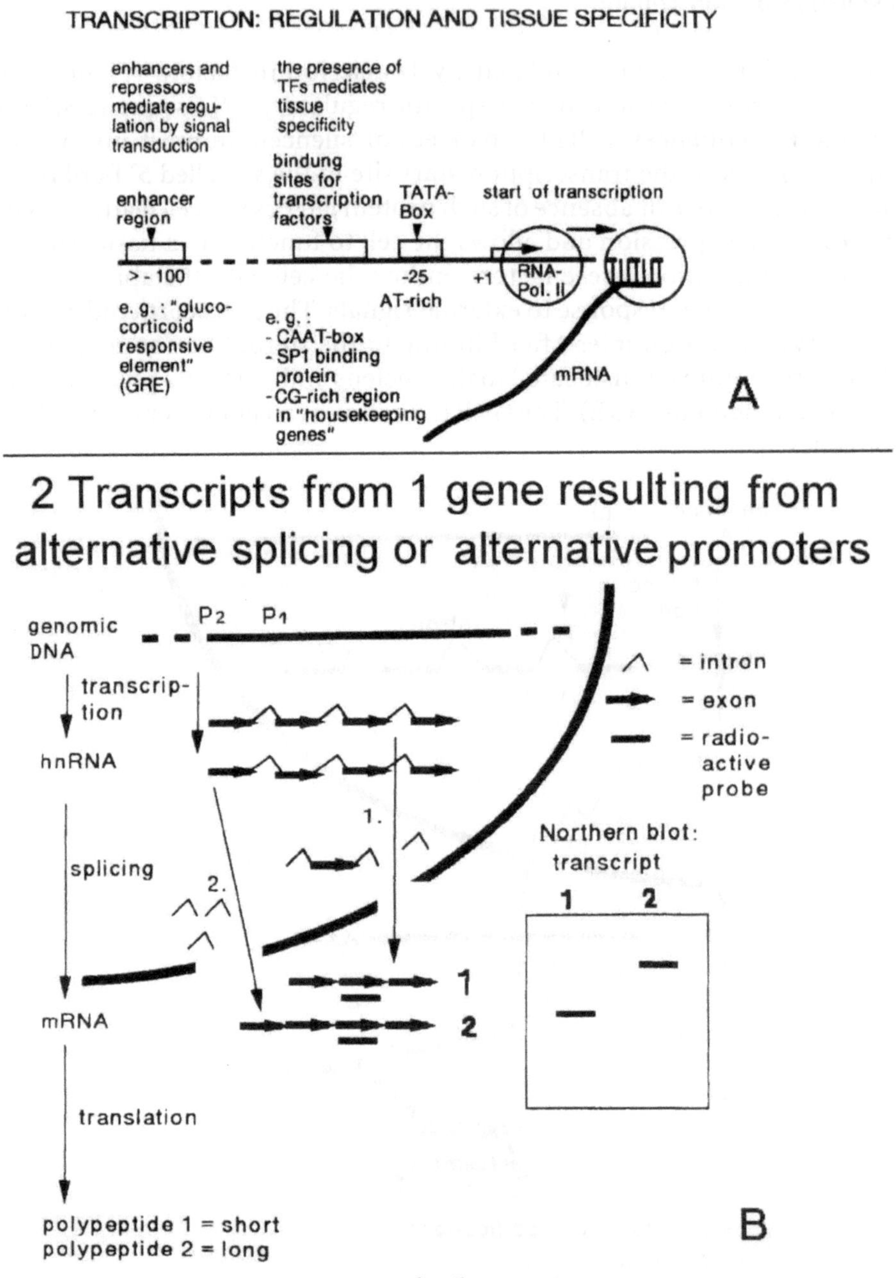

Fig. 11. Regulation of transcription. For details see text.

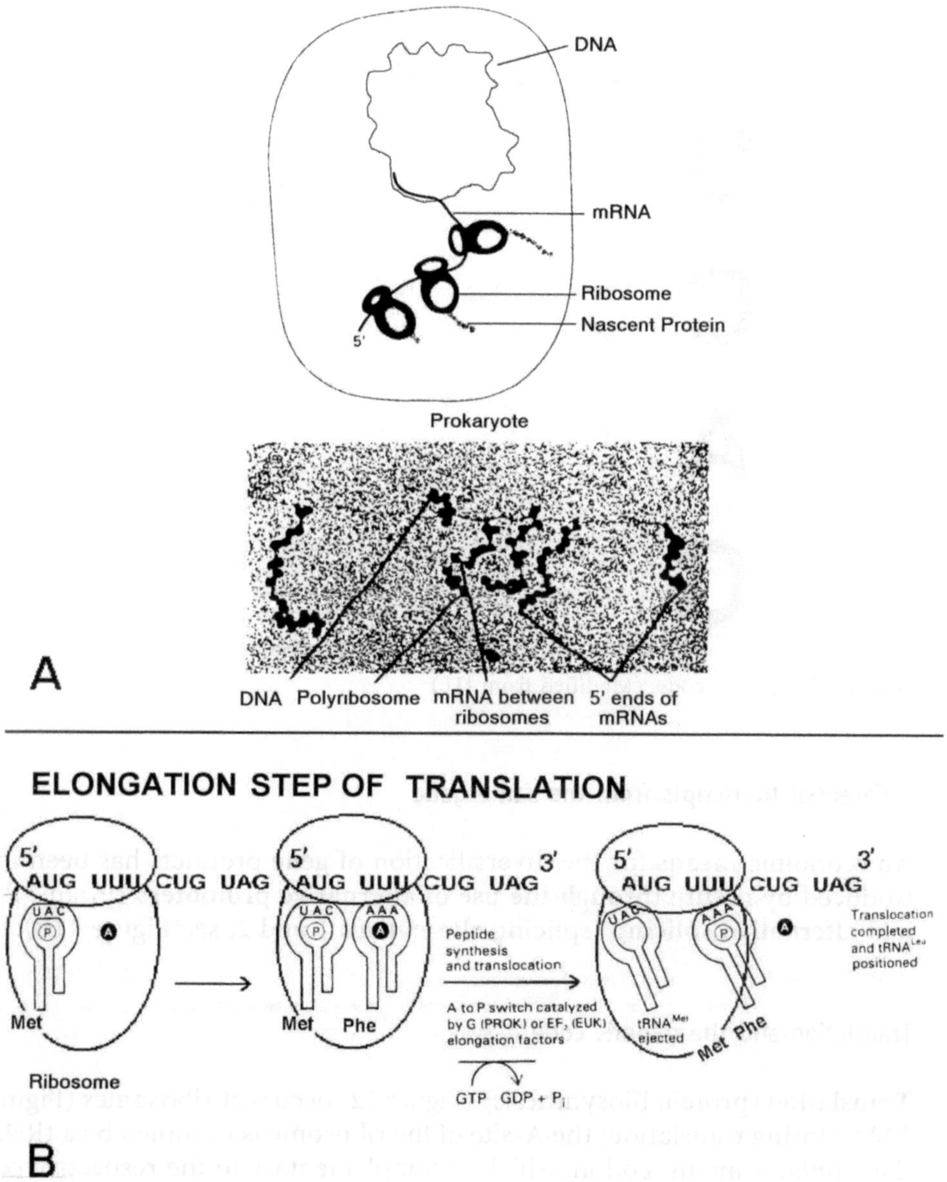

Fig. 12. Mechanisms of translation. For details see text. (Modified from [2].)

1st position (5' end) ↓	2nd position: U	C	A	G	3rd position (3' end) ↓
U	Phe	Ser	Tyr	Cys	U
	Phe	Ser	Tyr	Cys	C
	Leu	Ser	STOP	STOP	A
	Leu	Ser	STOP	Trp	G
C	Leu	Pro	His	Arg	U
	Leu	Pro	His	Arg	C
	Leu	Pro	Gln	Arg	A
	Leu	Pro	Gln	Arg	G
A	Ile	Thr	Asn	Ser	U
	Ile	Thr	Asn	Ser	C
	Ile	Thr	Lys	Arg	A
	Met	Thr	Lys	Arg	G
G	Val	Ala	Asp	Gly	U
	Val	Ala	Asp	Gly	C
	Val	Ala	Glu	Gly	A
	Val	Ala	Glu	Gly	G

Fig. 13. The genetic code. (Modified from [1].)

Differential transcripts from the same gene

An economic means for the diversification of gene products has been introduced by nature through the use of **alternative promoters** (P_1 and P_2) and **alternative splicing** (splicing alternatives 1 and 2, see Figure 11B).

Translation and the genetic code

Translation (protein biosynthesis) (Figure 12) occurs at ribosomes (Figure 12A). During translation, the A-site of the ribosome is occupied by a tRNA that contains an anticodon, which is complementary to the respective codon on mRNA. A peptide bond is formed between the amino acid carried by the tRNA and the nascent polypeptide chain, which occupies the P-site of the ribosome (Figure 12B). Translation always starts at the **start codon** AUG (methionine) and ends at one of the **3 stop codons** UAA, UAG, UGA.

The **genetic code** (Figure 13) is called **degenerate**, since more than one triplet may encode the same amino acid. There is a correlation between the second base of the codon and the chemical nature of the encoded amino acid (non-polar, ionized polar, non-ionized polar). The charging of the tRNA with the amino acid corresponding to the anticodon occurs by the respective tRNA synthetase. This process has been termed "the second genetic code", since an erroneous charging would lead to an error in translation.

References

1. Alberts B, Bray D, Lewis J, Raff M, Roberts K, et al. (eds) Molecular biology of the cell. Garland Publishing, New York and London 1983
2. Darnell J, Lodish H, Baltimore D (eds) Molecular cell biology. Scientific American Books, W. H. Freenmann and Company, New York 1990
3. Lewin B (ed) Genes III. John Wiley & Sons, New York 1987
4. Sippel AE and Nordheim A (eds) Erbsubstanz DNA. Spektrum-der-Wissenschaft-Verlagsgesellschaft 1986
5. Watson JD, Hopkins NH, Roberts JW, Steitz JA, Weiner AM (eds) Molecular biology of the Gene. The Benjamin/Cummings Publication Company, Inc., Reading 1987
6. Wood WB, Wilson JH, Benbow RM, Hood LE. Biochemistry – a problems approach. Benjamin/Cummings Publication Company, Inc., Reading 1981
7. Beebe TP, Wilson TE, Ogletree DF, Katz JE, Balhorn R, et al. (1989) Direct observation of native DNA structures with the scanning tunneling microscope. Science 243: 370–372
8. Landegreen U, Kaiser R, Caskey CT, Hood L (1988) DNA diagnostics – molecular techniques and automation. Science 242:229-237

Chapter 2

Enzymatic Modification of DNA and RNA

KAI-OLAF NETZER

Introduction

Enzymes represent the tools of the molecular biologist. They serve to manipulate DNA and RNA in vitro. Today, more than 200 different enzymes with various uses are offered by commercial manufacturers. Figure 1 shows a rough classification of enzymes by their principal actions. This chapter will provide examples from each group and protocols for their use.

Subprotocol 1
Restriction Endonucleases

The first observations that bacteria such as *E. coli* contain site-specific DNA degrading enzymes date back to the late 1960s. The so-called restriction endonucleases (or restriction enzymes) are part of a system that protects resident DNA against contamination by sequences of foreign origin. None of the early restriction enzymes lived up to the hopes of potential laboratory use. They recognized the DNA at specific sites, but cleaved at random sites far removed. Today, restriction enzymes have been isolated that fall into three different groups according to their action (Figure 2). Both restriction enzymes with site specific recognition and cleavage are routinely used in molecular biology.

Kai-Olaf Netzer, Krankenhaus Köln-Merheim, Medizinische Klinik I, Ostmerheimer Str. 200, Köln, 51109, Germany (*phone* +49-221-8907-2200; *fax* +49-221-8907-2335; *e-mail* kai.netzer@uni-koeln.de)

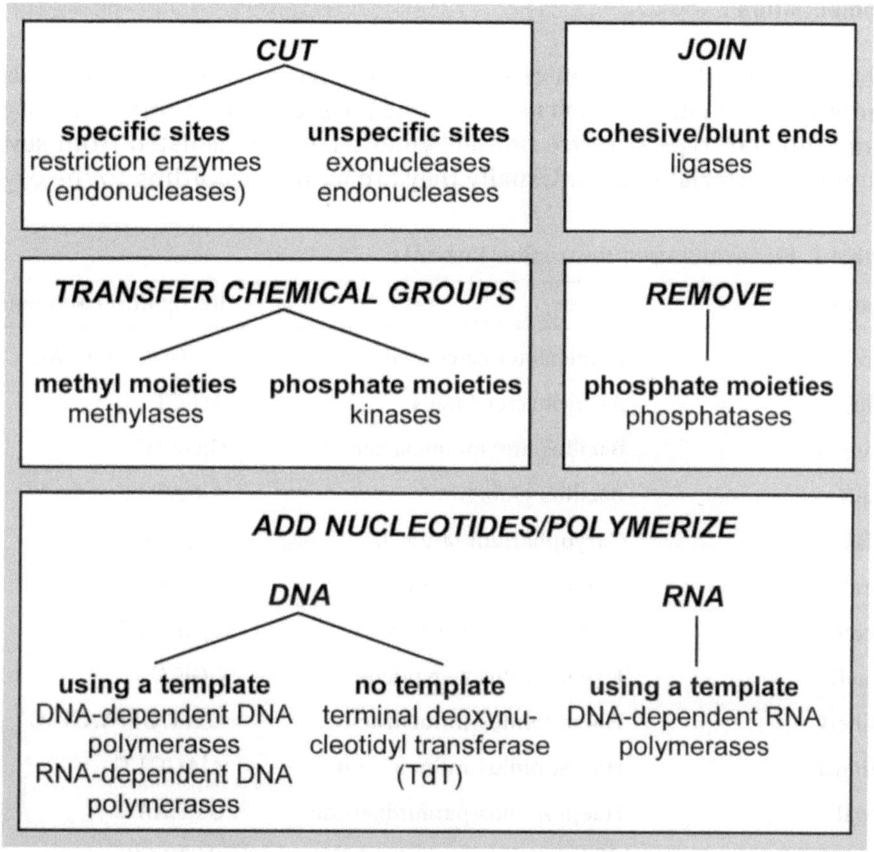

Fig. 1. Enzyme Classification by Principal Action on DNA/RNA

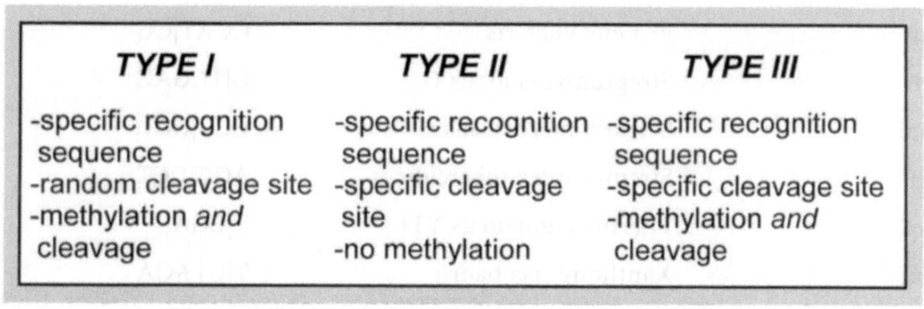

Fig. 2. Classification of Restriction Endonucleases

Nomenclature

The first type II restriction enzyme was discovered in the bacterium Haemophilus influenzae. It was named HindII based on the name of the organism. Thus far, type II restriction enzymes have been isolated from several hundred bacterial strains. Usually they are named according to the organ-

Table 1. Frequently used Restriction Enzymes

Name	Source	Recognition Sequence
AccI	Acinebacter calcoaceticus	GT|(A/C)(T/G)AC
AluI	Arthrobacter luteus	AG|CT
BamHI	Bacillus amyloliquefaciens H.	G|GATCC
BglII	Bacillus globigii	A|GATCT
ClaI	Caryophanum latum L	AT|CGAT
DraI	Deinococcus radiophilus	TTT|AAA
EcoRI	Escherichia coli RY13	G|AATTC
HaeIII	Haemophilus aegypticus	GG|CC
HincII	Haemophilus influenzae Rc.	GT(T/C)|(A/G)AC
HindIII	Haemophilus influenzae Rd.	A|AGCTT
HpaI	Haemophilus parainfluenzae	GTT|AAC
KpnI	Klebsiella pneumoniae OK8	GGTAC|C
MspI	Moraxella species	C|CGG
NcoI	Nocardia corallina	C|CATGG
NotI	Nocardia otitidis-caviarum	GC|GGCCGC
PstI	Providencia stuartii	CTGCA|G
PvuI	Proteus vulgaris	CGAT|CG
SalI	Streptomyces albus G.	G|TCGAC
SmaI	Serratia marcescens	CCC|GGG
StuI	Streptomyces tubercidicus	AGG|CCT
TaqI	Thermus aquaticus YTI	T|CGA
XbaI	Xanthomonas badrii	T|CTAGA
XhoI	Xanthomonas holcicola	C|TCGAG

ism of origin (three initial letters, e.g. Hin), the bacterial strain (one letter, e.g. d), and by roman numerals (e.g. II) if more than one enzyme has been isolated from a bacterium. More than 150 different cleavage sites have been found. Table 1 contains a list of commercially available, frequently used type II restriction endonucleases.

Recognition sequence

The recognition sequences are usually 4 to 8 bp long and frequently display twofold symmetry. They represent palindromic sequences, i.e. they have the same sequence when read forward on one DNA strand and backward on the complementary strand. A given 4 bp site is calculated to occur every $4^4 = 256$ bp, a 6 bp site every $4^6 = 4096$ bp, and an 8 bp site every $4^8 = 65536$ bp. The site distribution in native DNA, however, is nonrandom and depends in part on the base composition, so that a specific DNA region may be cut more or less frequently than the statistical average. For example, NotI has an 8 bp recognition sequence which includes CpG dinucleotides that occur very infrequently in mammalian DNA. It produces fragments of 1 to 1.5 million bp ("rare cutter") and is therefore used for long range physical mapping of DNA.

Isoschizomers and compatible ends

Several enzymes isolated from different organisms have identical recognition sequences. These enzymes are called isoschizomers. As shown in Figure 3, identical recognition sequences can be cleaved differently by isoschizomers. Cleavage of the recognition sequence GGTACC by Asp718, and KpnI, generates two fragments that cannot be joined ("ligated") together. Enzymes may have different recognition sites but produce identical or compatible ends. In this example, DNA fragments cleaved by MboI and BamHI, could be directly ligated together.

Mechanism

How do restriction endonucleases achieve their high specificity? The mechanism of action has been studied extensively for EcoRI (McClarin et al. 1986). It has been shown that the symmetry of the palindromic target sequence (GAATTC) is matched by the symmetry of the enzyme. The enzyme

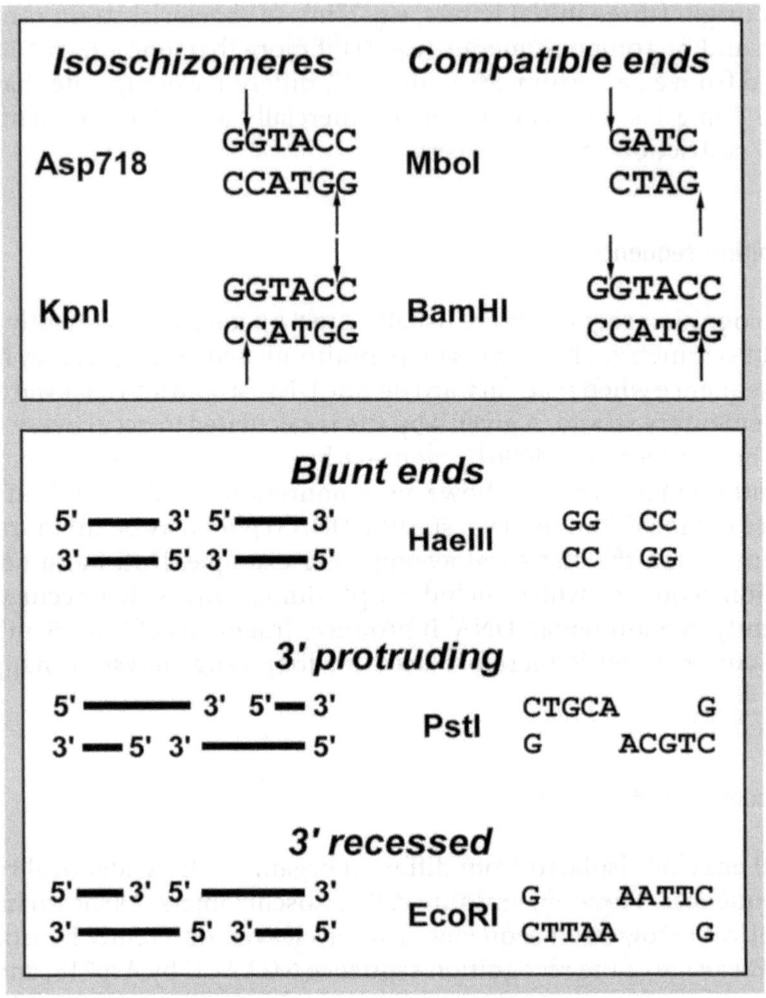

Fig. 3. Cleavage by Isoschizomeres

is a dimer comprised of two identical 31 kD subunits. Contact of the enzyme with the major groove of B-DNA (the classical double helix) leads to kinking of the double helix in the center of the hexanucleotide target. The DNA is unwound by 25 degrees, the major groove is widened 0.4 nm and the enzyme enters the groove. A total of 12 hydrogen bonds is formed between the alpha-helices of the enzyme and the two trinucleotides of GAA. The amino ends of the alpha-helices interact with the negatively charged phosphate backbone of the DNA and hydrolytic cleavage is initiated.

Cleavage of a DNA fragment may result in blunt ends, 3' protruding ends, or 3' recessed ends. Figure 3 shows examples for each group. Blunt ended DNA fragments can be ligated to any other blunt ended fragment, regardless of the enzymes they were generated with. Blunt ends have no inherent tendency to stick together, since they are base paired out to their ends. In contrast, PstI and EcoRI produce short single stranded tails. Complementary single stranded tails tend to associate by base pairing. They are therefore called cohesive, or sticky ends. Cohesive ends can be permanently joined together by the action of DNA ligase (see below).

Uses

Restriction enzymes are used to cut purified preparations of genomic, plasmid, phage, phagemid, or cosmid DNA. Digestion of genomic DNA results in many fragments of different lengths which appear as a smear on agarose gel electrophoresis, whereas digestion of cloned DNA results in distinct bands. The lengths of these bands in base pairs (bp) are predictable from the sequence. Digested DNA may be used for Southern blotting, in which specific fragments are identified using a labeled DNA probe. Restriction endonuclease-digested DNA is also used for molecular cloning. Since all digested fragments have cohesive or blunt ends of predictable sequence, DNA from different sources may be ligated together. For example, a PstI fragment of a human gene may be ligated into a bacterial plasmid vector that has been digested with the same enzyme, thus creating a molecule of "recombinant DNA".

Protocols for restriction endonuclease digestion

Most restriction enzymes are available as purified proteins from bacterial cultures. However, some have been cloned and are available as in vitro expressed recombinant proteins. Most restriction enzymes are shipped and stored in glycerol containing buffers at –20°C. Their activity is given in U/μl. One unit is defined as the amount of enzyme that cuts 1 μg of lambda phage DNA to completion in one hour.

If cloned DNA is to be cut, the amount of enzyme needed can be calculated from the following equation:

$$[U] = \frac{48000\ \text{bp}_{\text{lambda}}}{\text{number of sites}_{\text{lambda}} \times 1\mu\text{g}} \times \frac{\text{X}\mu\text{g}_{\text{pDNA}} \times \text{number of sites}_{\text{pDNA}}}{\text{Y bp}_{\text{pDNA}}}$$

Four variables have to be known:

- the number of restriction enzyme sites in lambda phage DNA,
- X, the amount of cloned DNA to be cut in μg,
- the number of sites within the cloned DNA, and
- Y, the length of the cloned DNA fragment.

Example: A digestion is to be set up using an enzyme with 6 recognition sites in lambda. 1 μg plasmid DNA shall be digested. The fragment contains two sites for the enzyme, and is 1000 bp long. 16 U of enzyme would be needed to completely digest the DNA in one hour. If more time, e.g. three hours, is allowed, the amount of enzyme may be reduced to approximately one third, 5 U. If preparations of genomic DNA are to be digested, a good rule of thumb is to add 3 U per μg DNA, and to allow very long incubation times of several hours to over night (16 hours). If the DNA is impure and contains contaminating protein, it may be necessary to add 1-2 U/μg of fresh enzyme after several hours and to incubate an additional three to five hours.

Materials

Supplies

- Heating block or water bath (temperature range 37 to 65°C)
- 0.5 or 1.7 ml reaction tubes
- Micropipettes, 1-10 μl and 10-100 μl
- Autoclaved pipet tips

Reagents and solutions

- DNA, 0.1-0.2 μg/μl in water or TE buffer (10 mM Tris-HCl pH8.0, 1 mM EDTA)
- Enzyme (should not exceed 1/10 of the total volume)
- Buffer (usually supplied as 10x reaction buffer by the manufacturer. It may contain NaCl 10-150 mM, Tris-HCl pH7.5-8.0 50-150 mM, $MgCl_2$ 0-10 mM, KCl, and DTT in varying concentrations depending on the enzyme and manufacturer)
- Ultrapure water
- Acetylated bovine serum albumin (BSA, Sigma)

Procedure

Plasmid DNA

This protocol is used for a standard 20 µl reaction, e. g. digestion of a plasmid miniprep. It can easily be scaled down to 10 µl or scaled up to 100 µl.

1. On ice, pipet the following into an autoclaved microcentrifuge tube:

DNA	5 µl	0.5-1 µg
10x buffer	2 µl	
Enzyme	1 µl	usually 5 U
Water	12 µl	

Note: If only one digest is prepared, add water and buffer first. The enzyme should be removed from the freezer briefly before addition and should be kept on ice all the time. If many different DNA samples are to be digested with the same enzyme, prepare a master mix by pipetting n+1 times the volume for one reaction (without DNA) in a reaction tube, mixing, and dispensing the volume for one reaction in each of n tubes. Subsequently add the DNA.

2. After addition of enzyme and DNA, mix briefly by vortexing or by tapping the tube.
3. Place reaction tube in a heating block or water bath preheated to the appropriate temperature.

Note: Usually this will be 37°C. However, some enzymes such as TaqI require higher temperatures (65°C). See the manufacturer's instructions for optimal incubation temperature.

4. Incubate for 1 to 3 hours.

Note: If the DNA preparations are pure, the reactions may be left overnight without the risk of degradation. If DNA purity is limited, rather choose a greater reaction volume in order to dilute the sample, and keep incubation times short.

Genomic DNA

This protocol is used for a standard 100 µl reaction. It can easily be scaled up or down.

1. On ice, pipet the following into an autoclaved microcentrifuge tube:

DNA	50 µl	5-10 µg
10x buffer	10 µl	
Enzyme	6 µl	usually 30 U
Water	34 µl	

Note: The same basic rules for plasmid DNA apply to genomic DNA. However, greater care should be applied when pipetting and mixing genomic DNA due to its increased fragility and viscous character. Genomic DNA should not be vortexed when large intact restriction fragments are desired.

2. Mix by tapping the tube. Place the tube in a heating block or water bath preheated to the appropriate temperature (usually 37°C).
3. Incubate for 12 to 16 hours.

Note: Completion of the digestion may be checked by removing 5 µl from the reaction mix, and analyzing the aliquot by agarose gel electrophoresis (0.8% agarose). If the digest is incomplete, add approximately 10 U of enzyme to each reaction and incubate another 3 to 5 hours. In general, impure preparations of DNA require greater amounts of enzyme.

Double digestion

If a DNA sample is to be digested with two different enzymes, there are several approaches.

- The two enzymes may be added simultaneously to the reaction mixture. This is possible if buffer conditions and incubation temperature are identical or very similar. Tables listing enzyme activity in different buffers are supplied by several manufacturers (e. g. Gibco-BRL, Promega). Sub-optimal buffer conditions may lead to reduced activity or "star activity", i. e. unspecific cleavage. We tend to optimize double-digest conditions for the less robust, more expensive of the two enzymes. For the second enzyme, activity losses up to 75% due to sub-optimal buffer conditions seem to be acceptable. Allow extended incubation time.
- Alternatively, if the only difference in the buffers is the NaCl concentration, digestion with one enzyme in the lower NaCl concentration is done first. Then the NaCl concentration is raised by addition of NaCl and digestion with the second enzyme is performed.
- The two enzymes are added one after another in two separate reactions. Between the two incubations, the DNA has to be ethanol precipitated or

purified by using a chaotropic buffer (NaI)/silica-gel spin column system (e. g. Qiagen) to remove all traces of buffer.

Subprotocol 2
Exonucleases, Other Nucleases and Ligases

In contrast to restriction enzymes, nucleases cleave DNA and RNA at non-specific sites. Substrate for nucleases could be double stranded or single stranded polynucleotides. They are degraded either from the termini (exonucleases) or from within the sequence (endonucleases). In this section, three examples are given for enzymes that are of some use in molecular biology.

Deoxyribonuclease I (DNase I)

This enzyme is isolated from bovine pancreas. DNase I is an endonuclease that digests both double-stranded and single-stranded DNA. It displays two modes of action depending on the ions added to the reaction.

- In the presence of Mg^{++} ions, each DNA strand is attacked independently with cleavage sites distributed randomly. In this way, the enzyme is used for introducing random nicks (i. e. discontinuities in one strand of a double-stranded molecule) into intact DNA prior to radiolabeling (nick translation).
- In the presence of Mn^{++} ions, DNase I cleaves the complementary strands of double stranded DNA at approximately the same sites to yield very short, mostly blunt-ended fragments.

Nuclease S1

Nuclease S1 is isolated from *Aspergillus oryzae*. It specifically degrades single-stranded DNA or RNA, whereas double-stranded molecules are resistant to cleavage. If the amount of enzyme in a reaction is increased, double-stranded DNA is nicked. This enzyme is used to remove single-stranded tails or loops from DNA fragments, and to analyze the structure of DNA: RNA hybrids.

Ribonuclease (RNase A)

RNase A is isolated from bovine pancreas. It is an endonuclease specific for RNA. It hydrolizes the phosphodiester bound on the 3' side of pyrimidine residues. RNase works in the absence of cofactors and divalent cations. The enzyme is used after hybridization experiments to remove unhybridized RNA probes. Another use is to cleave single base mismatches in DNA:RNA hybrids in order to map single base mutations. Contamination with RNase is often a problem when working with RNA. Its action can be inhibited by placental RNase inhibitor (RNasin™, Promega) or by vanadyl-ribonucleoside complexes.

Ligases

Ligases are isolated from *E. coli* (*E. coli* DNA ligase) or from bacteriophage T4-infected bacteria (T4 DNA ligase).

T4 ligase is used more frequently since it is able to join blunt-ended DNA fragments. It is a 68 kD enzyme that catalyzes the formation of phosphodiester bonds between adjacent 3'-hydroxyl and 5'-phosphate termini in DNA (two dephosphorylated fragments cannot be ligated). ATP (contained in the reaction buffer) supplies energy for the two step reaction. Joining DNA fragments with cohesive ends is very efficient. Joining DNA fragments with blunt ends, however, is much slower and requires more enzyme.

Materials

For T4 ligase reaction

Supplies

- Cooling waterbath (16°C)
- Alternatively: Thermal cycler, programmed to 16°C/hold

Reagents

- DNA, purified and dissolved in water at approximately 0.1 μg/μl; Vector DNA should be dephosphorylated prior to ligation in order to prevent autoligation. The presence of EDTA should be avoided since it may reduce enzyme activity.
- Buffer: 10x T4 DNA ligase buffer (usually supplied by the manufacturer) is 0.5 M Tris-HCl, pH7.5, 100 mM $MgCl_2$, 100 mM DTT, 10 mM ATP, 250 μg/ml bovine serum albumine; for ligation of blunt ends, use 10x blunt

end ligation buffer: 0.66 M Tris-HCl pH7.6, 50 mM MgCl2, 50 mM DTT, 10 mM hexaminecobalt chloride, 2 mM ATP, 5 mM spermidine-HCl, 250 µg/ml bovine serum albumine

- T4 DNA ligase (1-5 Weiss units/µl, available from various suppliers)

Procedure

T4 ligase reaction

The protocol gives an example for a standard ligation reaction of cohesive or blunt DNA ends, e. g. the ligation of a 1000 bp DNA insert into a 3000 bp plasmid vector.

1. Pipet the reaction components into a 0.5 ml reaction tube. Keep all stock solutions and the microcentrifuge tube on ice.

DNA (vector)	2 µl	200 ng
DNA (insert)	1 µl	100 ng
Water	5 µl	

2. Melt any cohesive termini that have reannealed by heating the reaction to 45°C for 5 min. Place on ice. Add:

10x buffer	1 µl	
Ligase	1 µl	0.5 Weiss units

Note: The enzyme is used in some excess in this protocol. If blunt ends are to be ligated, however, it may be necessary to increase the concentration of T4 DNA ligase.

3. Mix by tapping the tube.
4. Incubate for 1-4 hours at 16°C or overnight at 12°C

Note: The low reaction temperature helps to minimize autoligation of DNA fragments and increases the yield of intermolecular ligation. Incubation time may be increased to overnight, i. e. 16 hours, with no adverse effects.

5. Check the ligation by agarose gel electrophoresis of 1/10 of the reaction products.
6. Use 1 (to 5) µl to transform 50 to 100 µl of competent bacteria.

Subprotocol 3
Phosphatases - CIP Reaction

Two enzymes, bacterial alkaline phosphatase (BAP) and calf intestine alkaline phosphatase (CIP) are used in molecular biology. The enzymes catalyze the removal of 5' phosphate residues from DNA, RNA, rNTPs, and dNTPs. The bacterial enzyme BAP is more active but is also far more resistant to inactivation procedures at the end of dephosphorylation reactions.

Phosphatases are used to dephosphorylate vector DNA after restriction enzyme digestion to prevent self ligation. It is essential, however, to completely remove the phosphatases so that subsequent ligations work efficiently. Another use is to remove 5' phosphates from DNA or RNA prior to labeling 5' termini with ^{32}P.

Selection of phosphatase is mainly dependent on individual preference. Both enzymes CIP and BAP work well. We prefer calf intestinal phosphatase since it is easier and less time consuming to remove after the reaction is complete.

Materials

Supplies

- Heating block (37°C)
- Microfuge
- Spin columns

Reagents

- DNA to be dephosphorylated (0.1-0.2 µg/µl in water, TE, or any restriction enzyme buffer containing 50 mM Tris-HCl)
- 10x CIP buffer (1 mM $ZnCl_2$, 10 mM $MgCl_2$, 500 mM Tris-HCl pH 8.0)
- CIP enzyme (1 U/µl; if purchased at higher concentration, dilute an aliquot in storage buffer to 1 U/µl, store at +4°C)
- EDTA 0.5 M pH 8.0
- Chaotropic buffer (e. g. Qiaex™)

Procedure

CIP reaction

The following gives an example for the dephosphorylation of 1 µg vector DNA, i. e. approximately 1 pmol DNA. Use at least 1 U CIP per pmol DNA. If more than 10 µg DNA is to be dephosphorylated, the reaction should be scaled up to 100 µl.

1. On ice, pipet the following into a 1.7 ml microcentrifuge tube (if using a heating block or waterbath) or 0.5 ml tube (if using a thermal cycler):

DNA	10 µl	1 µg
10x buffer	5 µl	
CIP	1 µl	1 U
Water	34 µl	

Note: Restriction enzyme digests may be used directly for CIP reactions if 50 mM Tris-HCl is present. In this case, add 1 U CIP per pmol DNA and proceed to step 2.

2. Incubate at 37°C for 30 min.
3. Add 1 µl CIP and incubate for another 30 min. at 37°C.
4. Purify the DNA by using a chaotropic buffer/silica-gel spin column system (e. g. Qiagen).
5. Elute the DNA in 30 µl water.

Subprotocol 4
Kinases

Kinases catalyze the transfer of γ-phosphate from ATP to the 5' terminus of DNA or RNA. A frequently used enzyme, T4 polynucleotide kinase, has originally been isolated from phage T4 infected E. coli. Today, it is available as a recombinant enzyme. There are two types of reactions in which kinases are used:

- In a forward reaction, the γ-phosphate is transferred to dephosphorylated DNA. If the γ-phosphate is radioactive, the DNA is thereby endlabeled. In this way, the reaction is used to label oligonucleotides for sequencing reactions.

- In an exchange reaction, the substrate is phosphorylated DNA. An excess of ADP, however, first leads to dephosphorylation of the DNA which is then rephosphorylated by transfer of ^{32}P γ-phosphate.

T4 polynucleotide kinase (forward reaction)

The following gives a standard protocol for kinase mediated end-labeling of a DNA oligonucleotide primer. The amount of 50 pmole labeled primer would be sufficient to perform e.g. 50 direct, PCR mediated sequencing reactions (4 reactions [A, G, C, T] each).

Materials

Supplies

- Heating block (37°C)
- Alternatively: Thermal cycler

Reagents

- Dephosphorylated primer (1-50 pmol in approx. 1-20 µl water); the solution should be free from unincorporated nucleotides and low molecular weight DNA which may decrease the efficiency of the T4 reaction; in addition, care should be taken that ammonium ions which strongly inhibit T4 polynucleotide kinase are absent from the primer solution.

Primer length	ng of primer/pmol
17	5.6
19	6.3
21	6.9
23	7.6
25	8.2
27	8.9
29	9.6

- 10x buffer (500 mM Tris-HCl pH 7.6, 100 mM $MgCl_2$, 50 mM DTT, 1 mM spermidine)
- γ-^{32}P ATP (>3000 Ci/mmol, 10µCi/µl, 3.3 µM)

- T4 polynucleotide kinase (usually 10 U/μl)
- Ultrapure water
- EDTA 0.5 M pH8.0

Procedure

1. On ice, pipet the following into a 1.7 ml or 0.5 ml reaction tube:

T4 kinase

Primer	20 μl	50 pmol
10x buffer	5 μl	
γ - ^{32}P ATP	15 μl	150 μCi, 50 pmol
T4 kinase	2 μl	20 U
Water	8 μl	

Note: The reaction can be scaled down if less primer is needed. Labeling of 1 pmol primer can be performed in as little as 5 μl reaction volume. The primer should then be added as 1 pmol/2 μl solution. Use only 1 μl 10x buffer, 1.5 μl γ-^{32}P ATP, 10 U T4 polynucleotide kinase.

2. Incubate at 37°C for 30 min.

Note: If small volumes are used, the reaction may directly be used for subsequent sequencing reactions without further purification. In this case, skip the following steps.

3. Add 2 μl EDTA solution to stop the reaction.
4. Purify the DNA by using a chaotropic buffer/silica-gel spin column system (e. g. Qiagen).
5. Elute the DNA in 30 μl water.

Note: Unincorporated nucleotides can also be eliminated by small scale gel filtration (e. g. QuickSpin™, Boehringer).

Subprotocol 5
DNA-Dependent DNA Polymerases – E. Coli DNA Polymerase I and Klenow Enzyme

Many protocols in molecular biology involve the in vitro synthesis of DNA or RNA. These reactions are catalyzed by DNA and RNA polymerases, respectively. DNA polymerases synthesize new DNA strands complementary to template strands which may be either DNA (DNA polymerase I, Klenow enzyme, T4/T7 polymerase, Taq polymerase) or RNA (reverse transcriptase).

In vivo, both prokaryotic and eukaryotic cells contain multiple DNA polymerase activities. Three DNA polymerase enzymes have been characterized in E. coli. However, only one enzyme, DNA polymerase III, performs the replicase function. The others play subsidiary roles in replication or participate in repair of damaged DNA sequences. All bacterial DNA polymerases share the same fundamental type of synthetic activity: DNA is extended in a 5'→3' direction by adding nucleotides to the end containing a free 3' OH. The fidelity of the replication is guaranteed by a postsynthetic proofreading function performed by a 3'→5' exonuclease activity. The error rate in bacteria appears to be 10^{-8} to 10^{-10}.

The most abundant DNA polymerase in *E. coli* is DNA polymerase I. Apart from polymerase and proofreading activity, the enzyme possesses a 5'→3' exonucleolytic activity. Therefore, the holoenzyme is able to start DNA replication at nicks, which is of great practical use for labeling DNA molecules in vitro (nick translation).

Proteolytic cleavage of DNA polymerase I with subtilisin results in two fragments, the larger of which is called Klenow fragment. The Klenow fragment carries polymerase and proofreading activity but lacks the 5'→3' exonuclease activity. Therefore, with the use of Klenow enzyme there is no risk of template degradation. Klenow enzyme has several applications in molecular biology. One is to generate labeled DNA probes in a random hexamer primed reaction. The random primers bind to denatured, single stranded DNA and initiate the DNA synthesis. Radiolabeled nucleotides are incorporated into the newly synthesized strand (random primed labeling). Another application is the filling in of recessed 3' termini created by digestion of DNA with certain restriction enzymes (for a protocol, see subprotocol 1). Filling of 3' ends generates blunt ends which can be ligated or to which synthesis linkers can be added.

Materials

Supplies

- water bath 75°C
- Alternatively: Thermal cycler, programmed to 22°C/30 min, 75°C/10 min, maximum temperature slope

Reagents and solutions

- DNA (0.1-5 µg in 20 µl) which has been digested with restriction enzymes that create recessed 3' ends. Klenow enzyme works well in virtually all restriction enzyme buffers. Therefore, the filling in reaction can be performed after DNA digestion in the same reaction mix.
- 10x buffer (if necessary) is 0.5 M Tris-HCl pH7.6, 0.1 M $MgCl_2$
- dNTP solution: 1 mM dATP, 1 mM dCTP, 1 mM dGTP, 1 mM dTTP in water. This solution should be stored in small aliquots at -20°C.
- Klenow enzyme (various suppliers; usually 1 U/µl)
- TE buffer (10 mM Tris-HCl pH8.0, 1 mM EDTA)
- Chaotropic buffer, silica-gel spin columns

Procedure

Filling in recessed 3' ends with Klenow enzyme

Klenow enzyme

1. After completion of restriction enzyme digest, add the following to the reaction:

DNA (RE digest)	20 µl	0.1-5 µg
dNTP	1 µl	1 nmol each
Klenow enzyme	1 µl	1 U

2. Incubate at room temperature for 30 min.
3. Inactivate the enzyme by heating to 75°C for 10 min.
4. Increase volume to 100 µl with TE.
5. Purify the DNA by using a chaotropic buffer/silica-gel spin column system according to the manufacturer's instructions.

Subprotocol 6
T4 and T7 DNA Polymerase

Bacteriophage T4 and T7 encoded DNA polymerase are similar to the Klenow fragment of E. coli polymerase in that they possess 5'→3' polymerase activity and 3'→5' exonuclease (proofreading) activity. However, the T4 and T7 enzymes are 200 fold and 1000 fold, respectively, more active (faster) than the Klenow fragment. One of the uses of T4 polymerase is the conversion of double stranded DNA with protruding 3'-OH ends to blunt ended molecules (see below). T4 polymerase can also convert protruding 5'-OH ends to blunt ended molecules by using 5'→3' polymerase activity.

The primary use of T7 DNA polymerase is in primer extension reactions that require the copying of long stretches of template, such as sequencing reactions. For this purpose, a modified T7 DNA polymerase (Sequenase™) lacking the 3'→5' exonuclease activity is commercially available.

Removing protruding 3' termini

The 3'→5' exonuclease activity of T4 DNA polymerase removes protruding 3' termini until it reaches the second strand. Further exonucleolytic activity is then balanced by incorporation (5'→3' polymerase activity) of dNTPs which have to be added to the reaction. T4 DNA polymerase is preferred over Klenow enzyme due to its 200-fold higher activity.

Materials

Supplies

- Cooling water bath 12°C, water bath 75°C
- Alternatively: Thermal cycler, programmed to 12°C/30 min, 75°C/10 min, maximum temperature slope

Reagents and solutions

- DNA (0.1-5 µg in 20 µl) which has been digested with restriction enzymes that create protruding 3' ends. T4 polymerase has about 50% of its maximum activity in all commonly used restriction enzyme buffers. Therefore, the reaction can be performed after DNA digestion in the same reaction medium. Alternatively, restriction enzyme digests may be purified with chaotropic buffer/silica-gel spin columns before the reaction.

- 10x buffer (if necessary) is 0.33 M Tris acetate pH 8.0, 0.66 M potassium acetate, 0.1 M magnesium acetate, 5 mM DTT, 1 mg/ml bovine serum albumin. This buffer should be stored in small aliquots at –20°C.
- dNTP solution: 1 mM dATP, 1 mM dCTP, 1 mM dGTP, 1 mM dTTP in water. This solution should be stored in small aliquots at –20°C.
- T4 polymerase (various suppliers; usually 1 U/µl)
- TE buffer
- Chaotropic buffer, silica-gel spin columns

Procedure

Protruding 3'termini

1. After completion of restriction enzyme digest, combine the following:

DNA (restriction digest)	20 µl	0.1-5 µg
dNTP	2 µl	2 nmoles each dNTP
T4 DNA polymerase	2 µl	2 U

2. Incubate at 12°C for 30 min.
3. Inactivate the enzyme by heating to 75°C for 10 min.
4. Add TE buffer to 100 µl total volume.
5. Purify the DNA by using a chaotropic buffer/silica-gel spin column system according to the manufacturer's instructions.

Taq DNA polymerase

In the late 1970s, the first thermostable DNA polymerase was isolated from *Thermus aquaticus*, a bacterium dwelling in hot springs. The enzymes lacks 3'→5' proofreading activity. This results in a rather high error rate of 10^{-4}. Nevertheless, Taq DNA polymerase found widespread use in the polymerase chain reaction due to its. Another use is in direct sequencing of DNA. The enzyme is a 65 kD DNA-dependent DNA polymerase that works best at 70°C to 80°C. It requires Mg^{+} in mM concentrations. Several thermostable polymerases from other microorganisms, as well as recombinant enzymes with modifications for increased stability and decreased error rates are now commercially available for a number of specific PCR applications.

Subprotocol 7
RNA-Dependent DNA Polymerases

Reverse transcriptase

Enzymes which preferably copy RNA templates into DNA are called reverse transcriptases (RT). These RNA-dependent DNA polymerases were discovered in retroviruses independantly by Temin and Baltimore in 1970. The reverse transcriptases currently used in molecular biology are purified from avian myeloblastosis virus (AMV) or Moloney murine leukemia virus (MMLV). The latter is available as a recombinant enzyme. Both enzymes have 5'→3' polymerase activity and RNaseH activity. RNaseH activity is a bidirectional exoribonuclease activity on RNA:DNA hybrids. However, both enzymes lack 3'→5' proofreading activity which explains their considerable error rate of 10^{-4} in vitro. Reverse transcriptase is used to transcribe mRNA into single-stranded or double stranded cDNA (complementary DNA). Oligo (dT), random primers, or specific oligo nucleotides can serve to prime the reaction. While first strand cDNA synthesis is performed, the RNA template is removed by the RNaseH activity of RT. Alternatively, when using genetically engineered reverse transcriptase lacking RNaseH activity, the RNA template is digested after reverse transcription in a second reaction with RNaseH. Second oligonucleotides or self priming initiate the second strand synthesis which is necessary when a cDNA library is to be generated. Second strand synthesis is not necessary if subsequent PCR is planned.

Protocol for the generation of first strand cDNA by using MMLV reverse transcriptase

The quality of the RNA template is the limiting factor in cDNA synthesis. Therefore, great care should be taken to avoid RNA degradation during preparation. Our experience is that for first-strand synthesis with subsequent cDNA amplification, total cellular extracts of RNA are sufficient (e. g. using RNAzol™). Successful reactions have been performed with RNA from as little as 10^4 cells. With nested PCR, even fewer cell numbers may suffice. A useful amount of total cellular RNA to start with is 5 µg.

Materials

Supplies

- Thermal cycler programmed to 37°C/60 min, 65°C/5 min, 4°C/hold for reverse transcription. For the subsequent PCR amplification, choose optimal cycling conditions according to the principles in chapter 12.

Reagents and solutions

- total cellular RNA (0.5-1 μg/μl) in DEPC-treated water. It is best to check the integrity of the RNA preparation in a formaldehyde agarose gel.
- Primer: There are three options to prime first strand synthesis.
  - $(dT)_{12-18}$ (0.5 μg/μl in water) which hybridizes to the 3' poly(A) tail of mRNA molecules
  - Random hexamers, which hybridize to all RNA molecules present and are used when a particular mRNA molecule is difficult to copy in its entirety, 50 ng/μl in water
  - Specific antisense primers which hybridize only to the target mRNA (10 pmol/μl in water; for 21-mer, this is approximately 70 ng/μl)
- 10x buffer, 200 mM Tris-HCl, pH8.4, 500 mM KCl, 25 mM $MgCl_2$, 1 mg/ml bovine serum albumine
- DTT, 0.1 M
- dNTP mix, 10 mM each dATP, dCTP, dGTP, dTTP in water
- RNase inhibitor (RNasin®, Promega)
- MMLV reverse transcriptase, 200 U/μl. Use approximately 200 U/μg RNA
- PCR primers 1 (forward) and 2 (reverse), each at 20 pmol/μl
- Taq DNA polymerase
- DEPC-treated water

Procedure

MMLV reverse transcriptase

1. On ice, pipet the following into a 0.5 ml microcentriguge tube:

Total RNA	12 μl	1μg in DEPC treated water
Primer	1 μl	In the above concentration

2. Heat to 70°C for 10 min.

3. Place on ice for 2 min. Centrifuge at 12000 rpm in a microcentrifuge for 30 sec.
4. On ice, add the following to the reaction mixture:

10x buffer	2 µl	
dNTP mix	1 µl	10 nmol each dNTP
DTT	2 µl	0.2 µmol
RNasin™	1 µl	
MMLV-RT	1 µl	200 U

5. Mix gently. Collect the reaction mix by brief centrifugation.
6. Incubate at 37°C for 1 hour.
7. Increase the temperature to 65°C for 5 min. Place on ice for 2 min. Collect the reaction by brief centrifugation.
8. Remove 5 µl of the first strand synthesis reaction to a new 0.5 ml tube on ice. Store the remainder at −20°C.

Note: The newly synthesized cDNA may be directly amplified by PCR using specific amplification primers. If maximum sensitivity is required, use 20 µl of the first strand reaction as template for a 100 µl PCR amplification.

9. Add

10x buffer	4.5 µl	The same buffer as above may be used.
PCR primer 1	1 µl	20 pmol
PCR primer 2	1 µl	20 pmol
dNTP mix	6 µl	10 nmol each dNTP
Taq polymerase	0.5 µl	2.5 U
Water	32 µl	

10. Overlay the reaction with 50 µl mineral oil.
11. Perform 20 to 40 rounds of PCR with the appropriate temperature/time profile.
12. Analyze the amplified cDNA product on a 1.5-2.0% agarose gel.

Subprotocol 8
DNA – Dependent RNA Polymerases

In addition to the DNA polymerases mentioned above, several RNA polymerases are available. They are encoded by bacteriophage (SP6, T7, and T3) and synthesize RNA on double stranded DNA templates carrying the appropriate bacteriophage-specific promotor. The enzymes are used in vitro to generate large quantities of RNA complementary to one strand of foreign DNA which has been cloned immediately downstream from the promotor in appropriate plasmids. The RNA can be used for in vitro translation experiments, or, if a radioactive label is incorporated as probe in Northern blots or in situ hybridizations.

Protocol for in vitro transcription (T7, SP6 RNA polymerase)

RNA polymerases recognize the respective bacteriophage promotor sequence with high specificity. Plasmid vectors (e.g. the pGEM series, Promega) containing a multiple cloning site (MCS) with adjacent T7 or SP6 promotor sequences can be used for the transcription from either strand of a double-stranded DNA sequence cloned into the MCS. The protocol allows radiolabeling of RNA to high specific activity for in situ hybridization.

Materials

Supplies

- Heating block (37°C)

Reagents and solutions

- Plasmid vector containing a T7 or SP6 promotor upstream (i. e. 5') of the DNA strand to be transcribed into RNA. The template must be linearized by digestion with a restriction enzyme cutting downstream (i. e. 3') of the transcribed strand. Avoid restriction enzymes that create protruding 3' termini. Protruding 3' ends may aberrantly prime transcription of long RNA molecules. If necessary, refer to the protocol for removal of protruding 3' ends, subprotocol 2.6. Plasmid concentration is 0.1 pmol/µl, i.e. 200 ng/µl for a 3 kb plasmid.
- 10x buffer: 400 mM Tris-HCl pH 7.5, 60 mM $MgCl_2$, 20 mM spermidine HCl, 50 mM NaCl. The buffer should be stored in small aliquots at –20°C.
- rNTP mix: 5 mM each rATP, rCTP, rGTP
- [^{35}S]rUTP

- T7 RNA polymerase
- Rnasin™ 20 U/μl
- DTT, 0.2 M
- DNase I, RNase-free, 1mg/ml
- tRNA (yeast, DNase-free, RNase-free; e.g. Sigma) 1 μg/μl, which works as carrier in the precipitation reaction
- Ammonium acetate 5 M
- Ethanol 100%, 70%

Procedure

In advance

1. Linearize the DNA template with the appropriate restriction enzyme.
2. Purify the DNA by using a chaotropic buffer/silica-gel spin column system (e. g. Qiagen).
3. Check an aliquot (0.1-0.2 μg) of the DNA digest by agarose gel electrophoresis.
4. Dilute the DNA to 0.1 pmol/μl (i.e. 200 ng/μl for a 3 kb plasmid) in DEPC treated water.

In vitro transcription

5. At room temperature, mix the following components in a reaction tube:

DNA template	2 μl	0.2 pmol
10x buffer	2 μl	
rNTP mix (w/o rUTP)	2 μl	
DTT	1 μl	0.2 μmol
Rnasin™	1 μl	
DEPC water		10 μl
T7 RNA polymerase	1 μl	
[^{35}S]rUTP	1 μl	12.5 μCi

Note: If RNA is to be synthesized for in vitro translation or as control in cDNA reactions, no radiolabeling is necessary. Replace the [^{35}S]rUTP with nonradioactive rUTP in the rNTP mix (5 mM). If working with radioactivity, follow the safety regulations of your laboratory and discard pipet tips and contaminated solutions properly.

6. Incubate at 37°C for 1 hour.
7. Add 1 µl RNasin™ and 1 µl DNase I. Mix gently.
8. Incubate at 37°C for 10 min.
9. Purify the mixture by using a guanidine isothiocyanate/silica-gel spin column system (e. g. RNeasy™, Qiagen).
10. Elute the RNA in 20-100 µl DEPC treated water. Determine the counts per minute in 1 µl if the RNA is radiolabeled. Check 1 µl in a denaturing polyacrylamide or agarose gel (section 6).

Note: For some applications, it may be sufficient to remove unincorporated nucleotides and small molecules by small scale gel filtration (e. g. Quick-Spin®, Boehringer). If the RNA is not immediately used, it should be stored under ethanol at -70°C. Add 2 volumes of ethanol. To recover the RNA, add 0.1 volume ammonium acetate 5 M, mix, store at -20°C for 30 min, and centrifuge in a microfuge for 10 min. at 12000 rpm at 4°C.

Subprotocol 9
Terminal Deoxynucleotidyl Transferase (TdT)

TdT is a DNA polymerase which does not copy a template. It adds nucleotides preferentially to the 3' termini of existing DNA molecules. In vivo, TdT is found only in prelymphocytes and during early stages of lymphoid differentiation. In vitro, the enzyme is used to add homopolymeric tails to vector and cDNA, and to 3' end-label DNA.

Procedure

For a protocol, see manufactures recommendations.

References

McClarin JA, Frederick BC, Wang P, Greene H, Bryer W, Grable J, Rosenberg JM (1986). Structure of the DNA-Eco RI endonuclease recognition complex at 3 A resolution. Science 234: 1526-1541.

Purification of Nucleic Acids From Eukaryotic Cells

HANS GERD NOTHWANG AND FRIEDHELM HILDEBRANDT

Introduction

Essential to most molecular biology techniques is the analysis of DNA and RNA. In general, it is difficult to prepare both RNA and high-molecular-weight DNA from the same extract, because optimum conditions are mutually exclusive.

Ultimately, recognition of the optimum conditions for a particular purification procedure is dependent on the amount of starting material, the nature of the nucleic acids to be isolated and on their subsequent use. This chapter starts with the description of the most common purification method for nucleic acids, the extraction with phenol/chloroform. Subsequently, modifications to improve yield, purity and concentration are discussed. The second part covers methods for isolation of genomic DNA and RNA from complex mixtures of molecules such as cell lysates.

Subprotocol 1
Extraction of Nucleic Acids

Principles of phenol extraction. The standard way to prepare nucleic acids for use in enzymatic reactions is to extract first the DNA or RNA sample with phenol:chloroform:isoamyl alcohol (PCA) (25:24:1). Phenol leads to denaturation of proteins, which will subsequently accumulate at the interphase, whereas the hydrophilic nucleic acids will remain in the aqueous (phenol containing) phase. The mixture of different organic solvents makes deproteinization more efficient. Furthermore, phenol alone does not com-

Hans Gerd Nothwang, *Correspondence to* Friedhelm Hildebrandt, Universitäts-Kinderklinik, Mathildenstrasse 1, Freiburg, 79106, Germany (*phone* +49-761-270-4301; *fax* +49-761-270-4533; *e-mail* hildebra@kkl200.ukl.uni-freiburg.de)

pletely inhibit RNAse activity, and it is a solvent for RNA molecules that contain long poly(A) tails (Sambrock et al. 1989; Brawerman et al. 1972). Chloroform reduces the amount of aqueous solution retained in the organic phase and its high density enhances phase separation. Isoamyl alcohol inhibits foam formation at the interphase. The phenol/chloroform extraction is followed by one extraction with chloroform alone to remove any traces of phenol from the nucleic acids, which might otherwise inhibit subsequent

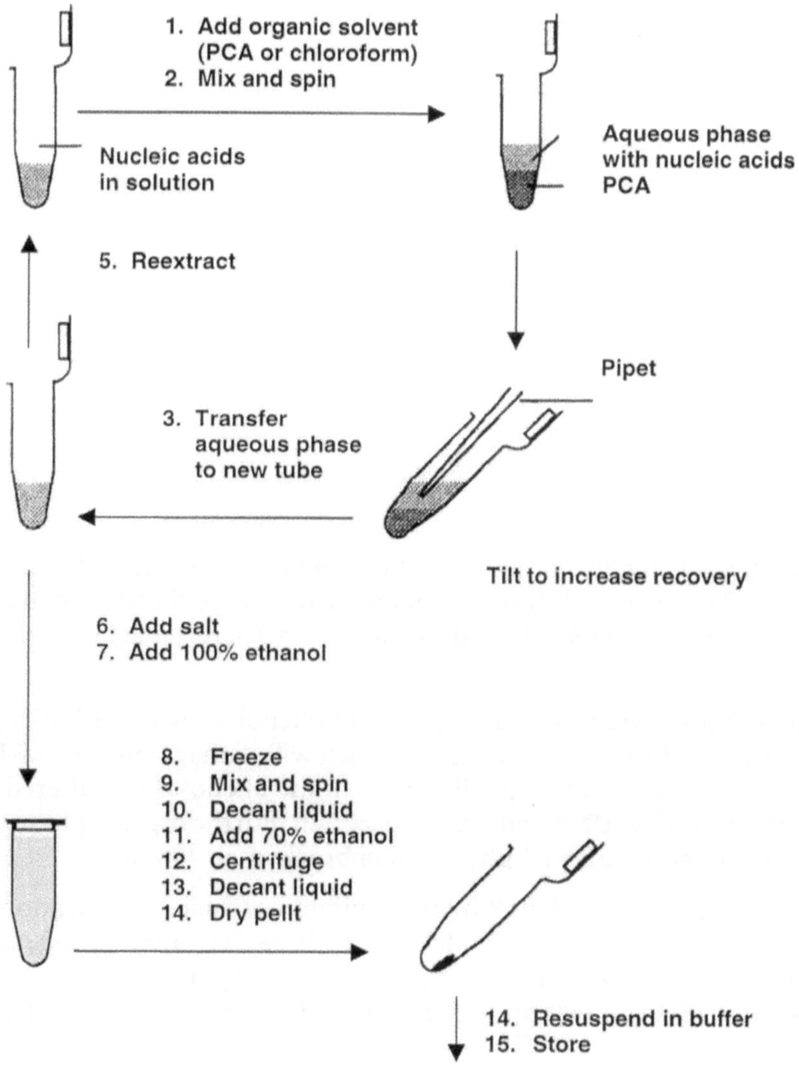

Fig. 1. Extraction of nucleic acids

enzymatic reactions. The nucleic acid is finally recovered from the aqueous phase by precipitation (Figure 1).

Materials

Reagents

- salt-saturated phenol (SS phenol)
- chloroform
- isoamyl alcohol
- 70% and 100% ethanol, ice-cold
- 3M sodium acetate (pH 5.2)
- TE-buffer (pH 7.6) (10 mM Tris-HCl (pH7.6), 1 mM EDTA (pH8.0))

Supplies

- 1.5 ml sterilized polypropylene tubes
- microcentrifuge
- chemical hood
- spectrophotometer

Procedure

In advance

Note: Phenol is highly corrosive and can cause severe burns. Wear gloves, gogges and protective clothing. All manipulations should be carried out in a chemical hood. Dispose of used phenol in a glass receptacle, do not pour down the drain.

1a) Before use, check the color and pH of phenol. Pink color indicates the presence of oxidation products, which will damage nucleic acids. The phenol must therefore be discarded. If the phenol is unbuffered or the pH is still <7.8, DNA will tend to partition into the organic phase. In this case, prepare buffered phenol (Sambrock et al. 1989).

1b) Add 0.5 g of antioxidant 8-hydroxyquinoline (final concentration 0.1%) to a 2-liter glass beaker containing a stir bar. Gently pour 500ml of liquified phenol or melted crystals of redistilled phenol (melt at 65°C in a water bath) and 500 ml of 0.5 M Tris-HCl (pH 8.0) into the beaker.

1c) Stir the mixture on a magnetic stirrer for 15 min and then turn off the stirrer. Allow phases to separate and decant as much as possible of the top (aqueous) phase into waste receptacle, removing remainder with a pipet and suction bulb.

1d) Add an equal volume of 0.1 M Tris-HCl (pH 8.0) to the phenol. Proceed as described in 1c). Repeat the extraction, until phenol phase reaches pH 7.8 (as checked with pH paper).

1e) After the phenol is equilibrated and the final aqueous phase has been removed, add 0.1 vol of 0.1M Tris-HCl (pH 8.0) containing 0.2% ß-mercaptoethanol. The phenol solution may be stored up to 2 months at 4°C in light-tight bottles. Phenol for high-molecular-weight RNA extractions should be made freshly melted and the excess should be discarded.

1f) Mix 50 ml phenol, 48 ml chloroform and 2 ml isoamyl alcohol. This mixture may be stored under 100 mM Tris-HCl (pH 8.0) in a light-tight bottle at 4 °C for periods of up to 1 month.

Note: If RNA is to be extracted, treat all water used directly or for buffer preparation with diethyl pyrocarbonate (DEPC) to inactivate RNAses (see Subprotocol 3, Precautions)

Phenol extraction

2. Add an equal volume of phenol:chloroform:isoamyl alcohol to the nucleic acid sample (prefentially 200 µl) in a polypropylene microcentrifuge tube (only these are resistant to the organic solvents).
3. Mix the contents of the tube until an emulsion forms. When isolating small DNA molecules (<10 kb), this can be done by vortexing, or by gentle shaking for DNA of moderate size (10-30 kb). When isolating large DNA molecules, the two phases should be mixed by rotating the tube slowly (20 rpm) on a wheel.
4. Spin at 12000 rpm, room temperature (RT), until the two phases are separated (15 sec-2 min).
5. Transfer top (aqueous) phase to a fresh tube. Avoid carry-over of the interphase.

Note: Normally, the aqueous phase forms the upper phase. However, if the aqueous solution is dense because of salt (>0.5 M) or sucrose (>10%), it will form the lower phase. The organic phase is easily identifiable because of its yellow colour.

6. Repeat steps 2-5, as long as white flocculent interface is present.
7. Add an equal volume of chloroform and repeat steps 3-5.
8. Add 0.1 vol of 3 M sodium acetate (pH 5.2). Mix by vortexing briefly or by flicking the tube several times. Add 2.5 vol of ice-cold 100% ethanol.
9. Keep at 4 °C for 10 min.
10. Spin at 15000 rpm for 10 min.
11. Remove the ethanol carefully.
12. Add 1 ml 70% ethanol to remove precipitated salt, vortex briefly.
13. Repeat steps 10 and 11.
14. Dry pellet in a desiccator under vacuum or in a Speedvac® evaporator.

Note: Genomic DNA and large DNA fragments should not be dried completely because rehydration may take days. Air dry genomic DNA under sight and dissolve pellet when still moist and shiny.

15. Resuspend pellet in TE-buffer at <1mg/ml and store at –20 °C or at 4 °C, when used frequently. RNA is stored at –70 °C under 70-80% ethanol.

Note: High-molecular-weight genomic DNA may require one to several days to dissolve and should be shaken gently to avoid shearing.

16. Analyse the quantity and quality of the nucleic acid content by spectrophotometric determination of an aliquot at wave-lengths of 260 nm and 280 nm.

Note: The ratio between the readings at 260 nm and 280 nm provides an estimate of the purity of the nucleic acids and should be between 1.8 and 2.0. Contamination by phenol or protein will result in ratios significantly less than 1.8. An OD of 1 at 260 nm corresponds to approximately 50 µg/ml for double-stranded DNA, 40 µg/ml for single-stranded DNA and RNA and ~20 µg/ml for single-stranded oligonucleotides.

Comments

Optimizing extraction conditions

- Technical considerations.
  Of fundamental importance for successful phenol extraction of intact nucleic acids is the performance of all steps in one session (especially

when dealing with RNA) and as quickly as possible. Extractions are best carried out on ice with refrigerated buffers to minimize nuclease activity. Only when SDS (see below) is present would a higher temperature (10 °C) be required to prevent precipitation of the detergent.

- Purity
  Detergents are employed to promote the dissociation of proteins from nucleic acids. The most commonly used are the ionic detergents sodium dodecylsulfate (SDS) (0.1 to 2%) or sodium desoxycholate, which in addition enhances the dissolution of lipid membrane structures (Wallace 1987). The dissociation can further be improved by incubation of the sample with proteinase K (nuclease free) (50-200 μg/ml, final concentration; 37 °C for 30 min) in the presence of SDS. To increase RNA over DNA content, solutions can be made to pH 5-6. DNA is then selectively retained in the organic phase, leaving RNA in the aqueous phase. The low pH also reduces the activity of many nucleases.

- Yield
  For maximum recovery, re-extract all organic phases by adding 100 μl of TE-buffer to the first organic phase and proceed with this aqueous phase through all subsequent organic phases. Then combine this second aqueous phase with the first and concentrate by preciptation.
  After 1-15 min at 0 °C in the absence of carrier, nucleic acid concentrations as low as 20 ng/ml will form a precipitate that can be quantitatively recovered by centrifugation in a microcentrifuge (Sambrock et al. 1989). To recover nanogram quantities of nucleic acids, add 10 μg of 10 μg/μl carrier tRNA prior to extraction. If this is not possible due to interference with subsequent experiments, add 0.1 vol of 1 M spermine tetrahydrochloride (Hoopes and McClure 1981) or apply more rigorous ultracentrifugation like in the technique of Shapiro (1981). To short DNA fragments (2 to 10 mM final concentration), increase ice-cold ethanol to 3 vol, and precipitate at −70 °C for 30-60 min. Large volumes or nucleic acid samples of unknown concentration should also be precipitated for 30-60 min at −70 °C to permit temperature equilibration to obtain a maximum precipitate.

- Precipitation
  The choice of salt used for precipitation is determined both by the nature of the sample and by the intended use of the nucleic acid. Substitution of sodium acetate (pH 5.2) by ammonium acetate (2.0-2.5 M, final concentration) prevents the precipitation of dNTPs and therefore is useful for removing unreacted triphosphates from the products of reverse tran-

scriptase, DNA polymerase, or terminal transferase-catalyzed reactions. Two sequential precipitations result in the removal of 99% of the dNTPs and greater than 90% DNA recovery (Okayama and Berg 1982). However, ammonium acetate should not be used, when the precipitated nucleic acid is to be phosphorylated or tailed, since ammonium ions inhibit the enzymes required for these processes.

LiCl (0.8 M, final concentration) is frequently used for precipitation at −70°C, since LiCl is very soluble under this condition. Li^+ should be avoided when the precipitated RNA is to be reverse transcribed. No chloride should be used when translation in cell-free systems is intended, since Cl^- interferes with the initiation of protein synthesis. Should the sample contain SDS, sodium chloride (0.2 M) should be used. The detergent then remains soluble in 70% ethanol. Samples containing phosphate or 10 mM EDTA should be dialysed prior to precipitation, as these materials will precipitate along with the nucleic acid.

- Isopropanol is useful for precipitation of nucleic acids, if smaller volume of alcohol are needed, e.g., when precipitating DNA or RNA from volumes >0.5 ml in a microcentrifuge tube. Substitute 1 vol isopropanol for 2.5 vol 100% ethanol. But salts are less soluble in isopropanol than in ethanol.

Subprotocol 2
Isolation of Genomic DNA From Mammalian Cells

Principles of DNA isolation. Genomic DNA is mainly used for restriction enzyme digestion and mutation analysis, Southern blots, PCR and library construction. Isolation from tissue culture cells is particularly convenient, but preparation from blood or whole organs is also possible. The major difficulties encountered during preparation are the high viscosity of genomic DNA, which makes it difficult to redissolve after alcohol precipitation and its high-molecular weight (average human chromosome: 100-150 Mb) which makes it sensitive to shear forces (Birnboim 1992). Care should also be taken to avoid any contamination by recombinant DNA. A variety of commerically available purification kits exists (e.g. Qiagen®). But there are also rather easy laboratory protocols. The one provided here, allows for preparation of genomic DNA from different sources, i.e. cultured cells, tissue samples or blood samples. (Recently, a single-tube genomic DNA isolation method from blood was reported by Gruber and Zingales, 1995.)

The DNA is isolated by lysing cells in the presence of high concentration of EDTA (chelates divalent cations which are essential for the action of most

DNAses and endonucleases), SDS (to solubilize membranes) and proteinase K (a protease which destroys most contaminating DNAses and other proteins). This is followed by an extraction with phenol (Sambrock et al. 1989).

Materials

Reagents

- mammalian cells
- 25:24:1 phenol:chloroform:isoamyl alcohol (with Tris-buffered phenol, see Subprotocol 1, Materials)
- citric acid
- 0.5 M EDTA (pH 8.0)
- glucose
- phosphate-buffered saline (PBS) or Tris-buffered saline (TBS)
- Proteinase K (20 mg/ml)
- 10mg/ml RNAse (DNAse free)
- 10% SDS
- sodium citrate
- 5 M sodium chloride
- TE-buffer (pH8.0) (10 mM Tris-HCl (pH 8.0), 1 mM EDTA (pH8.0))
- 1 M Tris-HCl (pH 8.0)
- liquid nitrogen
- sterilized H_2O

Supplies

- microcentrifuge tubes
- polypropylene tubes (50 ml)
- Erlenmeyer flask
- microcentrifuge
- cell culture centrifuge
- water bath
- mortar and pestle
- chemical hood

Procedure

In advance

1. In advance prepare the following solutions on ice:

Extraction buffer:

Component	Stock Concentration	Final Concentration	Amount
a. Tris-HCl (pH 8.0)	1 M	10 mM	0.5 ml
b. EDTA (pH 8.0)	0.5 M	100 mM	10 ml
c. RNAse A[a]	10 mg/ml	20 µg/ml	0.1 ml
d. SDS	10 %	0.5 %	2.5 ml
e. H_2O			36.9 ml

[a] DNAse free. To prepare RNAse A free of DNAse, dissolve RNAse A in TE-buffer at 10mg/ml, and heat between 75–100 °C for 15 min. Allow to cool down slowly. Store aliquots at -20 °C.

If using blood samples, prepare the following anticoagulant acid citrate dextrose solution (ACD):

Component	Amount
a. citric acid	0.48 g
b. sodium citrate	1.32 g
c. glucose	1.47 g
d. H_2O	to 100 ml

DNA isolation

2a) Cells growing in monolayers: Wash the monolayers twice with ice-cold TBS. Add 0.5 ml TBS and scrape cells. Transfer cell suspension to a microcentrifuge tube, stored on ice. Wash the Petri dish with 1ml of TBS and add the washing to the centrifuge tube. Spin at 1500 g for 10 min, 4°C. Preheat water bath to 37 °C. Wash the cells in 5-10 volumes of ice-cold TBS and repeat centrifugation. Resuspend cells in TE-buffer (pH 8.0) at a concentration of 5×10^7 cells/ml. Transfer the solution to an Erlenmeyer flask (for n ml of cell suspension, use a 50 × n ml flask). Add 10ml of extraction buffer for each milliliter of cell suspension. Break pellet to 2-3 pieces with a pipet. Incubate the solution for 1h at 37°C in a water bath, and then proceed to step 3.

2b) Cells growing in suspension: Recover the cells by centrifugation at 1500 g for 10 min, 4°C. Resuspend the cells in a volume of ice-cold TBS equal to the volume of the original culture. Recover the cells by centrifugation, repeat the washing and proceed as described above.

2c) Tissue samples: Excise and immediately mince tissue quickly and freeze in liquid nitrogen. Grind 200mg to 1g tissue with prechilled mortar and pestle, or crush with hammer to fine powder. Add little by little 10vol of extraction buffer in a beaker. Allow powder to spread over the surface of the extraction buffer, and then shake the beaker to submerge the material. When all of the material is in solution, transfer it to a 50ml centrifuge tube, incubate for 1h at 37°C, and then proceed to step 3.

2d) Blood samples: Collect approximetaly 20ml of fresh blood in tubes, containing 3.5 ml of ACD. The blood may be stored at 0°C for several days or at –70°C indefinitely before the DNA is prepared.

- Fresh blood (20 ml): Spin at 1300 g for 15 min and discard supernatant plasma. Carefully transfer the buffy coat (a broad band of white blood cells of heterogenous density) to a fresh tube and recentrifuge. Resuspend the buffy coat in 15 ml of extraction buffer. Incubate for 1h at 37°C, and proceed to step 3.
- Frozen blood (20 ml): Thaw in a water bath at room temperature, transfer 10 ml to a 50 ml centrifuge tube, and mix with 2 vol of PBS. Then spin the blood at 3500 g for 15 min at room temperature and discard the supernatant, which contains lysed red cells. Resuspend pellet in 15 ml of extraction buffer, incubate for 1h at 37°C, and proceed to step 3.

3. Add proteinase K to a final concentration of 100 μg/ml and incubate at 50°C for 3h or at 42°C overnight in a water bath. Pipette up and down the solution periodically with a wide-bore pipet 5 times.

Note: Only if sample is transparent, proceed to step 4. Otherwise, the granular cell fragments will trap most of the DNA during the subsequent extraction and centrifugation.

Note: Incomplete removal of all traces of proteinase K by subsequent phenol extraction can result in DNA that is refractory to digestion with resctriction endonucleases.

4. Cool the solution to room temperature, and add an equal volume of phenol:chloroform:isoamyl alcohol (phenol >pH 7.8) (see Subprotocol 1) and gently mix the two phases by slowly turning the tube end over end for 10 min or by rotating slowly (20 rpm) on a wheel.

5. Spin at 400 g for 15 min, RT.
6. Transfer the viscous aqueous phase with a wide-bore pipet (3 mm-diameter orifice) to a clean centrifuge tube and repeat steps 4-5 twice.
7. a) To isolate very-high-molecular-weight DNA (>200 kb), dialyze the aqueous phase at 4 °C four times against 4 l of a solution of 50 mM Tris-HCl (pH 8.0), 10 mM EDTA (pH 8.0) until the OD_{270} of the dialysate is less than 0.05. Allow room in the dialysis bag for the volume of the sample to increase 1.5 to 2-fold. Continue with step 15.

   b) To isolate smaller sized DNA add 1/12th of the volume of 5 M NaCl, mix by slowly turning the tube. Add 2.5 vol of ice-cold 100% ethanol.
8. Keep at 4 °C for 10 min.
9. Spin at 5000 g for 10 min.
10. Remove the ethanol carefully.
11. Add 50 ml 70% ethanol to remove precipitated salt, mix briefly.
12. Repeat step 9.
13. Store the pellet in an open tube at room temperature until the last visible traces of ethanol have evaporated.

Note: Do not allow the pellet to dry completely; otherwise, it will be very difficult to dissolve.

14. Add 1ml of TE-buffer (pH 8.0) for each 5×10^6 cells. Place the tube on a rocking platform and gently rock the solution until the DNA has completely dissolved. This usually takes 12-24 h.
15. Measure the optical density of the DNA at 260 nm and 280 nm.
16. Expect 200 µg from 5×10^7 cells or 20 ml of blood.

Subprotocol 3
Isolation of Cytoplasmic RNA

Principles of RNA isolation. The ability to isolate pure, intact RNA is essential in the study of gene expression, RT-PCR and in cDNA cloning. The difficulty in isolating RNA is that ribonucleases are ubiquitous, very stable and function under a variety of conditions unsuitable for most other enzymes. RNA is not shear-sensitive but it is very labile to alkali, so the pH must be controlled at the pH of maximal stability, particulary if RNA is to be heated. All protocols for RNA isolation use RNAse inhibitors or methods that disrupt cells and inactivate RNAses simultaneously. (A comparison of commercially available kits and different laboratory protocols is given in Liedke et al. 1994.) The RNA is than fractionated from the other cellular macromolecules. One important point to keep in mind is that inactivation of RNAses by chaotropic agents is reversible, and their removal without parallel elimination of RNAses will result in reactivation of the enzymes and subsequent degradation of the sample. First, a number of precautions that can be used to avoid problems with RNAses are listed. Subsequently, two common methods are presented. The first describes a rapid and quantitative preparation of cytoplasmic RNA from small numbers of cells. This results in RNA ready for Northern blot or RT-PCR. A second, more laborious protocol will provide for RNA in high yield, suitable for isolation of poly(A)$^+$ RNA, i.e. mRNA.

Precautions

- Sterile, disposable plastic ware is essentially free of RNAses and can be used after being autoclaved. General laboratory plastic ware should be rinsed with chloroform.
- Glassware and other items are filled with diethyl pyrocarbonate (DEPC) (0.1% in water), which is a strong, but not absolute inhibitor of RNAses. After 2 h at 37°C, the items are rinsed several times with sterile water and then autoclaved, if possible. These treatments remove traces of DEPC, which might otherwise modify purine residues in the RNA by carboxymethylation.

Note: DEPC is suspected to be carcinogen.

- All solutions (including 70% ethanol) should be prepared using RNAse free glassware and plastic ware, DEPC-treated, autoclaved water and chemicals reserved for work with RNA. Whenever possible treat all buffers with DEPC by adding 0.2 ml DEPC to 100 ml of the solution to be treated. Stir for 12 h at 37 °C. Autoclave the solution to inactivate the remaining DEPC.

Note: DEPC reacts rapidly with amines and cannot be used in solutions, containing buffers such as Tris. Reserve a fresh, unopened bottle of Tris for preparation of RNAse free solutions.

- Store and use all items to be used for RNA isolation at a designated place. Wear and change frequently disposable gloves throughout the purification. Proceed as fast as possible, in one session and keep solutions at 4°C.

- Several RNAse inhibitors are available, but they are expensive and are mostly used at the end of the purification process.

Rapid preparation of cytoplasmic RNA by urea/SDS/phenol/chloroform/extraction

This protocols makes use of cellular fractionation by differential centrifugation. Genomic DNA is removed by pelleting the cell nuclei. The cytoplasmatic lysate is then urea/SDS/phenol/chloroform extracted. The procedure is extremely rapid (30 min) and useful for processing of large numbers of samples (Gough 1988).

Materials

Reagents

- 5 x 10^6 cells
- phenol:chloroform:isoamyl alcohol (25:24:1)
- 1 M $MgCl_2$ (DEPC-treated)
- 5 M NaCl (DEPC-treated)
- NP-40
- 10% SDS
- 1 M tris-HCl (pH 7.5)
- urea
- 70% and 100% ethanol, ice-cold
- RNAse-free TE-buffer (pH 7.0) (10 mM Tris-HCl (pH 7.0), 1 mM EDTA (pH 8.0))
- DEPC-treated water

Supplies

- 1.5ml sterilized polypropylene tubes
- microcentrifuge at 4°C
- chemical hood

Procedure

1. In advance prepare the following solutions:

In advance

Lysis Buffer

Component	Stock Concentration	Final Concentration	Amount
a. Tris-HCl (pH 7.5)	1 M	10 mM	0.5 ml
b. NaCl (DEPC-traeted)	5 M	150 mM	1.5 ml
c. $MgCl_2$ (DEPC-treated)	1 M	1.5 mM	0.075 ml
d. NP-40	100%	0.65%	0.325 ml
e. H_2O (DEPC-treated)			46.925 ml

Extraction buffer

Component	Stock Concentration	Final Concentration	Amount
a. Tris-HCl (pH7.5)	1 M	10 mM	0.5 ml
b. NaCl (DEPC-treated)	5M	350 mM	3.5 ml
c. EDTA (DEPC-treated)	0.5 M	10 mM	1 ml
d. SDS	10%	1%	5 ml
e. urea	-	7M	21.02 g
f. H_2O (DEPC-treated)			to 50 ml

Note: For DEPC-treatment, see Subprotocol 3, Precautions. DEPC is suspected to be a carcinogen.

Preparation of cytoplasmic RNA

2. Transfer up to 5 x 10^6 cells directly from medium without washing in a microcentrifuge tube.
3. Spin for 5min at 1000 rpm (without brake).
4. Remove supernatant.
5. Resupend pellet in 200 µl ice-cold lysis buffer, vortex rigorously, until pellet is dissolved.

6. Spin for 5 min at 1000 rpm, 4°C, to pellet nuclei.
7. Transfer supernatant (cytoplasmatic lysate) to a fresh microcentrifuge tube containing 200 μl ice-cold extraction buffer and 400 μl phenol:-chloroform:isoamyl alcohol (25:24:1), vortex rigorously.

Note: Samples may be stored at this stage at –70°C for up to several weeks.

8. Spin for 2 min at 15000 rpm.
9. Transfer top (aqueous) phase to a new microcentrifuge tube.
10. Add 1 ml ice-cold 100% ethanol and chill for 20 min at –20°C.
11. Spin at 15000 rpm for 15 min.
12. Remove the ethanol carefully.
13. Add 1 ml 70% ethanol to remove precipitated salt, vortex briefly.
14. Repeat steps 11 and 12.
15. Dry pellet.
16. Resuspend RNA pellet in 200 μl TE-buffer.
17. Mix 10 μl of the RNA sample and 490 μl buffer and measure the absorbance of the solution at 260 nm.
18. If not using the remaining RNA sample immediately, add 500 μl ice-cold 100% ethanol and store at –70°C. RNA can be applied directly to formaldehyde-agarose gels or RT-PCR.

Subprotocol 4
Rapid Preparation of Total RNA by Acid Guanidinium Thiocyanate/ Phenol/Chloroform/Extraction

For principles and precautions see Subprotocol 3, Introduction.

Guanidinium thiocyanate is one of the most effective protein denaturants and facilitates the isolation of intact, functional RNA from a wide range of biological sources, even those with extraordinarily high levels of nuclease like pancreatic cells. Denaturation is enhanced by including ß-mercaptoethanol, a reductant, which breaks intramolecular protein disulfide bonds. This method, established by Chomczynski and Sacchi (1987), provides both high and pure yields of undegraded RNA and can easily be scaled up or down.

Materials

Reagents

- cells or tissue
- chloroform
- isoamyl alcohol
- phenol (water-saturated)
- ß-mercaptoethanol
- 0.5M EDTA (pH8.0)
- ethanol
- guanidinium thiocyanate
- 2 M sodium acetate (pH 4.0)
- 1 M sodium citrate (pH 7.0)
- 10% (w/v) sodium lauryl sarkosinate
- 70% and 100% ethanol, ice-cold
- RNAse free TE-buffer (pH 7.0) (10 mM Tris-HCl (pH 7.0), 1 mM EDTA (pH 8.0))
- DEPC-treated water

Supplies

- 4 to 15 ml disposable polypropylene tubes (e.g. Falcon, Cat.Nos. 2063, 2059)
- sterilized microcentrifuge tubes
- microcentrifuge at 4°C
- chemical hood
- spectrometer

Procedure

In advance

1. In advance prepare the following solution:
   - 49:1 (v/v) chloroform:isoamyl alcohol
   - solution D[a]:

Component	Stock Concentration	Final Concentration	Amount
a) guanidinium thiocyanate		4 M	47.2 g
b) sodium citrate (pH 7.0)	1 M	25 mM	1.25 ml
c) sodium lauryl sarkosinate	10%	0.5%	2.5 ml
d) H_2O (DEPC-treated)			to 50 ml
e) ß-mercaptoethanol[b]	14.3 M	0.1 M	0.35 ml

[a] Mix reagents a–d at 65°C. This stock solution can be stored at least 3 months at room temperature.
[b] Add just before use.

Preparation of total RNA

2a) For tissue, mince 100 mg freshly prepared tissue and homogenize with 1ml of solution D in a glass-Teflon homogenizer.

2b) For 10^7 cultered cells, either spin suspension cells and discard supernatant, or remove the culture medium from cells grown in monolayer cultures. Add 1ml solution D and pass lysate through a pipet 10 times.

Note: Carry out all steps at room temperature unless otherwise stated.

3. Transfer cells into a 4 ml polypropylene tube. Add 0.1 ml of 2 M sodium acetate (pH 4.0) and mix. Add 1ml water-saturated phenol, mix, and add 0.2 ml of chloroform:isoamyl alcohol (49:1). Shake vigoursly and cool on ice for 15 min.
4. Spin 20 min at 10000 g (9000 rpm in a SS-34 rotor), 4°C.
5. Transfer upper aqueous phase to a fresh tube. Precipitate RNA with 2.5 ml (2.5 vol) ice-cold 100% ethanol, place >30 min at –20°.
6. Spin 10 min at 10000 g, 4°C.

Note: For isolation of RNA from tissues with a high glycogen content (e.g. liver), wash out glycogen from RNA pellet by vortexing in 4 M LiCl after the ethanol precipitation. Sediment insoluble RNA 10 min at 5000 g and follow reminder of protocol.

7. Dissolve RNA in 0.3 ml solution D and transfer to a 1.5 ml microcentrifuge tube. Precipitate RNA with 0.75 ml ice-cold 100% ethanol >30min at –20°C.
8. Spin 10 min at 10000 rpm, 4°C.

9. Carefully remove the ethanol.
10. Add 1 ml 70% ethanol to remove precipitated salt, vortex briefly, and incubate 10-15 min at room temperature.
11. Repeat steps 8 and 9.
12. Dry pellet in a desiccator under vacuum or in a Speedvac® evaporator for 5-15 min, and dissolve in 100-200 µl DEPC-treated water.
13. Quantitate the nucleic acid content by spectrophotometric determination of an aliquot at wave-lengths of 260 nm and 280 nm.

Subprotocol 5
Preparation of Poly(A)$^+$ RNA

Principles of poly(A)$^+$ RNA preparation. The protocols for preparing RNA from eucaryotic cells yields total RNA, which contains primarily tRNA and rRNA. Since the vast majority of mRNAs in mammalian cells carry tracts of poly(A) at their 3'-termini, they can be separated from the bulk of cellular RNA by their capacity to bind to oligo(dT)-cellulose (Edmonds et al. 1971; Aviv and Leder 1972). This step is essential for construction of cDNA libraries and usually improves Northern hybridization or nuclease-S1 assays (Sambrock et al. 1989).

Materials

Reagents

- 0.1M NaOH
- 3 M sodium acetate (pH 5.2)
- 10% (w/v) sodium lauryl sarcosinate
- 1 x column buffer
- 2 x column loading buffer
- elution buffer
- 70% and 100% ice-cold ethanol
- RNAse-free TE-buffer (pH 7.0) (10 mM Tris-HCl, pH7.0, 1 mM EDTA (pH 8.0))

- methanol
- DEPC-treated water

Supplies

- oligo(dT)-cellulose (0.5-1 ml packed volume)
- glass wool and pasteur pipet or sterile small disposable column with a 2 ml capacity
- sterilized microcentrifuge tubes
- microcentrifuge at 4°C
- spectrophotometer

Procedure

In advance

1. In advance prepare the following solutions:
   - 2 x column-loading buffer[a]

Component	Stock Concentration	Final Concentration	Amount
Tris-HCl (pH 7.5)	1 M	40 mM	4 ml
LiCl (DEPC-treated)	5 M	1 M	20 ml
EDTA (DEPC-treated)	0.5 M	2 mM	0.4 ml
sodium lauryl sarcosinate[b]	10%	0.2%	2 ml
H_2O (DEPC-treated)			73.6 ml

[a] Do not sterilize solution by autoclaving, since it will froth extensively.
[b] Heat up sodium lauryl sarcosinate (10%) to 65°C for 30 min.

   - Elution buffer[a]

Component	Stock Concentration	Final Concentration	Amount
Tris-HCl (pH7.5)	1M	10mM	0.5ml
EDTA (DEPC-treated)	0.5M	0.1mM	0.1ml
DS	10%	0.05%	0.25ml
H_2O (DEPC-treated)			48.25ml

[a] Do not sterilize solution by autoclaving, since it will froth.

Note: For DEPC-treatment, see Subprotocol 5, Precautions; DEPC is suspected to be a carcinogen.

- Soak cuvettes for use with the spectrophotometer for 1h in concentrated HCl:methanol (1:1) and then wash extensively with DEPC-treated water.

Preparation of poly(A)$^+$RNA

2. Suspend 0.5-1.0 g of oligo(dT)-cellulose in 0.1N NaOH. Pour a column of oligo(dT)-cellulose (0.5-1.0 ml packed volume) in a sterile disposable column or a pasteur pipet, plugged with sterile DEPC-treated and autoclaved glass wool.

Note: Up to 10 mg of total RNA can be loaded onto 1 ml of packed oligo(dT)-cellulose. When using smaller quantities, the amount of oligo(dT)-cellulose should be reduced accordingly to avoid loss of poly(A)+ RNA.

3. Equilibrate the oligo(dT)-column with 10-20 ml 1 (loading buffer, until pH is <8.0.
4. Heat the RNA in water 5 min at 65°C. Cool quickly to room temperature, and add an equal volume of 2 x column-loading buffer.

Note: Heating the RNA disrupts regions of secondary structure that might involve the poly(A)+ tail.

5. Apply the solution to the column, and immediatly begin to collect the eluate in a sterile tube. Wash column with 1vol 1 x column-loading buffer and continue to collect.
6. Heat all the eluate to 65°C for 5 min and re-apply it to the column. Collect all eluate.
7. Wash the column with 5-10 column volumes of 1 x column-loading buffer.
8. Elute the poly(A)+ RNA with 2-3 column volumes of elution buffer. Collect fractions equivalent to 1/3 to 1/2 of the column volume.

Note: The oligo(dT)-column can now be stored at 4°C and reused many times. Between uses, regenerate column by sequential washing with NaOH, water, and 1 × column-loading buffer as described in steps 2-3.

9. Measure the absorbance of the fractions at 260 nm, using HCl:methanol treated cuvettes. Pool the fractions containg RNA.
10. Add 0.1 vol of 3 M sodium acetate (pH 5.2). Mix by vortexing briefly or by flicking the tube several times. Add 2.5 vol of ice-cold 100% ethanol, mix and store for at least 30 min on ice.

11. Spin at 15000 rpm for 15 min at 4°C.
12. Wash the pellet (which is often invisible) with 70% ethanol.
13. Redissolve the RNA in a small volume of water. Measure the absorbance at 260 nm, using HCl:methanol (1:1) treated cuvettes.

Note: Approximately 1% of the input RNA should be recovered as poly(A)+ RNA. The yield of poly(A)+ RNA from 10^7 cultured mammalian cells should be 1-5 μg.

14. Transfer the RNA to a polypropylene microcentrifuge tube, add 3 vol of 100% ice-cold ethanol, mix well, and store the preparation at -70°C. To recover the RNA, add 3 M sodium acetate (pH 5.2) to a final concentration of 0.3 M, mix well, and spin at 15000 rpm for 15 min at 4°C.

References

Aviv H, Leder P (1972) Purification of biologically active globin messenger RNA by chromatography on oligothymidylic acid-cellulose. Proc Natl Acad Sci USA 69:1408-1412

Birnboim HC (1992) Extraction of high molecular weight RNA and DNA from cultured mammalian cells. Methods Enzymol 216:154-60

Brawerman G, Mendecki J, Lee SY (1972) A procedure for the isolation of mammalian messenger ribonucleic acid. Biochemistry 11:637-641

Chomczynski P, Sacchi N (1987) Single-step method of RNA isolation by acid guanidinium thiocyanate-phenol-chloroform extraction. Analytical Biochem 162:156-159

Edmonds M, Vaughan Jr MH, Nakazato H (1971) Polyadenylic acid sequences in the heterogeneous nuclear RNA and rapidly-labeled polyribosomal RNA of HeLa cells: possible evidence for a precursor relationship. Proc Natl Acad Sci USA 68:1336-1340

Gough NM (1988) Rapid and quantitative preparation of cytoplasmic RNA from small numbers of cells. Analytic Biochem. 173:93-95

Gruber A and Zingales B (1995) One-base double-stranded DNA sequencing using commercial kits. BioTechniques 19:748-750

Hoopes BC, McClure WR (1981) Studies on the selectivity of DNA precipitation by spermine. Nucleic Acids Res 9:5493-5504

Liedke W, Battistini L, Brosnan CF, Raine CS (1994) PCR Methods Applic 3:185-187

Okayama H, Berg P (1982) High-efficiency cloning of full-length cDNA. Mol Cell Biol 2:161-170

Sambrook J, Fritsch, EF, Maniatis T (1989) Molecular cloning. A laboratory manual. Cold Spring Harbor Laboratory Press, Cold Spring Harbor

Shapiro DJ (1981) Quantitative ethanol precipitation of nanogram quantities of DNA and RNA. Analytical Biochem 110:229-231

Wallace DM (1987) Large- and small-scale phenol extraction. Methods Enzymol 152: 33-41

Chapter 4

Electrophoresis of DNA and RNA Fragments

FENG QIAN AND GREGORY G. GERMINO

Introduction

Characterization of DNA and RNA samples usually involves electrophoretic analysis. This chapter reviews basic principles and applications of electrophoresis, distinguishing between single field and pulsed field techniques. Also included is a review of the various matrices commonly used in molecular biology (agarose, acrylamide, Hydro-Link™). Methods of purification, detection and identification are discussed.

Electrophoresis plays an integral part of most strategies that involve characterization or manipulation of nucleic acids. The net negative charge of the phosphate groups along the DNA or RNA backbone results in migration of nucleic acids to the anode when exposed to a voltage gradient. In this section, we review many of the factors that affect the rate of electrophoretic migration of nucleic acids and discuss how some of these can be adjusted to optimize the separation, identification, purification or quantification of nucleic acids.

Parameters affecting the rate of DNA migration

Molecular size of the DNA

Molecules of linear double-stranded DNA are oriented in an end-on conformation in an electric field and migrate at rates that are inversely proportional to the logarithm of their molecular weight. Consequently, the molecular weight of a DNA fragment of unknown size can be determined by comparing its mobility to that of the molecular weight standard of known

Feng Qian, The Johns Hopkins University, School of Medicine, Division of Nephrology, 720 Rutland Avenue, Ross 958, Baltimore, Maryland, USA, *Correspondence to* Gregory G. Germino, The Johns Hopkins University, School of Medicine, Division of Nephrology, 720 Rutland Avenue, Ross 958, Baltimore, Maryland, USA (*phone* +01-410-614-1650; *fax* +01-410-614-5129; *e-mail* ggermino@welch.jhu.edu)

size. There are, however, a number of other factors that also affect rates of DNA migration.

DNA conformation

The relationship between size and migration rate is more complicated when the DNA molecules are not linear or are single-stranded. Closed circular (form I), nicked circular (form II), and linear duplex (form III) DNA of the same molecular weight migrate at different rates. The rate of migration for a circular molecule depends on the degree of its superhelicity. Closed circular DNA, which is typically negatively supercoiled, migrates faster than linear or nicked circular DNA. Conversely, a nicked circular molecule migrates more slowly than a linear one for any given length. Although the mobility of all the fragments is reduced when ethidium bromide, the intercalating dye, is incorporated into the gel, this effect is most pronounced with negatively supercoiled DNA. The dye reduces negative superhelicity in the form I molecule and the radii of the molecules increase, which results in decreased mobility. With increased concentration of ethidium bromide, all negative superhelical forms relax. As still more ethidium bromide is added, positive superhelical turns are generated, the DNA molecules become more compact and their mobility increases rapidly.

Single-stranded molecules (DNA and RNA) typically have considerable secondary structure in solution. The degree of secondary structure depends on a number of factors including temperature, base composition and buffer composition/concentration. The variability of secondary structure and its effects on electrophoretic mobility precludes accurate comparison of fragment lengths. Therefore, fragment lengths of single-stranded molecules (DNA or RNA) are determined by electrophoresis on denaturing gels (e.g., DNA sequencing gel or Northern blot gel) which eliminate secondary structure.

Applied voltage

At low voltage, the rates of migration of short (<2000 bp) and long linear DNA fragments are proportional to their lengths. This linear relationship is not valid for very long fragments subjected to a high voltage, however, and the resolution of larger fragments decreases when separated by high voltage. This effect is most pronounced in pulsed field gel electrophoresis (see below).

Matrix composition

The sieving properties of a gel depend on both the composition and concentration of its matrix.

Polyacrylamide gel electrophoresis (PAGE) is preferred for analysis of smaller fragments. A 6% gel can easily resolve differences of 1 bp in fragments of up to 300 bp in length. Polyacrylamide is usually stored as a 40% stock of the monomeric form mixed with bisacrylamide (a cross-linking agent) in a ratio of 19:1. TEMED and ammonium persulfate are added just prior to use to produce polymerization.

Advantages of PAGE:

- It can resolve molecules whose lengths differ by as little as 0.2% (i.e., 1 bp in 500 bp).
- Very small fragments can be resolved.
- Large quantities of DNA can be separated without loss of resolution: up to 10 µg of a single fragment of DNA can be loaded in a standard 1 cm x 1 mm slot.
- DNA samples isolated from polyacrylamide gels usually are free of contaminants and ready for use in the most demanding purposes (e.g., microinjection of mouse embryo).

Disadvantages of PAGE:

- The monomeric form is neurotoxic.
- The gels are a nuisance to prepare and to run.

Agarose is the matrix of choice for evaluating fragments from about 500 bp up to 10 Mb in length. Since DNA fragments of different sizes migrate at different rates through gels containing different concentrations of agarose, a concentration is selected which allows optimal separation of the range of sizes desired (Sambrook et al. 1989; Ausubel et al. 1987). Other favorable aspects of agarose include:

- Gels are sturdier and can be run in horizontal gel tanks allowing mid-run evaluation of progress.
- Agarose is non-toxic.
- Newer preparations of agarose are generally free of inhibitors, and some (e.g., NuSieve) can resolve fragments down to 10 bp in length.

Hydrolink™. AT Biochemical (USA) provides a new line of matrices with an interesting array of special properties. One matrix (proprietary information but probably a modified polyacrylamide), called the Long Ranger, has unique stacking properties that allow more base pairs to be separated per unit length than conventional acrylamide gels. Still others have features which allow better separation of molecules with conformational differences (i.e., for single-stranded conformation polymorphism analysis). Lastly, there is another matix which allows clearer resolution of very small DNA fragments (Hydrolink™).

Generally, the choice of matrix and its concentration depends on the range of sizes one hopes to resolve.

Electrophoresis buffers Of these, TBE (Tris-borate-EDTA) and TAE (Tris-acetate-EDTA) are the most widely used. Although these buffers have slightly different effects on DNA mobility and resolution, a buffer is usually selected based on its relative buffering capacity. TBE has significantly better buffering capacity than TAE and is the buffer of choice for extended pulsed field gel electrophoresis (PFGE) or high-voltage (sequencing gels) electrophoresis. For historical reasons, TAE is the most commonly used buffer for routine electrophoresis.

Quantity of DNA loaded The minimum that can be visualized by ethidium staining is about 5 ng. Up to 200 ng of a single DNA fragment can be loaded into a standard-size well (0.5 cm wide by 0.2 cm deep) before over-loading occurs. Over-loading results in trailing, band smearing and reduced mobility. These effects are particularly pronounced with DNA fragments above 10 kb (especially true for PFGE). The NaCl concentration of a sample has little effect. One advantage of acrylamide and Hydrolink® gels is that much larger quantities of DNA can be loaded into a lane without causing overloading.

Monitoring electrophoretic separation

Tracking dyes are commonly added to the loading buffer to monitor the progress of size separation during electrophoresis. The two most widely used (bromphenol blue and xylene cyanol) have very different electrophoretic mobilities. The fragments with which they co-migrate vary with the gel matrix and concentration (Sambrook et al. 1989; Ausubel et al. 1987). Two commonly used matrices are

- 1% agarose: xylene cyanol ~5 kb, bromphenol blue ~0.5 kb;
- 5% acrylamide: xylene cyanol ~260 bp; bromphenol blue ~60 bp.

Ethidium bromide is also commonly added to agarose gels as it allows direct visualization of DNA. This dye intercalates between the stacked bases and fluoresces red-orange when exposed to UV light (260–360 nm). The effect of ethidium on the mobility of linear DNA molecules is slight, and its addition to the gel offers the advantage of allowing monitoring of fragment separation during the run. Ethidium bromide can also be used to estimate the DNA quantity of a fragment by comparing its fluorescence to that of a known quantity of DNA.

Molecular weight markers

At least one lane containing a ladder of DNA fragments of known sizes is included on most gels; the choice of markers depends on the range of sizes to be resolved on the gel. A standard curve can be constructed from the relative positions of the known markers, and this curve can then be used to calculate the size of unknown DNA fragments. More often, however, the size of unknown fragments is simply estimated by comparing the distance they migrate relative to that of known fragments.

Subprotocol 1
The Recovery of DNA Fragments

Frequently, DNA samples are purified by electrophoresis. However, the samples must somehow be recovered from the gel. Therefore, a number of methods have been developed to maximize DNA recovery. Regardless of which method is selected, the purity of the final product critically depends on the quality of the preparative gel. Lane overloading results in significant trapping of other bands within the desired fragment, and re-electrophoresis may be necessary if a highly pure product is desired.

Procedure

Isolation of DNA fragments from agarose gels

Electroelution

In this method, one slices the fragment from a gel, places it in dialysis tubing with buffer and uses electrophoresis to drive the DNA out of the gel slice into the buffer. The chief advantage of this technique is that it can be used for recovery of a fragment of any size. The procedure is very inconvenient, however, and is usually used to isolate large fragments (>5 kb) which are inefficiently recovered by other methods.

Low melting point (LMP) agarose

A far simpler approach is to simply melt the gel slice containing the fragment of interest. Standard preparations of agarose, however, require boiling to melt the gel. This process denatures DNA and makes it unsuitable for many subsequent manipulations. This problem is circumvented by the use of a specially formulated agarose that melts at 65°C. There are a number of additional advantages of using this matrix. Since it remains fluid at 37°C and is very low in contaminants, enzymatic manipulations (such as ligation, labeling, restriction digestion, etc.) can be performed by adding a portion of the melted gel slice directly to the reaction mix. Furthermore, the gentle nature of the extraction procedure allows the isolation of intact high molecular weight DNA (though it works equally well for all sizes of DNA fragments).

Freeze-squeeze

An alternative method that is quick, inexpensive and efficient for recovery of molecules up to 5 kb in length can be used to isolate fragments from standard agarose gels. After excision, the gel slice is snap frozen in liquid nitrogen, thawed and then squeezed, expressing the buffer which contains the DNA.

Standard protocol

1. Cut out the desired fragment from an agarose gel.
2. Punch a hole in the bottom of an microcentrifuge tube with a needle.
3. Layer glass wool in the bottom of the tube. The glass wool should cover the hole.
4. After closing the cap, place the entire tube in liquid nitrogen for 30 sec to freeze the agarose slice within the tube.
5. Place the frozen tube into a collection microcentrifuge tube which is put below the frozen tube.
6. Centrifuge the entire device at 12,000 g for 5 min. The expressed liquid in the collection tube contains the DNA molecules, which are ready for ethanol precipitation.

Spin columns

Amicon offers a range of spin columns that can be used to separate larger DNA fragments from much smaller ones (i.e., >500 bp from <100 bp). These have the advantage of being very efficient, rapid, and useful for concentrating samples. The columns are especially useful for purifying PCR fragments from the amplification mix of primers, enzyme, buffer and nucleotides.

Agarose

Gel slices are dissolved when incubated with this enzyme. This is particularly useful when one is working with very small quantities of a fragment. This protocol is especially useful in purifying (without shearing) high molecular weight fragments sliced from pulsed field gel electrophoresis.

Silicon-based (glass-milk) purification systems

The number of methods that are commercially available increases each year. The two most commonly used systems, Gene Clean (Bio 101) and Prep-a-Gene (BioRad), differ primarily in the buffer they use to dissolve the agarose slice. Gene Clean (Bio 101) uses NaI whereas Prep-a-Gene (BioRad) uses Na perchlorate. Both bind the DNA to a silicon-based matrix which is then washed prior to DNA elution. Both produce yields of up to 80% when used for recovering fragments from 0.2 – 16 kb in size.

Isolation of DNA fragments from polyacrylamide gels

Purifying DNA fragments from polyacrylamide gels is much simpler than from agarose. The DNA is eluted from the slice by simply incubating the sample in 0.5 M ammonium acetate/1 mM EDTA buffer.

Subprotocol 2
Electrophoresis of RNA

Most commonly, RNA is electrophoresed for Northern blotting. The size and amount of a specific RNA molecule are then determined by hybridization of a labeled cDNA or anti-sense RNA probe to the blot. The quality of an RNA preparation can also be assessed by electrophoresis. For intact RNA samples, two sharp bands (28S and 18S ribosomal RNAs) should be readily visible when the gel is stained with ethidium bromide and exposed to UV light.

Many of the general principles discussed previously concerning DNA are applicable to the analysis of RNA. The migration rate of RNA molecules electrophoresed through agarose gels also is inversely related to the logarithm of the molecular weight. The size of unknown molecules can be determined by comparing their migration rates relative to those of RNA markers of known size, which are commercially available. The 28S and 18S bands are 5.0 kb and 1.9 kb respectively in humans, and are also useful for estimating size. A distinct feature of RNA molecules, however, is that they have considerable secondary structure since they are single-stranded, and this must be eliminated if their molecular weights are to be accurately determined. Therefore, RNA is usually denatured using one of the methods described below prior to loading on a gel, and the gel itself is also run under denaturing conditions.

Procedure

Denaturation of RNA using formaldehyde

RNA is denatured at 65°C in a formamide-formaldehyde buffer for 15 min, and then loaded onto a 0.8 to 1.2% agarose gel containing formaldehyde. Polyacrylamide gels can be used when smaller molecules are to be resolved.

Denaturation of RNA using glycoxal and DMSO

Samples are incubated in a glycoxal/DMSO buffer for 1 h at 50°C prior to loading onto the gel. The gel is prepared with 10 mM NaH_2PO_4, pH 7 running buffer. Recirculation of the buffer during the electrophoresis is required to keep the pH of the buffer at 7. At pH of 8 or greater, the glycoxal dissociates from the RNA. (Methyl mercuric hydroxide is also an excellent RNA denaturant, but its toxicity has limited its use.)

Note: An important point discussed in the section on RNA isolation needs to be re-emphasized in this section: *RNases are ubiquitous and stable, and RNA preparations are thus prone to be degraded.* Either an electrophoresis apparatus should be used that is solely dedicated to RNA electrophoresis, or it should be treated with 0.5 N NaOH and 10 mM EDTA solution prior to each use.

RNA protocols

Preparing an agarose/formaldehyde gel

The agarose concentration of a Northern gel is typically 0.8 – 1.2%. For 100 ml of a 1% gel, weigh out 1 g agarose, add 10 ml 10xMOPS buffer and 73 ml H_2O. Heat the mix in microwave to dissolve agarose. Pour the hot solution into a 100 ml graduated cylinder and make the volume to 83 ml, if necessary. Leave the solution in a fume cupboard until it reaches approximately 60°C, then add 17 ml of 37% formaldehyde, swirl to mix (avoid bubbles) and immediately pour into the gel tray.

Denaturation of RNA sample prior to loading

To the RNA samples (total RNA 10 – 20 µg; poly(A)$^+$ RNA 1 – 2 µg per lane) add the following solutions:

- 10 µl formamide (deionized)
- 3.5 µl formaldehyde (37%)
- 2 µl 10xMOPS buffer *
- 4.5 µl H_2O

Denature samples by heating at 65°C for 15 min, then cool them on ice. Parallel to RNA samples, run 5 µg RNA size markers treated in a similar manner (Gibco/BRL). RNA can be directly visualized in the gel by adding 1 µl of a 10 mg/ml ethidium bromide solution to the sample mix and denaturing in the presence of ethidium bromide. (Note: ethidium bromide solution must be prepared using DEPC-treated water.) Staining after the run is not as efficient as formaldehyde interferes with staining. Alternatively, RNA can be visualized after the transfer to the membrane by UV shadowing with an intense UV source.

* **10xMOPS**

- 0.2 M 3-morpholinopropanesulfonic acid, pH 7.0
- 50 mM sodium acetate
- 10 mM EDTA

Electrophoresis

Submerge gel in the running buffer (1xMOPS buffer with 3.7% formaldehyde) and load samples. To monitor progress of electrophoresis, add 5 µl dye mix (50% glycerol; 1 mM EDTA; 0.4% bromphenol blue and 0.4% xylene cyanol) to an unfilled well. Constant recirculation of the buffer is not necessary, but after 1 – 2 h the buffer from each reservior should be mixed.

Transfer of denatured RNA to the membranes

Remove the gel from the running buffer and wash two times with H_2O. If the blot will be probed for transcripts larger than 2 kb long, or if the gel is more than 1% agarose, perform the following steps:

1. Soak the gel in 50 mM NaOH; 0.15 M NaCl for 20 min.
2. Neutralize the gel in 0.1 M Tris (pH 7.5); 0.15 M NaCl for 30 min.

If the the blot will be probed for transcripts around 2 kb or smaller, rinse the gel in two changes of 20 x SSC to remove the formaldehyde. The gel is then placed in contact with the nitrocellulose filter or nylon membrane and the RNA is by capillary transferred to the membranes action in an ascending flow using 10 x SSC as the transfer buffer. Further details of this procedure can be found in the section describing Southern blots as the final set-up is identical.

Recently, a number of transfer devices have become commercially available which greatly decreases transfer times from 12 h to less than 1 h. The majority of transfer apparatuses use a negative pressure (vacuum) to draw the nucleic acids out of the gel onto the membrane. We, however, have excellent results using the PosiBlot® pressure blotter from Stratagene. This system uses a positive pressure to drive RNA (or DNA) molecules from the gel onto the membrane. Transfer is completed within 1 h and the efficiency of transfer is similar to that obtained with standard capillary technique. After blotting, the membrane is UV-crosslinked or baked at 80°C for 2 h.

Northern hybridization

The conditions for prehybridization, hybridization and washing of RNA immobilized on membranes are essentially the same as those used for DNA. We use the buffer described by Church and Gilbert (CG buffer).

1. Prehybridize the membranes for at least 2 h in CG buffer at the appropriate temperature.
   - 1 x CG buffer
   - 0.5 M Na_2HPO_4 (pH 7.2)
   - 1% BSA
   - 7% SDS
   - 1 mM EDTA
2. Add the denatured radiolabeled probe directly to the prehybridization solution. To detect low-abundance mRNAs, use at least 0.1 µg of probe whose specific activity exceeds 2 x 10^8 cpm/µg. Continue incubation for 16–24 h at the appropriate temperature.
3. Wash the membrane for 20 min at room temperature in 1 x SSC, 0.1% SDS, followed by 2–3 washes of 15 min each at 50–65°C in 0.2 x SSC, 0.1% SDS.
4. Expose the membrane to X-ray film (Kodak XAR-2 or equivalent) at –80°C with an intensifying screen.

References

Ausubel FM, Brent R, Kingston RE, Moore DD, Seidman JG, Smith JA, and Struhl K (eds) (1987) Current Protocols in Molecular Biology. Wiley Interscience
Sambrook J, Fritsch EF, and Maniatis T (1989) Molecular cloning - A laboratory manual. Cold Spring Harbor Laboratory Press, Cold Spring Harbor
Church GM and Gilbert W (1984) Proc Natl Acad Sci USA 81:1991-1995

Chapter 5

Single-Strand Conformation Polymorphism (SSCP) Analysis

FENG QIAN AND GREGORY G. GERMINO

Introduction

The detection of DNA sequence variation is important for the identification of disease-causing mutations, as well as for the detection of clinically silent DNA polymorphisms. Among the techniques capable of detecting a single base change, the single-strand conformation polymorphism (SSCP) analysis is most commonly used due to its simplicity and effectiveness (Orita et al. 1989a, 1989b; Sheffield et al. 1993). SSCP analysis permits the detection of mutations in very small amounts of genomic DNA or cDNA. This technique is rapid and inexpensive. It is based on the principle that a single-stranded DNA molecule adopts a sequence-specific conformation. A change in conformation resultung from a single base substitution can be detected as an alteration of its migration rate in a nondenaturing polyacrylamide gel (Figure 1).

The detection of conformation differences depends on a number of parameters. The optimal condition varies markedly from fragment to fragment and must be determined empirically for each DNA segment.

Procedure

Generating SSCP template by PCR reactions

The desired SSCP template is amplified using PCR. A total PCR reaction volume of 10 µl is sufficient. Add 0.1 µCi of ^{32}P-dCTP to each reaction.

Feng Qian, The Johns Hopkins University, School of Medicine, Division of Nephrology, 720 Rutland Avenue, Ross 958, Baltimore, Maryland, USA, *Correspondence to* Gregory G. Germino, The Johns Hopkins University, School of Medicine, Division of Nephrology, 720 Rutland Avenue, Ross 958, Baltimore, Maryland, USA (*phone* +01-410-614-1650; *fax* +01-410-614-5129; *e-mail* ggermino@welch.jhu.edu)

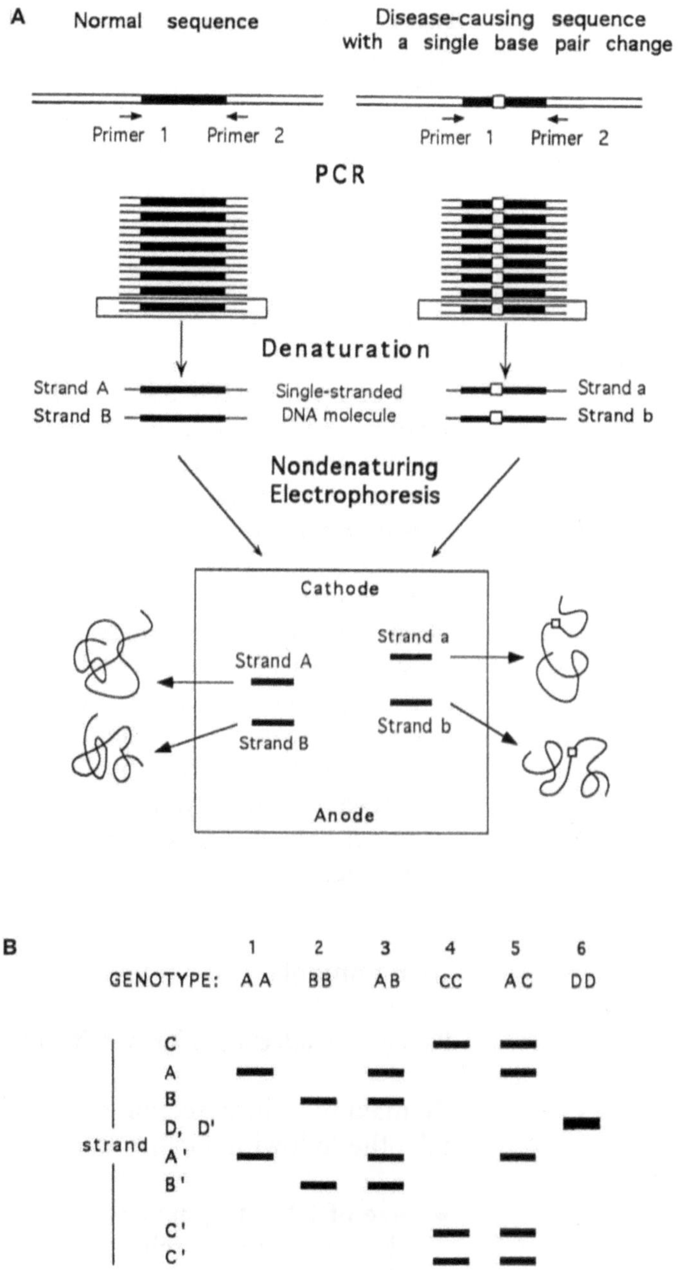

Fig. 1. A Schematic diagram of the SSCP principle. The target sequence (represented by a black block) of a normal (left panel) and a homozygous affected (right panel) individual is amplified by PCR using primer 1 and primer 2. The single base pair substitution in the affected individual is represented by an unfilled box in the target sequence. The PCR product is denatured prior to running in a nondenaturing polyacrylamide gel. The single base change causes altered conformations of the single-stranded DNA molecules. The conformational differences are detected by the altered migration rate of each single-stranded DNA molecule. **B A typical example of an SSCP pattern.** Generally, each strand of an allele runs as a single band representing a single conformation when electrophoresed through a nondenaturing gel (lanes 1, 2). PCR products amplified from individuals homozygous for such an allele will run as two bands. Heterozygous individuals having alleles A and B can be easily identified by the presence of four bands. On occasion, one strand of an allele will have 2 conformations at a given temperature, buffer, or matrix composition (lane 4, strand C' of allele C) and migrate as two bands. In this case, a homozygous individual will have three bands, and a heterozygous individual may have five (lane 5). Alternatively, the strands of an allele may not be distinguished under some running conditions, and all four strands run together as a single band (lane 6). It is important to keep in mind when interpreting SSCP gels that the band intensity of complementary strands often is not equal if α-[^{32}P]-dCTP is added to the nucleotide mix of the PCR reaction. This problem can be circumvented by using 5' end-labeled primers (see Chap. 7) in the PCR reaction.

This is best achieved by making a mastermix of all reagents and adding an appropriate amount of the radionucleotide. It is not necessary to reduce the concentration of cold dCTP in the reaction mix. After PCR amplification, the labeled PCR product is mixed with 5 times the volume of formamide loading buffer (95% formamide, 20 mM EDTA, 0.05% bromphenol blue and 0.05% xylene cyanol). The samples are denatured for 5 min at 95°C and placed directly on ice. 2 µl of each sample are loaded onto gels.

Electrophoresis

The electrophoresis in SSCP analysis is routinely carried out under two conditions. The first gel consists of 6% acrylamide (49:1 acrylamide:bis), 10% glycerol and 0.5 x TBE. Samples on this gel are electrophoresed for at least 2.5 h at 0.5 watt/cm) at room temperature in 0.5 x TBE running buffer. Constant temperature can be maintained using a cooling fan. The second gel condition has the same gel composition except it contains no glycerol and electrophoresis is performed at 4°C. Gels are then transferred to 3MM Whatman paper, dried and autoradiographed. The longer length of a sequencing gel apparatus (30 – 50 cm) offers the advantage of longer run time and less chance of running abberant bands off the bottom. We have had, however, excellent results using a shorter vertical gel system such as the Protean IIxi by BioRad. A major advantage of this system is its internal cooling chamber. Attached to a temperature-controlled circulator, gel temperature can be easily controlled to optimize band separation on the bench top without the need for a fan or a cold room.

Comments

Parameters affecting SSCP sensitivity

To maximize the detection of single base pair changes by SSCP, one should take the following factors into account.

- **Size of DNA fragments**
  There is a striking relationship between sensitivity and fragment length. The optimal size for relatively sensitive detection of single base changes is approximately 200 bp or less. For larger fragments, this limitation can be overcome by digesting the DNA with appropriate restriction endonucleases to reduce the fragment sizes prior to SSCP assay.

- **Gel matrix composition**
  The pore size of the acrylamide gel is an important determinant for sensitivity of SSCP. The ratio of acrylamide to bisacrylamide should be 49:1. Recently, a special gel matrix (MDE, AT Biochemical) reportedly allows better resolution of molecules with subtle conformational differences.

- **Electrophoresis conditions**
  Some single-stranded conformations can be remarkably sensitive to glycerol concentration and temperature in gels. The electrophoresis in SSCP analysis is thus routinely performed using gels comprised of different glycerol concentrations and run at different temperatures. There is almost an infinite combination of possible running conditions. Generally, one starts with two conditions:
  - 10% (v/v) glycerol and electrophoresis at room temperature
  - no glycerol and electrophoresis at 4°C

One can enhance the sensitivity by trying additional conditions which differ in the glycerol concentration of the gel and in the temperature of the buffer/ gel during the run.

Further parameters affecting SSCP sensitivity include the sequence of the fragment, the nature of the base pair change, and the base position of the nucleotide substitution with the fragment.

References

Orita M, Iwahara H, Kanazawa H, Hayashi K, and Sekiya T (1989a) Detection of polymorphisms of human DNA by gel electrophoresis as single-strand conformation polymorphisms. Proc Natl Acad Sci USA 86: 2766–2770

Orita M, Suzuki Y, Sekiya T, and Hayashi K (1989b) Rapid and sensitive detection of point mutations and DNA polymorphisms using the polymerase chain reaction. Genomics 5: 874–879

Sheffield VC, Beck JS, Kwitek AE, Sandstrom DW and Stone EM (1993) The sensitivity of single-stranded conformation polymorphism analysis for the detection of single base substitution. Genomics 16:325–332

Chapter 6

Denaturing Gradient Gel Electrophoresis

KAI-OLAF NETZER

Introduction

Denaturing gradient gel electrophoresis (DGGE) (Abrams and Stanton 1992; Myers et al. 1987) is a technique suitable for detecting point mutations as well as small deletions and insertions of several base pairs in double stranded DNA. The method is based on the analysis of strand separation (denaturation) of DNA molecules. The process of strand separation by disruption of hydrogen bonds is referred to as "melting" of the DNA, while the "melting point" is defined as the temperature or the denaturant concentration at which half of all possible hydrogen bonds are disrupted. It appears that melting occurs in domains, i. e. stretches of several tens of base pairs melt at the same temperature. Several factors influence the melting of a given DNA sequence: temperature, length, and base composition. GC-rich sequences have higher melting temperatures than AT-rich sequences. Moreover, since hybridization depends on hydrogen-bond formation, the presence of hydrogen-bond disrupting agents (denaturants) in the medium will also affect the melting point. Computer algorithms are available that can predict melting properties of DNA sequences (Lerman and Silverstein 1987). The resulting melt maps can be used to determine optimal experimental conditions.

Melting of a DNA fragment within a gel matrix greatly alters electrophoretic mobility. In polyacrylamide gels (PAG), mobility decreases exponentially with the length of the melted sequence. Introduction of a sequence change in a melting domain usually alters the melting temperature by several tenths of a degree Celsius. Therefore, sequence changes within a melting domain can be visualized by differential migration in denaturing gra-

Kai-Olaf Netzer, Krankenhaus Köln-Merheim, Medizinische Klinik I, Ostmerheimer Str. 200, Köln, 51109, Germany (*phone* +49-221-8907-2200; *fax* +49-221-8907-2335; *e-mail* kai.netzer@uni-koeln.de)

dient gels, as shown for **parallel** and **perpendicular** DGGE in Figure 1 and Figure 2. To improve electrophoretic detectability of mutations, heteroduplex molecules are generated by mixing normal and mutated DNA fragments prior to electrophoretic analysis. Subsequently, these mixtures are denatured and reannealed. The resulting heteroduplex molecules contain base mismatches at the site of the mutation. The reduced binding energy of mismatches leads to a reduced melting temperature and to greater migration differences upon electrophoresis.

The gradient needed for melting analysis can be generated by casting a PAG with a gradient for denaturing substances such as urea and formamide (DGGE), or by generating a physical temperature gradient along the gel using a heat exchanger attached to the gel (temperature gradient gel electrophoresis, TGGE).

Suitable DNA fragments for analysis can be generated by restriction enzyme digestion or PCR, the latter being used most commonly. The fragments should be in the size range of 100 to 400 bp. If PCR is used to generate fragments, the melting profile can be altered by
(i) varying the location of the primers, or
(ii) adding a GC clamp, i. e. a sequence of 40 nt of G and C to the 5' end of one primer. The GC clamp has a very high melting temperature (100°C) and acts as a clamp to hold together the double strand at one end, since it does not melt under normal gel conditions. This modifies the melting profile such that long domains (more than 100 bp) of uniform melting temperature are created. Theoretically, these domains are longer and easier to analyze than the naturally occurring domains in a given DNA fragment. However, in our experience, a drawback is that PCR with GC-clamped primers sometimes proves difficult with low yields of product.

Very recently, psoralen phosphoramidite has become commercially available. Psoralen phosphoramidite can be added to AA dinucleotide tails at the 5' ends of primers during synthesis in an oligonucleotide synthesizer. When used in PCR, the photo-reactive moieties are introduced into the PCR product. The two strands can then be covalently linked with 10 to 20 minutes of UV irradiation (360 nm). The covalent bond seems to replace the GC-clamp and exerts the same impact on the melting profile (Costes et al. 1993). Moreover, PCR with psoralen-labeled primers results in higher yields of product, and synthesis of primers is less costly.

The following protocol provides a general outline of the use of DGGE for analyses of PCR products. DGGE may be set up in two different ways: In perpendicular DGGE where the temperature gradient is perpendicular to the direction of electrophoresis whereas in parallel DGGE the gradient is parallel to the direction of electrophoresis. Perpendicular DGGE is useful

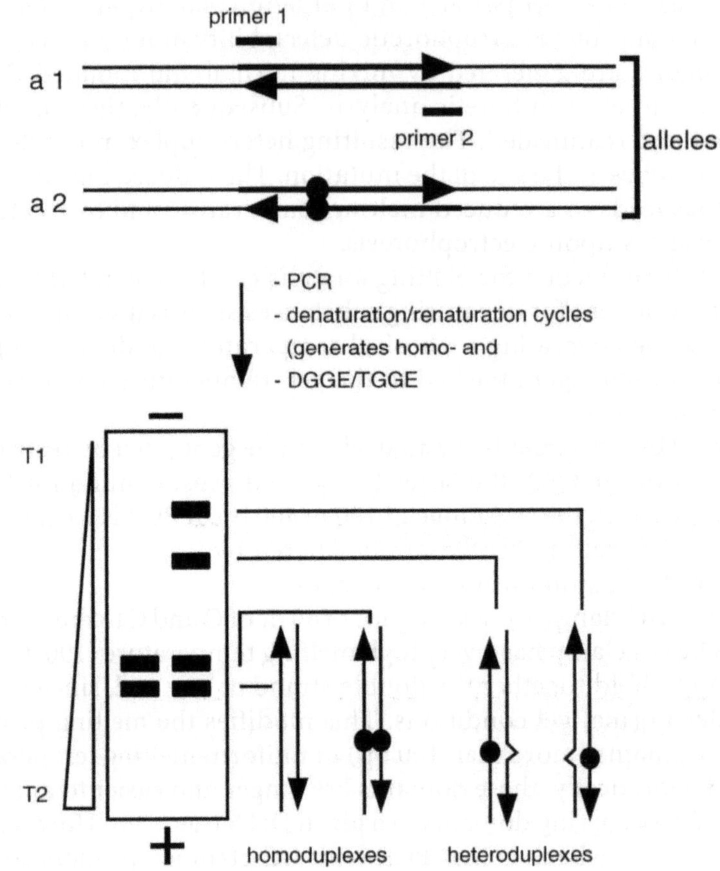

Fig. 1. Heteroduplex analysis in parallel DGGE/TGGE

Fig. 2. Heteroduplex analysis in perpendicular DGGE/TGGE

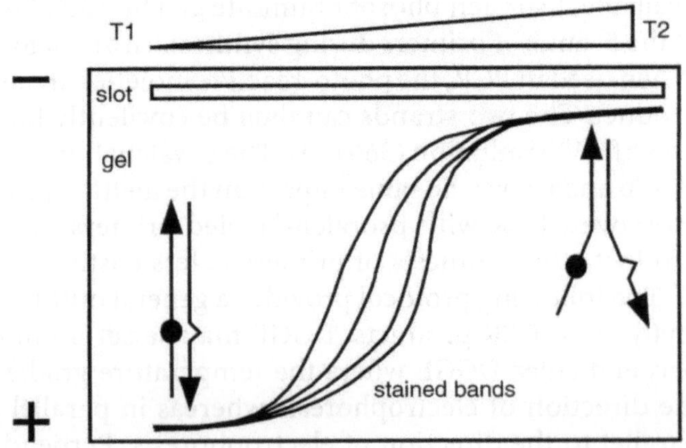

for experimentally elucidating melting domains within a given DNA fragment. Only one sample can be separated per gel. Parallel DGGE can separate many samples on one gel and is therefore the method of choice for mutation screening.

Materials

Supplies

- Gradient maker (15-25 ml capacity per side), with silicone tube (4 mm external diameter)
- Magnetic stirrer
- Gel-casting stand
- Vertical electrophoresis unit with heat exchanger (5 l buffer volume) to hold 16 x 18 cm gels, 0.75 mm thickness (e.g. Hoefer Scientific SE600)
- Heating water bath with external circulation, silicone tubes (12 mm external diameter)
- Power supply
- Peristaltic pump
- Variable clamp for flow adjustment in silicone tubes; 20 gauge needle

Reagents and solutions

- Polyacrylamide and bis-acrylamide 38.4:1.6 solution (e.g. Roth)
- Urea p.a. grade
- Formamide, deionized, p.a. grade; in our experience no deionization is necessary for up to 6 months if stored in the original vessel. If deionization is required, mix Dowex XG8 mixed-bed resin with formamide 1:20 w/v, stir gently for 30 minutes, carefully remove the formamide avoiding the resin beads and store from light protected at –70°C.
- 50x TAE buffer (2 M Tris, 50 mM EDTA, adjust pH with acetic acid to 8.0)
- 10% ammonium persulfate in water, prepared fresh daily
- TEMED
- Ultrapure water
- 6x gel loading buffer (0.25% bromophenol blue, 30% glycerel in water)
- DNA fragment in water or TE buffer (10 mM Tris-HCl, pH 8.0, 1 mM EDTA), or PCR reaction buffer. Best results are obtained with fragments

70-400 bp long. 1 µg/gel is required for perpendicular DGGE (i.e. product from a 100 µl PCR reaction), 50-100 ng/lane for parallel DGGE (i.e. approximately 1/10 of a 50 µl PCR reaction). Crude PCR reactions can be used, if no non-specific product is visible on agarose gels.

Procedure

In advance

1. Prepare the DNA fragment by PCR using the appropriate conditions. At the conclusion of 35 cycles, perform a 5 minutes/72 °C step to fill in single-stranded tails.
2. Cast a 1.8% agarose gel to check each PCR reaction for the presence of products of correct length and absence of non-specific products.
3. If mutation screening is intended, mix *half* of a patient's reaction mixture with the same amount of normal control mixture using a fresh 0.5 ml PCR tube. The remaining half of the patients' mixture is run in a parallel line. Alternatively, a mixture from a different patient may be substituted for control if distinct mutations are expected. Overlay with 50 µl mineral oil and place the tube in a temperature cycler. Use the following program for denaturing/reannealing to generate heteroduplex molecules: 95°C, 10 min; 55°C, 1 h, with a temperature slope of 0.1°C/sec in between.
4. Prepare two DGG stock solutions containing 0%, and 80% denaturing agents (100% corresponds to 40% formamide/7 M urea):

Component	Amount to give 100 mL	
	0 %	80 %
Urea	–	33.6 g
Formamide	–	32 ml
Acrylamide/Bis 38.4:1.6	17 ml	17 ml
50x TAE buffer	2 ml	2 ml
Ultrapure water	81 ml	21,7 ml

Store protected from light at 4°C. Solutions should be used within two weeks.

Casting a perpendicular denaturing gradient gel

5A. Clean glassplates thoroughly with detergent and water. Rinse with distilled water, then 100% ethanol, air dry.

6A. Assemble the glass plates and 0.75 mm spacers in the gel casting stand according to the manufacturer's instructions. Instead of inserting a comb, insert a third spacer. The length of this spacer should be 5 mm less than the width of the gel, leaving space to insert the needle for casting the gel. The set-up is shown in Figure 3. Add some silicone grease at the closed end to prevent leakage; fix the spacer with two steel clamps.

7A. Turn the gel casting apparatus with the gel by approx. 89° so that the 5 mm open space is at the top end, to prevent trapping of air (see Figure 3).

8A. Prepare the two gradient solutions in separate 15 ml tubes (e. g. Falcon). For a 16 x 18 x 0.075 cm gel, 12 ml of each gradient solution should be sufficient.

Note: The gradient range depends on the melt map of the fragment to be analyzed (see "Optimizing Conditions"). For perpendicular DGGE, it is ad-

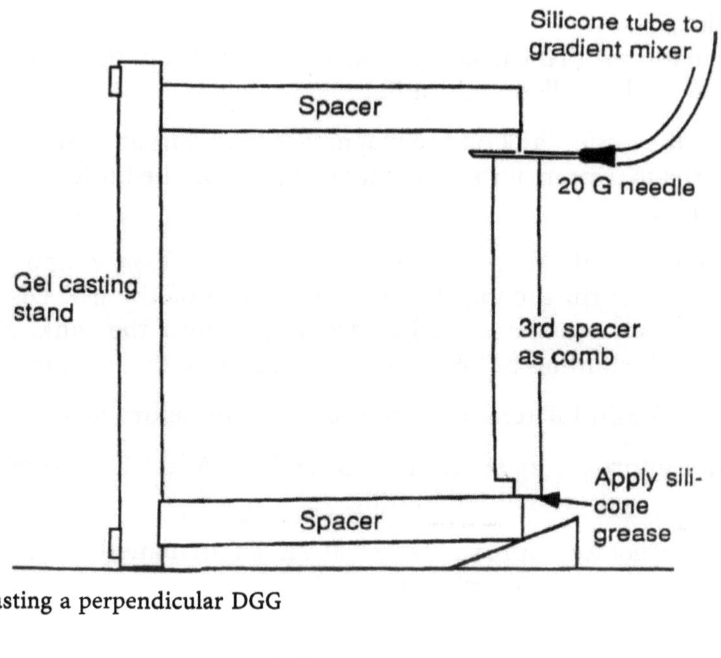

Fig. 3. Casting a perpendicular DGG

visable to first pour a gel with a wide gradient range (0-80%). Any other concentration range may be achieved by adjusting the composition of the two stock solutions (see step 7B. below).

9A. Add 70 μl 10% ammonium persulfate to each tube, mix gently, and add 7 μl TEMED. Mix gently.

10A. Carefully pour the higher concentration stock solution into the chamber of the gradient maker which is next to the exit. Open the stopcock connecting the two chambers so that the solution fills the closest passageway, then close again. Add the lower concentration stock solution to the second chamber. Place the gradient maker on a magnetic stirrer elevated 25 cm above the bench. Stir the chamber next to the exit at medium velocity. Mixing should not introduce bubbles into the solution. Insert the 20 gauge needle in the 5 mm opening at the top end of the gel. Carefully open the stopcock in between the two chambers and the exit of the gradient maker. Adjust the flow so that the gel is poured within 4 minutes.

11A. Allow one hour for polymerization at room temperature.

12A. Preheat approximately 5 l of 1x TAE buffer (for Hoefer Scientific SE600) to 60°C in a microwave oven.

Note: The buffer can be reused 3 times.

13A. Preheat the water bath to approximately 70°C. Connect the external circulation with the heat exchanger of the electrophoresis unit as soon as the tank is filled with buffer.

14A. Place the buffer tank on the magnetic stirrer. Stir at medium velocity. Insert a thermometer into the tank and adjust the buffer temperature to 60 °C.

15A. Remove the third spacer from the gel. If necessary, apply silicone grease to form a complete well at the top of the gel. Disassemble the casting apparatus and insert the gel into the tank. Flush the well with running buffer using a syringe and 20 g needle.

Note: Wait 10 min for temperature equilibration before loading the gel.

16A. Dilute approx. 1 μg DNA in 500 μl 1x TAE. Add 100 μl 6x gel loading buffer. Load the sample into the gel.

17A. Run the gel for approximately 500 Vh. Recirculate the buffer using a peristaltic pump (10 ml/min).

Note: Short runs at higher voltage produce sharper bands. Therefore, we prefer runs at 250 V for 2.5 hours.

18A. Disassemble the apparatus and stain the gel (see “Staining“).

Casting a parallel DGG

5B. See step 5A.

6B. Assemble the glass plates and 0.75 mm spacers in the gel casting stand according to the manufacturer’s instructions.

7B. Prepare the two gradient solutions. The gradient range depends on the melt map of the fragment to be analyzed (see “Optimizing Conditions“). Any denaturant concentration can be prepared by mixing the 0% and 80% stock solutions. For a 16 x 18 x 0.075 cm gel, 12 ml of each concentration should be prepared in 15 ml tubes (e. g. Falcon). Various concentrations can be prepared according to the following table:

Desired denaturant concentration (%)												
	10	15	20	25	30	35	40	45	50	55	60	70
0% Stock	10.5	9.7	9.0	8.2	7.5	6.7	6.0	5.3	4.5	3.7	3.0	1.5
80% Stock	1.5	2.3	3.0	3.8	4.5	5.3	6.0	6.8	7.5	8.3	9.0	10.5

8B. To each tube, add 70 µl 10% ammonium persulfate, mix gently, and add 7 µl TEMED. Mix gently.

9B. Carefully pour the higher concentration solution into the chamber of the gradient maker which is closest to the exit. Open the stopcock connecting the two chambers so that the solution fills the passageway, then close again. Add the lower concentration solution to the second chamber. Place the gradient maker on a magnetic stirrer elevated 25 cm above the bench. Stir the chamber next to the exit at medium velocity. Mixing should not introduce bubbles into the solution. Position the gel casting apparatus upright and insert the 20 gauge needle in between the two glass plates. Carefully open the stopcock between the two chambers, and the stopcock at the exit of the gradient maker. Adjust the flow so that the gel is poured within 4 minutes.

10B. Remove the 20 g needle. Insert a suitable comb 10 mm deep. Allow one hour for polymerization at room temperature.

11B. Preheat approximately 5 l of 1x TAE buffer (for Hoefer Scientific SE600) to 60°C in a microwave oven.

Note: The buffer can be reused 3 times.

12B. Preheat the water bath to approximately 70°C. Connect the external circulation with the heat exchanger of the electrophoresis unit. Fill the buffer tank with preheated buffer.

13B. Place the buffer tank on the magnetic stirrer. Stir at medium velocity. Insert a thermometer into the tank and adjust the buffer temperature to 60 °C.

14B. Carefully remove the comb from the gel. Disassemble the casting apparatus and insert the gel into the tank. Flush the wells thoroughly with running buffer using a syringe and a 20 gauge needle.

Note: Wait 10 min for temperature equilibration before loading the gel.

15B. Add 1/5 volume gel loading buffer to each DNA sample (approximately 100 ng in 10 µl, 1/10 of a 50 µl PCR reaction). Load the samples into the wells of the gel.

16B. Run the gel for approximately 500 Vh (see "Optimizing Conditions"). Recirculate the buffer using a peristaltic pump (10 ml/min).

17B. Disassemble the apparatus and stain the gel (see "Staining").

Comments

Optimizing conditions

Gradient range

To find optimal conditions, it is advisable to first analyze a PCR product in a *perpendicular gel* with a wide gradient range, i. e. 0-80%. This helps to elucidate the melting temperature of domains within the DNA fragment. The curve obtained for a molecule with one domain accessible for DGGE will be of sigmoid shape: High electrophoretic mobility in the region of low denaturant concentration, reduced mobility in the region of high denaturant concentration. A point of inflection in the curve (see Figure 2) occurs at the denaturant concentration that is equivalent to the **"melting temperature"** of a melting domain (i. e., the denaturant concentration at which 50% of the melting domain is denatured and 50% is non-denatured). The "melting temperature“ of this domain can be calculated from the concentration of denaturants at the point of inflection of the curve. As a rule of thumb, in a gel heated to 60°C, 3% denaturants equal an increase in temperature of 1°C. Therefore, a gradient gel with 0-80% denaturants will cover a temperature range from 60°C (i. e. buffer temperature) to 87°C. Once the melting temperature has been determined, parallel DGGE may be used to screen large numbers of samples. In our experience it is best to have a range of denaturants of 30% from top to bottom, i. e. ±15% flanking the melting temperature of the domain to be analyzed.

Running time

It is essential that the DNA fragment reaches the region of the gel that contains enough denaturant to melt the domain of interest. If the gradient is narrow and the domain of interest is the first to melt, this may be achieved with 500 Vh. Gels can be run at low voltage overnight, e.g. at 25 V for 16 hours. Short runs during the day time, e.g. 200 V for 2.5 hours, may be preferable since everything is under immediate control. In addition, bands will be sharper. It may be necessary to perform test runs in which one fragment is applied to the gel at 0, 1, 2, 3, etc. hours after initiation of the run. The ideal running time is achieved when the fragment reaches its melting temperature and gives a well-focused band in the gel.

Staining

We have used two different methods for staining of DGG. A very simple method is staining with ethidium bromide. The gel is immersed in 1 µg/ml ethidium bromide in water for 15 to 20 min, rinsed with water, then photographed using a UV transilluminator. Since ethidium bromid is carcinogenic, it must be discarded properly (follow the rules of your laboratory's safety guidelines). Silver staining is more time consuming but also more sensitive. There are good kits available commercially (e. g. BioRad). Follow the manufacturer's instructions.

References

Abrams ES, Stanton VP (1992). Use of denaturing gardient gel electrophoresis to study conformational transitions in nucleic acids. Meth Enzymol 212: 71-104.

Myers RM, Maniatis T, Lerman LS (1987). Detection and localization of single base changes by denaturing gradient gel electrophoresis. Methods Enzymol 155: 501-527.

Lerman LS, Silverstein K (1987). Computational simulation of DNA melting and its application to denaturing gradient gel electrophoresis. Meth Enzymol 155: 482-501.

Costes B, Girodon E, Ghanem N, Chassignol M, Thuong NT, Dupret D, Goossens M (1993). Psoralen-modified oligonucleotide primers improve detection of mutations by denaturing gradient gel electrophoresis and provide an alternative to GC-clamping. Hum Mol Genet 2: 393-397.

Chapter 7

Introduction to Pulsed-Field Gel Electrophoresis (PFGE)

FENG QIAN AND GREGORY G. GERMINO

Principle

An agarose gel is made up of pores of various sizes with net pore size varying with concentration of the matrix. The mobility of a DNA molecule depends on how easily it can pass through these pores. Larger molecules must change their conformation to get through the smaller pores and thus migrate more slowly. DNA molecules larger than a certain size must squeeze to get through even the largest pores, and molecules exceeding this size all migrate at about the same rate. For ordinary agarose gels run in a uni-directional electric field, limiting mobility occurs between 20 kb and 40 kb. The significance of this limitation becomes clear when it is realized that a single genetic locus may occupy up to 1 Mb of DNA and that DNA molecules in the individual chromosomes of lower eukaryotes may be 7 Mb or more in length. However, Schwartz and Cantor (1984) showed that it was possible to resolve much larger fragments by varying the orientation of the electric field in the gel during the run. Each time the direction of the electric field is switched, the molecule is forced to change to a new conformation that will allow it to move in the new direction. The time required for this reorientation is very sensitive to molecular weight: larger molecules take more time to realign themselves than do smaller ones.

Feng Qian, The Johns Hopkins University, School of Medicine, Division of Nephrology, 720 Rutland Avenue, Ross 958, Baltimore, Maryland, USA, *Correspondence to* Gregory G. Germino, The Johns Hopkins University, School of Medicine, Division of Nephrology, 720 Rutland Avenue, Ross 958, Baltimore, Maryland, USA (*phone* +01-410-614-1650; *fax* +01-410-614-5129; *e-mail* ggermino@welch.jhu.edu)

Parameters affecting PFGE resolution

The PFGE resolution of DNA is dependent upon several critical parameters, which are discussed below.

Voltage

Resolving fragments greater than ~1 Mb in size requires reducing the voltage from the customary 6 – 10 V/cm. Why large molecules fail to be resolved with higher voltages is not known.

Temperature

Temperature has a dramatic effect on DNA mobility in PFGE (unlike in single-field electrophoresis). Although mobility is considerably increased with increased temperature, gels run at higher temperatures have diminished resolution. Most investigators maintain their buffer at 4 – 14°C during electrophoresis.

Agarose

As standard electrophoresis, the concentration of agarose will affect the separation of molecules. Although bands are sharper with higher concentrations of matrix, the rate of migration is considerably reduced.

Reorientation angle

In PFGE, for any fixed voltage, the migration rate of DNA molecules varies depending on the angle between the electric fields (the reorientation angle); reducing the reorientation angle increases fragment mobility. This effect is particularly pronounced for very large molecules (>1 Mb). In most systems, the angle is fixed. An advantage of the Programmable Autonomously-Controlled Electrodes (PACE) system is that this parameter can also be altered, allowing rapid resolution of fragments of any desired size (see “PACE“).

Switch interval (pulse times)

This is the most important determinant of relative mobilities for DNA molecules. The best resolution is obtained by using the shortest switch interval that permits separation of the complete size range of the fragments to be examined. The duration of the pulse time loosely correlates with the size of the molecules it resolves best. This relationship is not linear, however. Furthermore, there is a relatively narrow range of sizes for which each interval is best, and PFGE separations carried out using a single switch interval have regions of compressed and expanded resolution. Molecules longer or shorter than the ideal range will have reduced resolution. This problem is circumvented by programming the pulse time to linearly change during the course of an electrophoretic run, a technique called ramping. The duration of the run directly correlates with the pulse time. An electrophoretic run designed to separate yeast chromosomes greater than 3 Mb may take up to a week to complete.

DNA preparation

A preparation of high molecular weight DNA is essential to PFGE analysis. Usual DNA preparation protocols result in considerable shearing such that the average fragment size is about 100 - 200 kb. With special care this range can be extended up to 400 - 500 kb. Consequently, DNA that is to be analyzed by PFGE is generally prepared in agarose blocks, as discussed in other sections.

Finally, DNA markers of known size are as essential to PFGE as for standard electrophoresis. Those in common use include:

- lambda concatamers (monomer ~48 kb): Usual range of resolution is from ~50 - 1100 kb.
- *Saccharomyces cerevisiae*: Chromosomes of this yeast range from ~200 kb up to 3 Mb (in some strains, the upper limit is 1.6 Mb).
- *Candida albicans:* Chromosomes range from about 1 Mb to 3 Mb.
- *Schizosaccharomyces pombes:* Chromosomes range from 3 to 6 Mb.

All of these are now commercially available (Clontech, BRL, Biorad, FMC BioProducts, etc.). Protocols for preparation of these can be found in Current Protocols in Molecular Biology (Ausubel et al., 1987).

PFGE apparatuses

A number of different apparatuses have been developed over the past few years:

- OFAGE (Orthogonal Field Agarose Gel Electrophoresis)
  In this apparatus, the electrodes are arranged around a square box and the orientation of the electric field is periodically switched 90 degrees. One problem with this system is that the electric field within the gel box is not homogeneous. A consequence of this is that DNA molecules migrate at different rates depending on their location in the gel.
- FIGE (Field Inversion Gel Electrophoresis)
  Size separation can also be produced by simply periodically inverting a uniform field 180 degrees (Carle et al. 1986). There are several advantages to FIGE:
  - The lanes are straight, allowing accurate lane-to-lane comparisons and size estimations. This is especially advantageous for preparative gels.

  - No special gel tank or electrode system is required. A standard electrophoresis system can be easily adapted by adding a switching device and a buffer circulator.

  Disadvantages:
  - the upper limit of resolution is ~1000 kb;
  - DNA molecules of different sizes may exhibit the same mobility at any given switch time. This occurs when the reverse time is insufficient to adequately disrupt the conformation of larger molecules; hence, they may actually migrate faster than shorter molecules. Therefore, FIGE works best if the reverse interval is roughly equal to the time it takes for a molecule to change conformations, and a single reverse interval separates only a relatively narrow range of sizes. This problem is minimized by progressively changing the switching interval during the run. The range of sizes resolved by various sets of switching angles has been empirically determined.

- CHEF (Contour – Clamped Homogeneous Electric Fields)
  Chu et al. (1986) calculated the voltages needed to generate homogeneous electric fields using multiple electrodes arranged around a closed contour. This system, like the OFAGE, is capable of resolving very large molecules (>10 Mb); in addition, it also has the advantage of producing straight lanes. This system is currently the most widely used.

- PACE (Programmable Autonomously-Controlled Electrodes)
  This is the most recent modification of the pulsed field gel electrophoresis (Clark et al. 1988). In this system, the ability to alter the reorientation angle between the fields permits increased speed of separation for large DNA molecules. DNA fragments as large as 6 Mb can be resolved in 24 h. This is in contrast to the CHEF where a similar separation would take around 7 days.

Application of PFGE in gene mapping

In gene mapping, a restriction map expressed in base pairs is required to guide cloning strategies and precisely define localization of a sequence. However, standard single-field electrophoresis is only able to resolve fragments up to 40 kb. Genetic markers typically are too far apart to be linked by this technique. Likewise, yeast artificial chromosomes (YACs), bacterial artificial chromosomes (BACs), and P1 clones have inserts that are far larger than can be analyzed using single-field electrophoresis.

PFGE, with its ability to resolve fragments of up to 10 Mb, is a useful tool in most mapping strategies. DNA generally is prepared in agarose blocks and then digested with one of the rare cutting enzymes. The infrequency of sites for these enzymes is based either on the length of their recognition sequence (8 bp for Sfi I) or the presence of 1 or 2 CpGs within the recognition sequence. (The frequency of the dinucleotide CpG is generally reduced within the mammalian genome.) After electrophoresis, the gel is blotted in the usual manner and probed with radiolabeled fragments. There are a number of features characteristic of PFGE mapping.

Partial digests Single-copy sequences often detect multiple bands when hybridized to Southern blots of pulsed field gels. This occurs because most of the infrequent cutter enzymes will not cleave at sites where cytosine is methylated. Since different DNA molecules from the same DNA prep may have different methylation patterns, a labeled probe may hybridize to several bands. This frequently can be useful in linking markers that map to adjacent complete digest fragments.

Successive hybridizations Genetic markers are said to be physically linked when they hybridize to at least some of the same fragments. Since there is considerable gel-to-gel variation in the rate of migration of DNA samples relative to that of the molecular weight markers (as well as considerable differences in resolution), probes must be successively hybridized to the same filter. Only identical bands are considered co-migrating. Unfortunately, most blots lose signal after several hybridizations, making it difficult to link up multiple loci.

Co-migration One of the challenges in constructing a map is determining whether bands detected by different probes that appear identical on an autoradiograph represent the same fragment vs. co-migrating of separate fragments. Figure 1 illustrates this principle. In this example, all three probes hybridize to *Nru* I and *Pvu* I fragments of identical size. Indeed, even the partial digestion pattern with *Cla* I is consistent with the hypothesis that all three probes are on the same *Nru* I fragment. The *Not* I hybridization data does not disprove this since it is possible that there could be a *Not* I site between them. However, the *Not* I/*Nru* I double digest disproves the hypothesis. The different *Not* I fragments detected by Probes A and B/C cannot be simply due to a single site between them because the *Nru* I fragment of Probe A shifts down when digested with *Not* I whereas the *Nru* I fragment of Probes B and C remains unaltered. If all three were on the same *Nru* I fragment, each would have detected a shift in size.

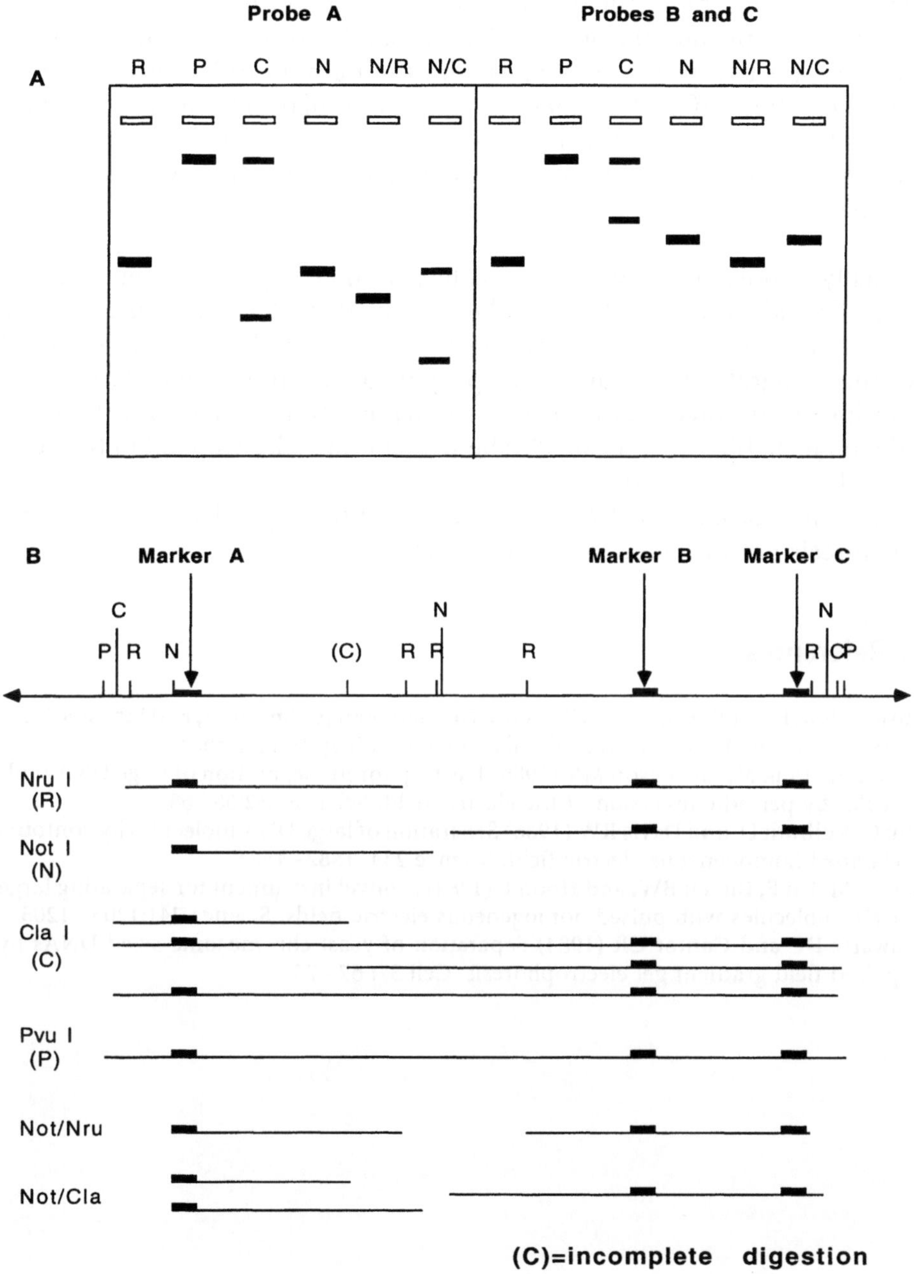

Fig. 1. A Schematic autoradiographs of the same Southern blot probed with marker A (left) and marker B and C (right). All three probes hybridized to the same *Nru I*, *Cla I* partial, and *Pvu I* bands. **B** The top half of the figure is a map summarizing the data from A. Below it is a schematic representation of the various molecules resulting from digests listed on the left. The *Not I/Nru I* double-digest proves that marker A can not be on the same *Nru I* fragment as Marker A and B. (C) = incomplete digestion.

Partial digestion, by virtue of increasing the number of bands detected in each lane, complicates the process. The probability of a chance co-migration occurring for two probes increases with the number of bands compared. Thus the pattern of partial digestion must be similar for two probes before they can be said to be physically linked. One other factor which affects the probability of co-migration is the resolution of the gel from which the blot was made.

The zone of limiting mobility

Virtually all pulsed field gels have a zone near their top where there appears to be a "cloud" of very high molecular weight DNA. This region is called the zone of limiting mobility because DNA fragments in this portion of the gel no longer migrate in a linear fashion proportional to their length. Fragments differing by millions of base pairs in length migrate at the same rate (and are indistinguishable) in "the zone". The size of molecules trapped in the zone depends on the pulse interval.

Specific protocols for DNA fragment resolution by pulsed field gel electrophoresis are provided in the next section.

References

Ausubel FM, Brent R, Kingston RE, Moore DD, Seidman JG, Smith JA, and Struhl K (eds) (1987) Current Protocols in Molecular Biology. Wiley Interscience

Carle GF, Frank M, and Olson MV (1986) Electrophoretic separation of large DNA molecules by periodic inversion of the electric field. Science 232:65 – 68

Chu G, Vollrath D, and Davis RW (1986) Separation of large DNA molecules by contour-clamped homogeneous electric field. Science 234: 1582 – 1585

Clark SM, Lai E, Birren BW, and Hood L (1988) A novel instrument for separating large DNA molecules with pulsed homogeneous electric fields. Science 241:1203 – 1205

Schwartz DC and Cantor CR (1984) Separation of yeast chromosome-sized DNAs by pulsed field gradient gel electrophoresis. Cell 37: 67 – 75

Chapter 8

Pulsed-Field Gel Electrophoresis: Protocols

GUDRUN A. RAPPOLD, KARIN RIED, ALBRECHT KLINK, ERCOLE RAO AND BIRGIT WEISS

Introduction

Pulsed-field gel electrophoresis (PFGE) techniques in combination with the cloning of large fragments of DNA into yeast artificial chromosomes (YACs) have revolutionized physical mapping in molecular genetics (Barlow and Lehrach 1987; Burke et al. 1987). DNA contigs of megabase length and physical maps of chromosome bands or entire human chromosomes are now at the stage of feasibility.

Conventional gel separation techniques use a single pair of electrodes to generate the electric field and cannot effectively resolve DNA fragments much larger than 50 kb. PFGE, however, represents a method for separating high molecular weight DNA in the range from 10 kb up to several Mbp by introducing new electrode configurations that generate electrical fields in alternating orientations (Schwartz and Cantor 1984). This periodic switching (pulsing) of fields forces high molecular weight DNA molecules through the gel in a "zig-zag" path. Based on the principle that large DNA fragments reorient more slowly than smaller fragments do, high molecular weight DNA can be separated by using longer pulses and small DNA molecules by using shorter pulses. Different PFGE systems have been designed and are used, e.g., OFAGE (orthogonal field-alternating gel electrophoresis (Schwartz and Cantor 1984), FIGE (field inversion gel electrophoresis (Carle et al. 1986), CHEF (contour-clamped homogeneous electric fields (Chu et al. 1986), ROFE (rotating field inversion gel electrophoresis (Southern et al. 1987[SV1]), OD-PFGE and 2D-PFGE (one- and two-dimensional pulsed-field gel electrophoresis). Besides its use for analytical purposes for con-

---

Correspondence to Gudrun A. Rappold, Institute of Human Genetics, Im Neuenheimer Feld 328, Heidelberg, 69120, Germany (*phone* +49-(0)6221-565068; *fax* +49-(0)6221-565332; *e-mail* gudrun_rappold@krzmail.krz.uni-heidelberg.de), Karin Ried, Albrecht Klink, Ercole Rao, Birgit Weiss

structing long-range restriction maps of chromosomes, PFGE has been also used as a preparative tool (Michiels et al. 1987).

Additional information on PFGE procedures can be found in books on PFGE methodology (e.g., Burmeister and Ulanovsky 1992; Birren and Lai 1993; Anand 1992).

Materials

Components of a pulsed-field gel experiment

- DNA agarose blocks made from specific cells, digested with infrequently cleaving restriction enzymes
- DNA size markers: Yeast chromosomes and lambda concatemers
- Agarose gel
- PFGE buffer
- OFAGE, FIGE, CHEF or ROFE apparatus with switching unit and cooling supply

Subprotocol 1
Preparation and Processing of High Molecular DNA in Agarose Blocks

To protect DNA from shearing, high molecular weight DNA is prepared from cells embedded in agarose blocks. In principle, DNA can be derived from single cell suspensions of all different tissues or cell lines; most commonly blood is used.

Materials

Reagents

- Whole blood, solid organ, cell line
- Agarose (LMP and normal agarose)
- PMSF (phenylmethylsulfonylfluoride)
- Proteinase K
- EDTA
- Sodium lauroyl sarcosinate

Solutions

- Lysis buffer (155 mM NH_4Cl, 10 mM $KHCO_3$, 0.1 mM EDTA (pH 7.4))
- PBS (8 g NaCl, 0.2 g KCl, 1.44 g Na_2HPO_4, 0.24 g KH_2PO_4 in 1 l, adjust to pH 7.4)
- PMSF stock solution (40 mg/ml in isopropanol)
- Proteinase buffer (1% sodium lauroyl sarcosinate (Sigma), 0.5 M EDTA (pH 8))
- TE (10 mM Tris base (pH 8.0), 1 mM EDTA)
- Distilled water
- Proteinase K

Supplies

- Falcon tubes
- Neubauer chamber
- Block formers
- Glass plate
- Sterile plastic loops

In advance

- Prepare a 2% agarose solution in PBS and keep at 45°C.
- Pre-cool ethanol-cleaned block formers on a glass plate sitting on ice.
- Have all solutions handy described above.

Time required

Day 1: 2-3 h
Day 3: 2 h

Procedure

1. Collect 10 ml of whole blood and add 30 ml of lysis buffer. Place on ice for at least 20 min until red blood cells lyse.
2. Spin at 2000 rpm for 10 min and decant off clear red supernatant; wash cells once more in lysis buffer. Then resuspend cells well in PBS.
3. Count the cells in a Neubauer chamber by diluting an aliquot of the single cell suspension.
4. Cells are resuspended in PBS such that 1 million cells are contained in 40 µl PBS (1 million diploid mammalian cells contain roughly 10 µg of genomic DNA).

5. A 2% solution of LMP agarose in PBS is melted and kept at 50°C.
6. An equal volume (1 ml) of cell suspension and agarose is mixed well at room temperature and immediately dispensed into the block formers.
7. Allow 20 min for the agarose to solidify; then push blocks through molds using the sterile disposable plastic loops (usually used to streak out bacteria) directly into proteinase buffer. Add proteinase K at 2 mg/ml.
8. Keep blocks in proteinase K buffer 2-3 days at 50°C. Up to 100 blocks can be placed in 50 ml proteinase buffer in a Falcon tube.
9. After proteinase K digestion, blocks can be stored in this buffer or in 0.5 M EDTA at 4°C.
10. Alternatively, continue the procedure by rinsing the blocks several times with sterile TE buffer.
11. Put blocks in a Falcon tube containing TE plus 0.04 mg/ml PMSF to inactivate remaining proteinase K.

Note: Caution! PMSF is poisonous! Incubate 2 x 30 min at 50°C. Replace 1 x TE and PMSF after the first incubation.

12. Rinse blocks afterwards with several changes of TE at room temperature; return blocks to a clean tube; use directly for enzyme digestion or store blocks in 0.5 M EDTA, (pH 8.0) at 4°C.
13. After storage in EDTA, rinse block 2 x 30 min in TE at room temperature.

Subprotocol 2
Preparation of Size Markers

Materials

Lambda concatemers

Time required Day 1: 1 h
Day 3: ready for use

1. Suspend **clean** concatemeric lambda DNA (Boehringer MA) in TE buffer at 4 μg per 40 μl.

2. Mix with an equal volume of 2% LMP agarose (kept at 45°C) in TE.
3. Pipet mixture into precooled block formers.
4. Incubate in TE + 100 mM NaCl for 2 days at room temperature

Yeast chromosomes

Reagents

See chapter on YACs

Solutions

See chapter on YACs

In advance

Pre-cool block formers on ice

Time required

- Day 1: <1 h
- Day 2: 2 h
- Day 4: ready for use

Procedure

1. Pick single colony from a YPD plate into 10 ml YPD preculture broth; grow with vigorous shaking 24h at 30°C. Then add 200 ml YPD broth and shake for 24-48 h (yields approx. 100 blocks).
2. Spin 4K/10 min, then resuspend in 50 mM EDTA/10 mM Tris-HCl (pH 7.5).
3. Spin 4K/10 min and resuspend in SCE (1 M sorbitol + 0.1 M Na citrate, pH 5.8 + 10 mM EDTA (pH 7.5)).
4. Count the cells in a Neubauer chamber by diluting an aliquot of the cell suspension.
5. Cells are resuspended (after centrifugation) in SCE such that 5×10^7 cells are contained in 40 µl SCE (corresponds to 5×10^7 cells/block of 80 µl).
6. The solution is mixed with 0.1 mg/ml Zymolyase 100T and 100 mM b-mercaptoethanol and kept at 37°C for 15-30 min.
7. Mix an equal volume of the cell suspension with 2% LMP agarose (kept at 50°C) in SCE.

8. Pipet mixture into precooled block formers as described above.
9. Blocks are then incubated in SCE with 10 mM DTT at 37°C for 1-2 h with shaking.
10. Transfer blocks into proteinase buffer, add proteinase K at 2 mg/ml and incubate at 50°C for 48h.
11. Wash blocks in 50 mM EDTA/10mM Tris HCl (pH 7.5), 3 x 20 min.
12. Blocks can be stored in this mixture at 4°C or loaded directly onto the gel without rinsing.

Subprotocol 3
Restriction Enzyme Digestion of DNA in Agarose Blocks

Materials

Reagents Infrequently cleaving restriction enzymes

Solutions
- High salt buffer (100 mM Tris-Cl (pH 7.5), 100mM $MgCl_2$, 1000 mM NaCl - 10x conc)
- Medium salt buffer (100 mM Tris-Cl (pH 7.5), 100 mM $MgCl_2$, 10 mM DTT, 500 mM NaCl-10x conc)
- Low salt buffer (100 mM Tris-Cl (pH 7.5),100 mM $MgCl_2$, 10 mM DTT-10x conc)
- Spermidine (10 mg/ml in H_2O)
- BSA (bovine serum albumin) (10% in H_2O)

Supplies
- Falcon tubes
- Eppendorf tubes
- Sterile plastic loops

Time required Day 3: 2 h

Procedure

1. Use sterile solutions and gloves to avoid degradation of DNA.
2. Rinse agarose blocks in 1 x TE for 3 x 20 min to remove EDTA (in Falcon tubes).
3. Mix: Reaction buffer (high, medium or low salt buffer), 100 mM spermidine (for reactions using high salt buffer) and 10-20 units enzyme. 20 units of enzyme are plenty for a complete digestion of 10 μg of DNA overnight.
4. Include a control of a digest containing all ingredients except the enzyme to check for unspecific DNA degradation.
5. Add agarose block to reaction mix; always manipulate blocks with sterile scalpels or inoculating loops.
6. Digestion with two different enzymes can be done together if the buffers are compatible, or sequentially (first use buffer with lower salt concentration; then adjust salt concentration). If the first enzyme requires 50°C, then replace buffer for second enzyme.
7. If partial digests are needed, first try using 1:10 diluted enzyme at same temperature and reaction time.

Standard reaction

Prepare a reaction volume of 170 μl in an Eppendorf tube and mix before adding the agarose block; total volume will be 250 μl per reaction.

Component	Volume	Stock concentration	Final concentration
Enzyme buffer	25 μl	10x	1x
BSA	25 μl	10 mg/ml	1 mg/ml
Spermidine	12.5 μl	10 mg/ml	1 mg/ml
Distilled H_2O	105.5 μl	-	-
Enzyme	2 μl	10 U/μl	0.08 U/μl
Agarose block	80 μl	-	-

Note: Choice of infrequently-cutting restriction enzymes: Check catalogues from companies for restriction enzymes. Enzymes which are most likely to cleave within CpG islands (Bird 1986) often found at the 5'ends of genes are those which contain CpG dinucleotides in their recognition sequence and are sensitive to CpG methylation, e.g., NotI, BssII, SstII or EagI. Enzymes with recognition sequences containing also A and T's also cleave outside CpG islands, e.g., SalI, MluI, NruI and ClaI (Bickmore and Bird 1993).

Subprotocol 4
Gel Electrophoresis

Materials

Reagents
- Agarose
- LMP-agarose (low melting point agarose)
- EtBr (ethidium bromide)

Solutions
- TBE (10 x TBE is 108 g Tris base, 55 g boric acid and 40ml 0.5 M EDTA (pH 7,5) in 1 l)
- Denaturing solution (0,5 M NaOH + 1.5 M NaCl)
- Neutralization solution (50 mM Na phosphate (pH 6.5))

Supplies
- Gel tray, frame and comb
- Scalpels
- Syringe and needle

In advance
Pre-cool 1-3 l of 0.25 x TBE in the apparatus to 10°C, 1 or 2 h before gel run

Time required
- Day 4: 1-2 h
- Day 6: 3 h

Procedure

1. Melt 0.8% agarose in 0.25 x TBE and cool to 50-60°C before pouring into the frame. Insert comb immediately.
2. Remove comb carefully and load the DNA block by using 2 sterile scalpels. Flame (and cool) scalpel between samples if you have digested blocks with different enzymes. Load DNA marker on both sides of the gel.

3. Seal the slot which contain blocks with 1% liquified LMP agarose gel in 0.25 x TBE.
4. In case of bubbles, remove the air with a syringe.
5. Once the sealing LMP agarose has solidified (~10 min), place gel into chamber and cover with running buffer.
6. Set the appropriate voltage and switching times (see Table 1) and start the run. Current (in Amps) should be approximately equal in the two directions.
7. Remove gel after the run and stain in 0.25 x TBE containing EtBr (0.4 μg/ml).
8. Drain buffer from box with a pump and wash box with distilled water.
9. Photograph gel; DNA will get nicked during the exposure.
10. Rinse gel 30 min in 0.25 M HCl to depurinate DNA and facilitate transfer.
11. Denature gel 2 x 20 min in alkali (denaturing solution), followed by 1-5 min in neutralization solution.
12. Transfer DNA to nylon membrane using standard Southern blotting protocol. In general, blot PFGE gels longer than ordinary gels (48 h).

Table 1. Running conditions of pulsed-field gel electrophoresis. OFAGE analysis using the LKB-Pulsaphor Electrophoresis Unit. Specific pulse times select which molecules will be resolved (separation range) and which remain unresolved (limiting mobility of the gel). A more homogeneous resolution over the whole separation range can be achieved by increasing switching times continuously (lower part of table).

Pulse Time(s)	Run Time (h)	Voltage (V)	Separation Range (kb)
1	20	150	0 - 50
5	20	150	10 - 100
20	40	150	10 - 350
40	40	150	30 - 600
60	40	150	50 - 800
80	40	150	50 - 1000
100	40	150	50 - 1500

Table 2. Continuos

Pulse Time(s)	Run Time (h)	Voltage (V)	Separation Range (kb)
10 - 30	24	180	10 - 300
20 - 80	42	140	10 - 700
20 - 100	42	140	10 - 1150
40 - 150[a]	34	140	-
150 - 300[a]	8	140	10 - 1500

[a] These two pulse time gradients are run consecutively

Comments

Optimizing running conditions

To optimize conditions for separation of DNA fragments in the appropriate size range, several parameters can be changed:

- **Pulse time**
  The lower the selected pulse time, the smaller the resolved molecules (Table 1). Separation range may differ within a given pulse time, depending on the gel box and running system used.
- **Voltage/current**
  Generally, gels are run at constant voltage (10 V/cm). For separation of DNA larger than 2 Mbp, voltage must be lowered because of trapping effects. The higher the voltage gradient, the shorter the switching interval for separating the same range of size markers. Current will increase with running time.
- **Concentration of agarose in gel**
  Agarose concentration between 0.8 and 1% are optimal for most purposes. Use of 0.7% rather than 1% agarose can speed up long gel runs but decreases the resolution and broadens the band width.
- **Temperature**
  Constant temperature in the gel box is very important; preferred temperatures are between 10-15°C, cooled by a Kryostat.
- **Run time**
  Usually, running times are between 16-48 h; separation of fragments larger than 2 Mbp require running times longer than two days.

- **Buffer concentration**
  Standard is 0.25-0.5 x TBE; the higher the buffer concentration, the sharper the bands observed. For gel runs which take longer than two days, 0.5 x TBE has the advantage of optimal buffering.

- **Position of electrodes**
  Shape of the lanes and sharpness of the bands are affected by the position of the electrodes; in all OFAGE boxes, the angle between point electrodes should be >90°C.

These protocols are based on lab protocols originally developed in the laboratory of Hans Lehrach. We wish to gratefully acknowledge Denise Barlow and Margit Burmeister. We thank Jocelyn Faberman for assistance in preparing the manuscript and Ralf Keil for the helpful discussions on PFGE conditions.

References

Anand R (ed) (1992) Techniques for the analysis of complex genomes. Academic Press, London, UK

Barlow DP, Lehrach H (1987) Genetics by gel electrophoresis: The impact of pulsed-field electrophoresis on mammalian genetics. Trends Genet 3:167-171

Bickmore WA, Bird AP (1993) The use of restriction endonucleases to detect and isolate genes from mammalian cells. Methods Enzymol 216:224-244

Bird AP (1986) CpG-rich islands and the function of DNA methylation. Nature 321: 209-213

Birren B, Lai E (ed) (1993) Pulsed-field gel electrophoresis, a practical guide. Academic Press, London, UK

Burmeister M, Ulanovsky L (ed) (1992) Pulsed-field gel electrophoresis, protocols, methods and theories. Methods in molecular biology vol 12, Humana Press Inc, Totowa, NY, USA

Burke DT, Carle GF, Olson MV (1987) Cloning of large segments of exogenous DNA into yeast by means of artificial chromosome vectors. Science 236:806-812

Carle GF, Frank M, Olson MV (1986) Electrophoretic separation of large DNA molecules by periodic inversion of the electric field. Science 232:65-68

Chu G, Vollrath D, Davis RW (1986) Separation of large DNA molecules by contour-clamped homogenous electric fields. Science 234:1582-1585

Michiels F, Burmeister M, Lehrach H (1987) Derivation of clones close to met by preparative field inversion gel electrophoresis. Science 236:1305-1308

Schwartz DC, Cantor CR (1984) Separation of yeast chromosome-sized DNAs by pulsed field gradient gel electrophoresis. Cell 37:67-75

Southern EM, Anand R, Brown WRA, Fletcher DS (1987) A model for the separation of large DNA molecules by crossed field gel electrophoresis. Nucl Acid Res 15:5925-5943

Radioactive Labeling of DNA and RNA Probes

PETER IGARASHI

Subprotocol 1
End-Labeling Using T4 Polynucleotide Kinase

Description

End-labeling is most commonly utilized for synthetic oligodeoxyribonucleotides, although double-stranded DNA may also be labeled (particularly if protruding 5' ends are present). End-labeled oligonucleotides are used for hybridization probes, primer extension analysis, or DNA sequencing using the Maxam-Gilbert chemical cleavage method. End-labeled double-stranded DNA is used in nuclease protection assays and other applications requiring probes of defined length.

5' ends of DNA (or RNA) are labeled with $\gamma[^{32}P]$ATP using bacteriophage T4 **polynucleotide kinase** (see also Chap. 2). Polynucleotide kinase transfers the γ-phosphate group from ATP to the free 5' hydroxyl group of DNA (or RNA) (Chaconas and van de Sande 1980). For DNA containing a 5' phosphate group, dephosphorylation with calf intestinal alkaline phosphatase is required prior to labeling. Alternatively, an exchange kinase reaction (Berkner and Folk 1977) may be performed in which polynucleotide kinase transfers the 5' phosphate group from the DNA substrate to an excess of ADP. The dephosphorylated substrate is then rephosphorylated by transfer of γ-phosphate from $\gamma[^{32}P]$ATP.

3' ends of DNA are labeled with **terminal transferase** (terminal deoxynucleotidyl transferase) which transfers dNTPs to free hydroxyl groups at the 3' ends of DNA (Deng and Wu 1983). End-labeled probes generated with polynucleotide kinase have relatively low specific activity because only one mole of ^{32}P is incorporated per mole of template. Since terminal transferase can add many nucleotides to each molecule of template, this enzyme is frequently preferred when higher specific activity is required (e.g., in situ hybridization) (Biroc et al. 1993).

Peter Igarashi, Yale University Medical School, Division of Nephrology, 333 Cedar Street, New Haven, Connecticut, USA

In all protocols described in this chapter, ^{33}P can be substituted for ^{32}P in applications where isotope emissions that are less energetic than ^{32}P are desired (e.g., cycle sequencing or in situ hybridization) (McLaughlin and Margolskee 1993; Evans and Read 1992).

Materials

Substrate

Synthetic oligodeoxyribonucleotides should have a free 5' hydroxyl group (i.e., trityl group removed if synthesized using phosphoramidite chemistry). Dephosphorylate (see Chap. 2) or use the exchange kinase reaction (see below) if a 5' phosphate group is present. Double-stranded DNA may also be labeled, but protruding 5' ends are labeled much more efficiently than blunt or recessed 5' ends.

Other reagents

- Bacteriophage T4 polynucleotide kinase (10 units/μl).
- $\gamma[^{32}P]$ATP (3000 Ci/mmol, 10 mCi/ml)
- 10x reaction buffer is:
  - 0.5 M Tris-Cl (pH 7.6)
  - 0.1 M $MgCl_2$
  - 50 mM dithiothreitol
  - 1 mM spermidine
  - 1 mM EDTA (pH 8.0)

Supplies and equipment

- 1.5 ml microcentrifuge tube (autoclaved)
- Table top centrifuge with swinging bucket rotor (optional)
- 37°C water bath
- Liquid scintillation counter
- Sephadex G-25 spin columns (Boehringer-Mannheim) or alternative method for separating probe from unincorporated radionucleotide (optional)

Procedure

Labeling 5' ends with T4

1. Combine in order in a sterile 1.5 ml microcentrifuge tube on ice:
   - 50 pmol synthetic oligonucleotide
   - 2.5 μl 10x reaction buffer
   - 10 μl $\gamma[^{32}P]$ATP (33 pmol)

- 1.5 μl T4 polynucleotide kinase (15 units)
- to 25 μl sterile water

2. Incubate at 37°C for 30 minutes.
3. While the reaction is incubating, prepare a Sephadex G-25 spin column by centrifugation at 2000x g for 2 minutes (optional).
4. Stop the reaction by adding 2 μl of 0.5 M EDTA or heating to 68°C for 10 min. Load the reaction mixture on to the spin column and centrifuge at 2000x g for 4 min (optional). Count 1 μl of column eluate in a liquid scintillation counter.

Note: It is important to avoid contamination of substrates with ammonium ions which inhibit polynucleotide kinase (e.g., avoid prior ethanol precipitation with ammonium chloride).

Subprotocol 2
End-Labeling Using the Exchange Kinase Reaction

Materials

Substrate Synthetic oligodeoxyribonucleotide or double-stranded DNA containing a 5' phosphate group.

Other reagents

- Bacteriophage T4 polynucleotide kinase (10 units/μl).
- $\gamma[^{32}P]$ATP (3000 Ci/mmol, 10 mCi/ml)
- 5 mM ADP
- 10x reaction buffer: 0.5 M imidazole-Cl (pH 6.6), 0.1 M $MgCl_2$, 50 mM DTT, 1 mM spermidine, 1 mM EDTA

Procedure

1. Combine in a sterile 1.5 ml microcentrifuge tube on ice:
   - 50 pmol synthetic oligonucleotide
   - 2.5 μl 10x reaction buffer
   - 1.5 μl ADP (300 μM final)
   - 10 μl $\gamma[^{32}P]$ATP (33 pmol)

- 1.5 μl T4 polynucleotide kinase (15 units)
- to 25 μl sterile water

2. Incubate at 37°C for 30 min.

3. Stop the reaction by adding 2 μl of 0.5 M EDTA or heating to 68°C for 10 min. Remove unincorporated nucleotides by centrifugation through a Sephadex G-25 spin column (2000x g for 4 min) (optional). Count 1 μl of column eluate in a liquid scintillation counter.

Note: The exchange kinase reaction is much less efficient than the "forward" reaction but is satisfactory for labeling molecular weight standards, for example.

Subprotocol 3
Labeling 3′ Ends with Terminal Transferase (Forward Reaction)

For principle considerations see Subprotocol 1, Description.

Materials

Substrate

Synthetic oligodeoxyribonucleotide, double-stranded DNA (protruding 3' ends are preferred), or single-stranded DNA containing a free 3' hydroxyl group.

Other reagents

- Terminal transferase (15 units/μl).
- α[^{32}P]dATP (3000 Ci/mmol, 10 mCi/ml)
- 5x reaction buffer : 0.5 M potassium cacodylate (pH 7.2), 10 mM $CoCl_2$, 1 mM DTT

Supplies and equipment

- 1.5 ml microcentrifuge tubes (autoclaved)
- Table top centrifuge with swinging bucket rotor (optional)
- 37°C water bath
- Liquid scintillation counter
- Sephadex G-25 spin columns (Boehringer-Mannheim) or alternative method for separating probe from unincorporated radionucleotide (optional)

Procedure

1. Combine in a sterile 1.5 ml microcentrifuge tube on ice:
   - 1-50 pmol Synthetic oligonucleotide
   - 5 µl 5x reaction buffer
   - 10 µl α[^{32}P]dATP (33 pmol)
   - to 25 µl sterile water
   - 1 µl Terminal transferase (15 units)
2. Incubate at 37°C for 30 minutes.
3. Stop the reaction by heating to 65°C for 5 min. Remove unincorporated nucleotides by centrifugation through a Sephadex G-25 spin column (2000x g for 4 min) (optional). Count 1 µl of column eluate in a liquid scintillation counter.

Note: dATP is the preferred nucleotide when preparing hybridization probes in order to reduce non-specific binding.

Note: The number of dNTPs incorporated per molecule of substrate can be adjusted by altering the duration of the reaction or the molar ratio of radionucleotide to oligonucleotide. Alternatively, the reaction mixture can be "doped" with dideoxynucleotides (which are not extended).

Subprotocol 4
Random-Prime Labeling

The method of random primer extension has largely replaced the less efficient "nick-translation" as the preferred method for labeling double-stranded DNA (Feinberg and Vogelstein 1983). Template DNA is denatured and annealed to hexamers of random sequence. The hexamers anneal at various points along the template and are elongated by the **Klenow fragment** of *E. coli* DNA polymerase I in the presence of three unlabeled dNTPs and one ^{32}P-labeled dNTP. The Klenow fragment possesses the 5'→3' synthase activity but lacks the 5'→3' exonuclease activity so that newly synthesized probe is not rapidly degraded. The substrate DNA is not itself radiolabeled but serves as the template for DNA synthesis. The resultant products range in size from 400-600 nucleotides and have typical specific activities of 10^8-10^9 cpm/µg. Reagents are available in kit form from several manufacturers (e. g., Amersham).

Materials

Substrate

Double-stranded DNA (50-200 ng). Minimal length has not been determined, but we have labeled 135-bp fragments successfully. Plasmid DNA should be linearized prior to use. Usually restriction fragments or PCR products are isolated on agarose gels and purified with silica resins (Qiaex, GeneClean, Prep-a-Gene, etc.) prior to use. Plasmid DNA can also be used but non-specific hybridization will be higher.

Other reagents

- Klenow fragment of *E. coli* DNA polymerase I (2 unit/μl).
- α[^{32}P]dCTP (3000 Ci/mmol, 10 mCi/ml)
- 10x reaction buffer: 0.5 M Tris-Cl (pH 7.2), 0.1 M $MgCl_2$, 1 mM dithioerythritol, 2 mg/ml BSA, 2.5 mg/ml random hexamers
- 0.5 mM each of dATP, dGTP, dTTP

Supplies and equipment

- 1.5 ml microcentrifuge tubes (autoclaved)
- Table top centrifuge
- Microcentrifuge
- 37°C water bath
- 100°C heat block
- Liquid scintillation counter
- Sephadex G-50 spin columns (Boehringer-Mannheim) or alternative method for separating probe from unincorporated radionucleotide (optional)

Procedure

1. Combine in a sterile 1.5 ml microcentrifuge tube:
   - 10-200 ng template DNA (in 1-9 μl water)
   - Sterile water to 9 μl
2. Heat to 100°C for 5 minutes then place immediately on ice. Spin briefly in a microcentrifuge.
3. Add in order on ice:
   - 2 μl 10x reaction buffer
   - 1 μl 0.5 mM dATP (25 μM final)

- 1 μl 0.5 mM dGTP (25 μM final)
- 1 μl 0.5 mM dTTP (25 μM final)
- 5 μl α[^{32}P]dCTP (0.83 μM final)
- 1 μl Klenow fragment of *E. coli* DNA polymerase I (2 units)

4. Incubate at 37°C for 30 minutes.
5. While the reaction is incubating, prepare a Sephadex G-50 spin column by centrifugation at 2000x g for 2 min (optional).
6. Stop the reaction by adding 2 μl of 0.2 M EDTA (pH 8.0). Load the reaction mixture onto the spin column and centrifuge at 2000x g for 4 min (optional).
7. Count 1 μl of column eluate in a liquid scintillation counter. Prior to use in hybridization experiments, the probe should be denatured by heating at 100°C for 5 minutes, then placed immediately on ice for 5 min.

Comments

- The DNA template should be linear.
- Random hexamers are available commercially or may be synthesized on an automated DNA synthesizer using equimolar concentrations of all four bases at each position.
- The Klenow fragment of DNA polymerase I is sensitive to denaturation. Avoid vortexing the reaction mixture.
- Incubating longer than 30 min at 37°C may result in reduced specific activity of the probe. Alternatively, incubations may be performed at room temperature for 2 hours.
- ^{32}P-labeled probe that is prepared using this method is susceptible to radiolysis and should be used promptly.
- DNA fragments can also be labeled directly in low-melting-point agarose gels (e.g., NuSieve) (Feinberg and Vogelstein 1984). Gel slices containing template DNA are excised and diluted with three times their weight of water. The slices are melted at 95°C for 7 min, then placed in a 37°C water bath. 9 μl of the melted gel solution is then removed and labeled as described above (starting with step 3) except that the reaction mixture is kept at 37°C to prevent gelling.

Subprotocol 5 Riboprobes

Recently, methods have been developed for **in vitro transcription** of RNA from cloned cDNA inserts (Krieg and Melton 1987; Melton et al. 1984). These methods have been adapted to produce uniformly labeled, single-stranded RNA probes ("riboprobes") of high specific activity. Advantages of such probes include the avoidance of potential reassociation of double-stranded probes (which competes with hybridization to target). Moreover, RNA-DNA and RNA-RNA hybrids are thermodynamically more stable than DNA-DNA hybrids permitting experiments to be performed at higher stringency.

For preparation of riboprobes, the template cDNA is cloned into a plasmid containing a multiple cloning site (polylinker) flanked by promoters for two different bacteriophage RNA polymerases. For example, the multiple cloning site of the pBluescript plasmid (Stratagene) is flanked by promoters for T3 and T7 RNA polymerase. Other suitable plasmids include pUC19, pGEM, etc. The plasmid is linearized by restriction digestion at a site in the polylinker downstream from the cloned cDNA. Riboprobe is then synthesized using the appropriate **RNA polymerase** in the presence of three unlabeled rNTPs and one ^{32}P-labeled rNTP. The riboprobe may be purified from template DNA by gel electrophoresis or deoxyribonuclease treatment. Riboprobe prepared in this fashion is suitable for in situ hybridization, Northern or Southern blots, or nuclease protection assays.

Note: Either the coding or non-coding strand of the template cDNA will be transcribed (depending on which of the two RNA polymerases is selected). This permits generation of either "sense" or "antisense" riboprobes. Only the latter will hybridize to the cognate mRNA. The former can be used as a negative control, e.g., for in situ hybridization experiments.

Materials

Template

Plasmid DNA containing the cDNA insert of interest is usually prepared on silica ion-exchange columns (Qiagen) or CsCl gradients. Alkaline-lysis "mini-prep" DNA may also be used, but it is essential to avoid contamination with ribonuclease A. Prior to use, linearize the template by restriction digestion at a site in the polylinker downstream from the cloning site (or within the cDNA but near the distal end). Extract with phenol:chloroform:isoamyl alcohol and precipitate with ethanol. Resuspend to a concentration of 0.2-1 mg/ml. Verify completeness of the digestion by agarose gel electrophoresis.

Other reagents

- 5x transcription buffer: 200 mM Tris-HCl (pH 8.0), 40 mM $MgCl_2$, 250 mM NaCl, 10 mM spermidine
- Gel loading buffer (optional): 95% formamide, 20 mM EDTA, 0.05% bromphenol blue, 0.05% xylene cyanol
- Gel elution buffer (optional): 2 M ammonium acetate, 1% SDS, 25 µg/ml yeast tRNA
- Ribonuclease-free deoxyribonuclease I (RNase-free DNase, 1-10 unit/µl)
- 10 mM each of rATP, rGTP, and rCTP; 1 mM UTP
- RNAse Block II (1 unit/µl, Stratagene) or RNAsin (40 units/µl, Promega)
- $\alpha[^{32}P]$UTP (800 Ci/mmol, 10 mCi/ml) or α-^{35}S-UTP (1000 Ci/mmol, 10 mCi/ml)
- Bacteriophage T3, T7 or SP6 RNA polymerase (depending on the promoter present in the plasmid) (diluted to 10 units/µl in 1x transcription buffer)
- Tris-equilibrated phenol
- Chloroform:isoamyl alcohol (24:1, v/v)
- 7.5 M ammonium acetate
- Absolute ethanol (stored at –20°C)
- DEPC-treated water (add 0.1% diethylpyrocarbonate, then autoclave prior to use)

Supplies and equipment

- 37°C water bath
- Microcentrifuge
- Dry ice
- Sequencing gel electrophoresis apparatus and power supply (optional)
- X-ray film (X-Omat AR, Hyperfilm ß, or equivalent) (optional)
- Liquid scintillation counter
- Speed-Vac evaporator/concentrator
- Geiger counter

Procedure

1. Add in order in a 1.5 ml sterile microcentrifuge tube (at room temperature):
   - 5 µl 5x transcription buffer
   - 1 µg Template DNA
   - 1.2 µl 10 mM rATP (480 µM final)
   - 1.2 µl 10 mM rCTP (480 µM final)
   - 1.2 µl 10 mM rGTP (480 µM final)
   - 1 µl 1 mM UTP (40 µM final)
   - 1 µl 0.75 M dithiothreitol
   - 1 µl RNAse Block II or 0.5 µl RNAsin
   - 5 µl [α-^{32}P]UTP (2.5 µM final) or α-^{35}S-UTP (2 µM final)
   - to 25 µl DEPC-treated water
   - 1 µl T3, T7 or SP6 RNA polymerase (10 units)
2. Incubate at 37-40°C for 60 min.
3. If gel purification **will not** be done:
   a. Add 1 µl of ribonuclease-free deoxyribonuclease, and incubate 15 min at 37°C.
   b. Increase the volume to 100 µl and extract once with 100 µl of phenol: chloroform:IAA (25:24:1).
   c. Extract once with 100 µl of chloroform:IAA (24:1).
   d. Add 0.5 volume of 7.5 M ammonium acetate and 2 volumes of ethanol. Place on dry ice for 30 minutes.
   e. Centrifuge at 10,000x g for at least 5 min (drain supernatant and monitor with Geiger counter).
   f. Resuspend the pellet in 100 µl 1 M ammonium acetate and repeat steps d and e.
   g. Dry the pellet under vacuum (Speed-Vac) and resuspend in DEPC-treated water. Count 1 µl in a liquid scintillation counter.
4. If gel purification **will** be done:
   a. Omit treatment with RQ1 RNAse-free DNAse
   b. Prepare a sequencing gel (5% polyacrylamide/6M urea)
   c. Bring riboprobe volume to 100 µl with DEPC-treated water and extract once with 100 µl of phenol:chloroform:IAA (25:24:1). Extract once with 100 µl of chloroform:IAA (24:1).
   d. Add 0.5 volumes of 7.5 M ammonium acetate. Add 2 volumes of ethanol, and place on dry ice for 30 min.
   e. Centrifuge at 10,000x g for at least 5 min (drain supernatant and moni-

tor with Geiger counter). Dry pellet under vacuum, and resuspend in 5 μl DEPC-treated water.
f. Add 10 μl of loading buffer, and run the entire sample on the gel (heat at 85°C for 5 minutes and place on ice before loading). Electrophorese at 60 W for at least 1.5 hours.
g. Separate plates and cover the gel with plastic wrap. Expose to X-ray film for 30-60 seconds. Locate the riboprobe on the autoradiogram and excise the corresponding gel band with a razor blade.
h. Elute the gel slice in 400 μl of elution buffer at 37°C for 4 hours with continuous agitation. Remove the eluate to a new microcentrifuge tube. Monitor the eluate and the gel slice with a Geiger counter (there should be more counts in the eluate).
i. Add 1 ml of ice-cold ethanol and place on ice for 15 minutes.
j. Centrifuge at 10,000x g for 15 minutes. All counts should precipitate with ethanol. Re-dissolve the RNA pellet in DEPC-treated water and count 1 μl in a liquid scintillation counter.

Comments

- For linearization of plasmid, it is important to select a restriction endonuclease that generates a blunt or protruding 5' end since recessed 5' ends cause aberrant initiation of transcription. Alternatively, recessed 5' ends can be repaired by treatment with T4 DNA polymerase in the presence of excess dNTPs (Ausubel et al. 1993).
- Do not place the in vitro transcription reaction mixture on ice; spermidine in the buffer may precipitate the template DNA.
- Although digestion with deoxyribonuclease is sufficient for preparation of riboprobe for hybridization experiments, we have found that for ribonuclease protection assays, complete removal of template DNA by polyacrylamide gel electrophoresis is required.
- The specific activity of the riboprobe can be adjusted by altering the ratio of labeled and unlabeled UTP, although the total concentration should remain greater than 20 μM to avoid premature termination.

References

Ausubel FM, Brent R, Kingston RE, Moore DD, Seidman JG, Smith JA, Struhl K (1993) Current Protocols in Molecular Biology, pp. 3.5.11-3.5.12.

Berkner KL, Folk WR (1977) Polynucleotide kinase exchange reaction. Quantitative assay for restriction endonuclease-generated 5'-phosphoryl termini in DNAs. J Biol Chem 252:3176-3184.

Biroc SL, Murphy-Erdosh C, Fisher JM, Payan DG (1993) The use of ^{33}P-labeled oligonucleotides for in situ hybridization of vertebrate embryo frozen sections. BioTechniques 15:250-254.

Chaconas G, van de Sande JH (1980) 5'-^{32}P labeling of RNA and DNA restriction fragments. Method Enzymol 65:75-88.

Deng GR, Wu R (1983) Terminal transferase: Use in the tailing of DNA and for in vitro mutagenesis. Method Enzymol 100:96-116.

Evans MR, Read CA (1992) ^{32}P, ^{33}P and ^{35}S: selecting a label for nucleic acid analysis. Nature 358:520-521.

Feinberg AP, Vogelstein B (1983) A technique for radiolabeling DNA restriction fragments to high specific activity. Anal Biochem 132:6-13.

Feinberg AP, Vogelstein B (1984) Addendum: A technique for radiolabeling DNA restriction endonuclease fragments to high specific activity. Anal Biochem 137:266-267.

Krieg PA, Melton DA (1987) In vitro RNA synthesis with SP6 RNA polymerase. Method Enzymol 155:397-415.

McLaughlin SK, Margolskee RF (1993) ^{33}P is preferable to ^{35}S for labeling probes used in in situ hybridization. BioTechniques 15:506-511.

Melton DA, Krieg PA, Rebagliati MR, Maniatis T, Zinn K, Green MR (1984) Efficient in vitro synthesis of biologically active RNA and RNA hybridization probes from plasmids containing a bacteriophage SP6 promoter. Nucl Acid Res 12:7035-7056.

Chapter 10

Hybridization Methods (Southern and Northern Blotting)

KAI-OLAF NETZER

Introduction

Hybridization methods represent standard techniques in molecular biology. In general, they are used to detect particular sequences (target) within a complex mixture of DNA or RNA molecules. DNA or RNA are usually transferred and immobilized to nitrocellulose or, more commonly, to nylon membranes. Complementary single-stranded probes are labeled radioactively or non-radioactively. When hybridized to the filter, probes bind to their complementary target sequence via hydrogen bonds. Unhybridized probe is then washed away, and specifically-bound probe is detected by autoradiography or color reaction.

Hydrogen bonds produce a stable but reversible association of double-stranded DNA. For replication and transcription in vivo, it is most important that the two DNA strands can be separated and reassociated under physiological conditions without the need to break covalent bonds. This natural function represents the basis for the in vitro techniques described in this section. The term hybridization refers to the association of complementary single-stranded DNA or RNA molecules to produce hybrid double-stranded molecules.

Hybridization can occur

- in solution with the participating nucleic acid molecules being present as single strands,
- with one strand (target) immobilized to a filter membrane, and the other strand (probe) in solution, and
- in situ, with the target strand fixed in tissue sections, cells, or chromosome smears, and the probe present in solution.

Kai-Olaf Netzer, Krankenhaus Köln-Merheim, Medizinische Klinik I, Ostmerheimer Str. 200, Köln, 51109, Germany (*phone* +49-221-8907-2200; *fax* +49-221-8907-2335; *e-mail* kai.netzer@uni-koeln.de)

At the molecular level, hybridization begins as soon as the ends of two complementary molecules come together. Once pairing of bases is established at one end, it proceeds in a zipper-like fashion throughout the length of the molecule. The complexity, i.e. the sequence diversity, limits the speed of this process. The more repetitive sequence elements there are (i.e. the lower the complexity), the faster hybridization proceeds.

Factors influencing hybridization of nucleic acids

Table 1. Factors Influencing the Stability of Hybrid DNA Molecules

	Increased hybrid stability	Decreased hybrid stability
Molecular Properties		
Length of hybrid	long	short
GC-content	high	low
Number of mismatches	low	high
Buffer Conditions		
Temperature	low	high
Salt concentration (NaCl)	high	low
pH	low	high
Concentration of denaturants	low	high

Factors that influence the formation and stability of hybrid DNA and RNA molecules are listed in Table 1. Length, GC-content and number of mismatches depend on the primary structures of hybrid molecules. Longer hybrid molecules have higher total binding energy and stability. Since there are three hydrogen bonds between each pair of GC, and two between each pair of AT (AU in RNA), GC-rich hybrids are more stable than AT-rich hybrids. (To separate pure GC sequences, a temperature of approximately 100°C is necessary.) If two strands are not perfectly complementary, hybridization leads to mispaired bases (so-called mismatches). The total binding energy of mismatch-containing hybrid molecules is reduced; hence they are less stable.

Buffer conditions that influence hybridization are temperature, salt concentration, pH, and concentration of denaturants. The conditions are modified in order to achieve a certain **level of stringency**, i.e. specificity of hybridization. Low stringency permits hybridization of imperfectly matched sequences, whereas high stringency permits hybridization only of complementary strands. High temperature and low salt (NaCl) concentration create high stringency. This applies also to the post-hybridization washes.

Denaturation and melting point

The reverse of hybridization, **denaturation**, is accomplished by the disruption of double-stranded molecules. This is most simply achieved by boiling. Denaturation leads to strand separation. At the **melting point**, 50% of all possible double-stranded hybrids are formed. The melting point of a DNA-DNA hybrid in solution can be calculated according to the following estimation:

$$T_m[^\circ C] = 81.5 + 16.6 \log (M_{Na}) + 0.41 (P_{CG}) - P_M - B/L - 0.65 (P_F)$$

M_{Na} = molar concentration of monovalent cation (e. g., Na^+)
P_{CG} = percentage of guanine and cytosine in hybrid sequence
P_M = percentage of mismatches (on average, 1% mismatch reduces T_m by 1°C)
B = complexity (500 for hybrids longer than 100 nucleotides, 675 for hybrids shorter than 100 nucleotides)
L = length in base pairs (bp)
P_F = percent formamide in the hybridization solution

The calculated melting point of hybrids may help to establish optimal hybridization and washing conditions. In filter hybridizations, however, the kinetics of hybrid formation and the T_m of hybrid molecules are different from those in solution hybridzation reactions. As a consequence, hybridization time and stringency for filter hybridizations must be determined mostly empirically.

For more details on nucleic acid hybridization, see Hames and Higgins (1985).

DNA blot hybridization

There are two ways to blot nucleic acids:

- Direct spotting of DNA or RNA solutions on to nitrocellulose or nylon filter membranes (dot or slot blot), and
- Transfer of DNA or RNA to filter membranes after size fractionation by agarose gel electrophoresis (Southern or Northern blot). DNA is digested with restriction enzymes prior to electrophoresis.

RNA blot hybridzation

RNA blot hybridization is a technique that involves the detection of specific target sequences within total RNA or mRNA that has been immobilized to a filter membrane. The subprotocols 4, 5 and 6 describe direct spotting of RNA on to a membrane filter (dot blot).

Subprotocol 1
Preparation of Dot Blots for DNA

Dot blots are used to determine the presence or absence of specific sequences within the target DNA. In comparison with standard samples, the amount of target DNA may also be quantified. It is not possible, however, to determine the number and sizes of bands that belong to the target sequence. To achieve this, Southern blotting is the method of choice.

DNA dot blot hybridization has two components:

- preparation of DNA blots (filters), and
- hybridization of the filter to specific probes (see Subprotocol 3).

For the preparation of filters, commercial dot blot or slot blot manifolds may be used. In this case, follow the manufacturer's instructions. A simple and reliable protocol for direct spotting of DNA on to the membrane is given in the following.

Materials

- DNA (approximately 0.5 µg/µl in TE buffer)
- TE buffer (10 mM Tris-HCl, pH8.0, 1 mM EDTA)
- 20x SSC (3 M NaCl, 0.3 M sodium citrate)
- Membrane filter: nitrocellulose filters, uncharged and charged nylon filters are available. Nylon filters have better mechanical properties and are, therefore, preferable. Charged nylon filters have slightly better DNA binding capacity than uncharged nylon membranes; however, they tend to give more background problems upon hybridization.

Procedure

In advance

1. Sonicate genomic DNA at approximately 70 W for 1 minute on ice. Check the fragment size by agarose gel electrophoresis. The fragments should have an average size of 7 to10 kb.

Note: DNA that has been digested with an appropriate restriction enzyme may also be used. DNA should be purified by using a chaotropic buffer/silica-gel spin column system (e. g. Qiagen). If unfragmented genomic DNA is used, hybridization may be less efficient.

2. Dilute the DNA to 0.5 µg/µl in TE buffer.

Preparation of dot blots

1. Heat 5 µg DNA (in 10 µl TE buffer) to 95°C for 10 minutes, then chill on ice for 5 min. Collect the volume by brief centrifugation in a microfuge. Place tube on ice.
2. Cut a piece of filter membrane to the appropriate size. Mark the locations where the DNA will be spotted with a soft pencil. A one cm-grid may be helpful.
3. Prewet the membrane briefly in distilled water, then saturate it in 20x SSC. Place the membrane on a filter paper wad (Whatman 3MM) pre-saturated in 20x SSC.
4. Spot the DNA samples on to the filter in 5 µl aliquots. Allow each spot to dry before applying the second aliquot.

Note: The filter should remain damp but not too wet throughout the spotting of DNA samples.

5. Layer the filter membrane briefly on a sheet of dry filter paper (Whatman 3MM) to blot dry. Rinse in 5x SSC for 5 minutes.
6. Fix the DNA to the membrane. Nitrocellulose should be baked for 2 hours at 80°C under vacuum. Nylon filters should be wrapped in Saran-Wrap(or any comparable plastic wrap while still wet, placed DNA side down on a UV transilluminator and exposed to UV light (280 nm) for 3-5 minutes. The filter is then ready for hybridization.

Note: If not hybridized immediately, nitrocellulose filters should be stored under vacuum. Nylon filters should be sealed in plastic bags and stored at 4°C.

Subprotocol 2
Preparation of Southern Blots for Genomic DNA

The preparation of Southern blots involves DNA digestion with restriction enzymes, size fractionation by agarose gel electrophoresis, and transfer from the agarose gel to a filter membrane by capillary flow of buffer with high ionic strength, i.e. high sodium content (20x SSC). The technique was described by Southern (1975), hence the name Southern blot. (With reference to the Southern transfer of DNA, transfer of RNA to a filter membrane has been named Northern blot, and transfer of proteins to a filter membrane has been named Western blot.) Transfer of DNA from gel to filter membrane can be accelerated by the use of vacuum or positive pressure blot apparatuses which are commercially available from several suppliers.

Applications

Southern blots are used, for instance, to study genetic diseases. Genomic DNA for the analysis of the human genome can be extracted from any nucleated cell of an organism. Once blots are prepared from size-fractionated DNA samples, the presence of abnormal sequences or the absence of normal nucleic acid sequences can be verified by hybridization to specific probes. Bands of altered size, or missing bands, may indicate structural aberrations of a gene. On the other hand, variation in the size of restriction fragments may as well be caused by genetic polymorphisms (e.g. point mutations in the recognition sequence of restriction enzymes) that do not necessarily have a pathogenic effect. Polymorphisms are inherited according to mendelian principles. They result in restriction fragment length polymorphisms (RFLP) that can be utilized as markers for specific sites within the genome. Hence they can be used for linkage analyses (see Chap. 23).

The sensitivity of Southern blotting is sufficient to detect single-copy genes, i.e. genes that exist in only one copy per genome. For medical applications, the power of the method resides in the ability to analyze a small portion of the primary structure of human genomic DNA.

Materials

- Agarose gel containing size fractionated DNA which has been digested to completion with restriction enzymes prior to gel electrophoresis
- 20x SSC (3 M NaCl, 0.3 M sodium citrate)
- 0.25 M HCl
- Denaturing buffer (1.5 M NaCl, 0.5 M NaOH)
- Neutralization buffer (1.5 MNaCl, 0.5 M Tris-HCl, pH 7.5)
- Membrane filter: nitrocellulose filters, uncharged and charged nylon filters are available. Nylon filters have better mechanical properties and are therefore preferable. Charged nylon filters have slightly better DNA binding capacity than uncharged nylon membranes, however, they tend to give more background problems upon hybridization.

Procedure

In advance

1. Digest genomic DNA samples (10 μg) with the appropriate restriction enzyme according to the principles given in Chap. 2.
2. Run an agarose gel as described in Chap. 4. Usually, 0.7-1.0% agarose gels are used to separate genomic DNA samples. They give a good resolution over a size range of one to 15 kb. Either TBE or TAE buffer may be used. Gels should be run at a voltage of 1 V/cm. Large gels (20 x 25 cm) should be used if several bands of similar size are to be separated. Convenient size markers are HindIII-digested Lambda DNA, HaeIII-digested $\varphi \times 174$ DNA and 100 or 1000 bp size ladders. After electrophoresis, photograph the gel on a UV transilluminator with a ruler alongside the gel. Take good care not to drop or break the gel while moving it from the tray to the illuminator.

Preparation of Southern blots

1. Transfer the gel to a plastic box.

2. Add at least 4 gel volumes 0.25 M HCl to depurinate the DNA. Incubate for 15 minutes at room temperature on a rocker. The bromophenol blue from the gel loading buffer turns yellow during this procedure. If there is still blue color after 15 minutes, incubate for additional 5 minutes.

Note: Depurination facilitates capillary transfer of DNA bands of 10 kb and more. Hence, it may be omitted if only DNA bands of less than 10 kb are of interest.

3. Carefully drain the HCl from the plastic box. Rinse the gel once in distilled water.
4. Add at least 4 gel volumes of denaturing buffer (caustic; wear gloves and protective goggles!). Incubate for 20 minutes at room temperature on a rocker.
5. Repeat step 4 with fresh buffer. Carefully drain the denaturing buffer. Rinse the gel briefly with distilled water.
6. Add at least 4 gel volumes neutralization buffer. Incubate for 15 minutes at room temperature on a rocker.
7. Repeat step 6 with fresh buffer.
8. While treating the gel, cut a piece of filter membrane (nitrocellulose or nylon) to a size slightly larger than the gel. Prewet the filter membrane in distilled water, then soak in 20x SSC for at least 15 minutes.
9. Set up the transfer apparatus as shown in Figure 1. Fill a tray with blotting buffer (20x SSC). Make a platform above the buffer level, e.g. by inverting a gel tray, and cover it with a wick made from 3 sheets of filter paper (Whatman 3MM) saturated with 20x SSC. The wick should hang into the buffer on both sides of the platform. Use a 10 ml glass pipet to roll out air bubbles and to flatten the paper wick.

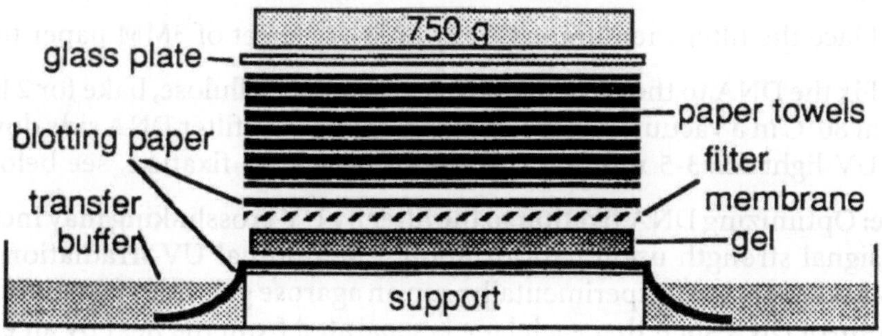

Fig. 1. Capillary transfer from agarose gels

10. Pour several ml of 20x SSC on top of the paper wick, then place the gel upside down on the wick. Carefully remove any air bubbles trapped in between the gel and paper wick. Surround the gel with cling film or plastic wrap to prevent buffer flow around the gel.

Note: The use of a glass pipet may be helpful to roll out air bubbles. Take care not to distort or squeeze the gel.

11. Cover the gel with several ml of 20x SSC. Place the filter membrane on top of the gel. Make sure that no air bubbles are trapped between the gel and the filter.
12. Place 3 sheets of 3MM paper cut to size and pre-wetted with 20x SSC on top of the filter membrane.
13. Place a stack of dry absorbent paper (blotting paper, or paper towels) on top of the 3MM paper (approximately 5-8 cm high).
14. Place a glass plate on top of the stack of paper, and put a 0.75-1 kg weight on top.
15. Allow transfer of DNA to proceed for 12-16 hours (e.g., overnight). Make sure that there is enough 20x SSC in the reservoir (1-3 l).
16. After completion of capillary transfer, carefully disassemble the blotting apparatus. Transfer the membrane together with the gel to a dry filter paper, gel side up. Mark the position of the gel and the wells on the filter membrane by using a soft pencil. Peel off the gel.

Note: To check efficiency of transfer, the gel may be stained with ethidium bromide (1 μg/ml) in water for 20 minutes, and then photographed on a UV transilluminator.

17. Wash the filter membrane briefly in 5x SSC to remove any traces of agarose.
18. Place the filter membrane briefly on a dry sheet of 3MM paper to dry.
19. Fix the DNA to the filter membrane. For nitrocellulose, bake for 2 hours at 80°C in a vacuum oven. For nylon, expose the filter DNA side down to UV light for 3-5 minutes (for optimizing DNA fixation, see below).

Note: Optimizing DNA fixation to the filter by UV crosslinking may increase the signal strength upon hybridization. The optimal UV irradiation time may be determined experimentally: run an agarose gel with 5 identical lanes of control DNA such that each lane is separated from the next by an empty lane. Treat and blot the gel as described above, and expose each lane to UV radiation for different times (e.g. 1, 2, 3, 4, 5 minutes). Hybridize the filter to the appropriate probe, autoradiograph, and determine the optimal signal strength.

20. The filter is now ready for hybridization or can be stored at 4°C. Nitrocellulose filters should be stored under vacuum. Nylon filters should be sealed in plastic bags.

Subprotocol 3
Hybridization of DNA Blots

Many different protocols have been described for the hybridization of DNA blots. In general, they fall into two categories: hybridization at 42°C in the presence of formamide, and hybridization at 65°C without formamide. The speed of nucleic acid reassociation is higher at 65°C, so that 16 hours hybridization are sufficient to detect single-copy genes in genomic DNA blots. In the following, a standard protocol for hybridization at 65°C is given. Normally, it is used with radioactively-labeled probes. With modifications, however, it may also work when non-radioactively-labeled probes are used. Refer to the manufacturer's instructions if you use non-radioactive label.

Materials

- DNA probe labeled to high specific activity (approximately 1 x 10^9 cpm/µg) with e.g. random primed labeling (for a protocol see Chap. 9)
- DNA filter
- 20x SSPE (3.6 M NaCl, 0.2 M NaH_2PO_4, 0.02 M EDTA, adjust the pH to 7.5 with 10 N NaOH [approximately 6 ml/l])
- 100x Denhardt's solution (2 g/100 ml bovine serum albumin, 2 g/100 ml Ficoll 400 [e.g. Pharmacia], 2 g/100 ml polyvinylpyrrolidine [PVP])
- Denatured, fragmented salmon sperm DNA (10 mg/ml). After complete dissolution pass 15 times through a 17-gauge hypodermic needle. Alternatively, sonicate at 70 W for 2 minutes on ice, then denature by boiling for 10 minutes. Chill quickly on ice.
- 10% SDS in water

- 20x SSC
- 20% (w/v) dextran sulfate in water (optional)
- Ultrapure water
- Prehybridization buffer. To make 10 ml, add

20x SSPE	2.5 ml	
100x Denhardt's solution	0.5 ml	
Salmon sperm DNA	0.1 ml	
Water	6.8 ml	(5.9 ml for nylon membranes)
10% SDS	0.1 ml	(1.0 ml for nylon membranes)

The solution should be prepared fresh from stock solutions before use. Add SDS as last component. Denhardt's and salmon sperm DNA stock solutions should be stored frozen at –20°C.

- Hybridization buffer (only required if dextran sulfate is to be used; otherwise, prehybridization buffer can be used; see note below). To make 10 ml, add add

20x SSPE	2.5 ml	
100x Denhardt's solution	0.5 ml	
Salmon sperm DNA	0.1 ml	
20% dextran sulfate	5.0 ml	(optional, see below)
Water	1.8 ml	(0.9 ml for nylon membranes)
10% SDS	0.1 ml	(1.0 ml for nylon membranes)

Note: Dextran sulfate may be added to increase sensitivity. However, increased background problems sometimes occur with dextran sulfate. Hybridization solution should be prepared fresh from stock solutions before use. Add SDS as last component.

- Wash buffer 1 (2x SSC, 0.1% SDS)
- Wash buffer 2 (0.25x SSC, 0.1% SDS)
- Hybridization bag
- Heat sealer (industrial standard is preferable over household plastic bag sealer)

- Water bath (65°C) or oven (65°C)
- Heating block (95°C)
- X-ray cassette with intensifying screen
- X-ray film (e.g. Kodak X-OMAT AR or Dupont Cronex)

Procedure

Hybridization

1. Wet the DNA filter in 5x SSPE.
2. Transfer the wet DNA filter into a plastic bag. Seal the bag at four sides and cut off one corner.
3. Add prehybridization solution (0.1 ml/cm[2)]) to the bag through the open corner. Try to avoid bubbles by careful pipetting. Squeeze out all air bubbles from the bag. Seal the bag completely.
4. Incubate the bag in a shaking water bath or between two glass plates in an oven for 2-4 hours at 65°C. Make sure that there are no air-bubbles on the filter surface.
5. Denature the probe for 10 minutes at 95°C. Immediately chill on ice for 5 minutes.
6. Cut off a corner of the hybridization bag. Empty the prehybridization buffer into a 15 or 50 ml Falcon tube. Add probe to the prehybridization buffer or to freshly prepared hybridization buffer (if the use of dextran sulfate is intended) at approximately 10-20 ng/ml (random primed labeling) or 50-100 ng/ml (nick translation). In general, 10^6-10^7 cpm/ml should suffice. Mix gently. Add the hybridization solution to the filter by using a disposable plastic pipet.

Note: If the use of dextran sulfate is necessary to improve sensitivity (see above), prepare fresh hybridization solution, preheat to 65°C, then add denatured probe. Mix gently, and add the solution to the filter. Do not add radioactive probe directly into the bag.

7. Squeeze out all air bubbles through the open corner of the bag. Use a paper towel to blot off traces of hybridization buffer that come out. Seal the bag carefully. Avoid leakage. If necessary, seal the bag in a second bag to prevent contamination.

Note: Wear gloves when working with radioactive probes, and work behind plexiglass shielding. Work according to your laboratory's safety rules. Dispose of radioactive waste properly.

8. Incubate at 65°C for 12-16 hours in a shaking waterbath or between two glass plates in an oven.
9. After hybridization, open the bag at one corner and squeeze out the hybridization mix into a 15 or 50 ml Falcon tube. Take care not to spill radioactive solution.

Note: The hybridization solution may be reused once.

10. Open the bag completely. Use blunt forceps to transfer the filter to a tray for the subsequent washes. Immediately rinse the filter with wash buffer 1.

Note: Take care that the filter does not dry.

11. Wash three times for 5 minutes with wash buffer 1 at room temperature.
12. Wash two times for 5 minutes with wash buffer 2 at room temperature.
13. Wash one to three times for 15 minutes with preheated wash buffer 2 at 65°C. Between the buffer changes, use a Geiger counter to check the radioactivity remaining in the center of the blot, where specific signal is, expected in comparison with the margins of the blot, where background signal is expected.

Note: The washes at 65°C containing decreased SSC concentration represent stringent washes. If higher stringency is required, increased time (30 minutes) or decreased salt concentration (0.1x SSC) may be used. Washes can be carried out in plastic boxes with tightly sealing lids in shaking water baths. Alternatively, washes can be carried out in a hybridization oven using roller bottles or a rocker.

14. After sufficient washing, **briefly** lay the filter on a sheet of 3MM paper to blot dry. Seal the damp filter in a thin walled plastic bag or plastic wrap.

Note: The filter should not be dried completely if rehybridization is intended. Probe cannot be stripped from the filter if the filter had been dried.

15. Autoradiography of the filter: place the blot (wrapped in plastic seal, DNA side up) on the bottom of the X-ray cassette, then lay the film on top. The intensifying screen can be conveniently affixed to the cover of the cassette. Close the cassette. Expose at –70°C for overnight or up to two weeks.

Note: One can save exposure time by placing two films on top of the blot. The film on top can be developed earlier to check the autoradiograph while the other film – which should be taped to the blot – is still exposed.

Subprotocol 4
Preparation of RNA Dot Blots

RNA dot blots are used to determine the presence or absence of specific transcripts. In comparison with standard samples, the amount of target RNA may also be quantified. It is not possible, however, to determine the sizes of mRNA transcripts. To achieve this, Northern blotting is the method of choice.

Materials

Use RNase-free reagents or DEPC-treated reagents only. Wear gloves when working with RNA.

- Formaldehyde, 37% solution (handle in a fume hood)
- Formamide (handle in a fume hood)
- 20x SSC (3 M NaCl, 0.3 M sodium acetate)
- Denaturing solution (500 µl formamide, 162 µl formaldehyde [37% solution], 100 µl 20x SSC)
- Membrane filter. The use of nylon membranes is recommended due to their superior handling properties.

Procedure

RNA blot hybridization

1. Cut a piece of filter membrane to the appropriate size. Mark the locations where the RNA will be spotted with a soft pencil. A one cm-grid may be helpful.
2. Soak the membrane in 20x SSC.
3. Denature 10-20 µg total RNA by incubation for 5 minutes at 65°C in the following solution:

Total RNA (10-20 μg)	2.0 μl
Denaturing solution	6.0 μl

Quickly chill on ice for 5 minutes. Spin briefly in a microfuge to collect the sample. Keep the tube on ice.

Note: If the desired amount of RNA is present in a volume greater than 2 μl, the reaction can be scaled up. Always add 3 volumes of denaturing solution.

4. Layer the membrane on a wad of filter paper (e.g. Whatman 3MM) soaked with 20x SSC.
5. Spot the RNA samples onto the filter in 4 μl aliquots. Allow each spot to dry between each application.

Note: The filter should remain damp but not too wet throughout the spotting of RNA samples.

6. Place the filter membrane briefly on a sheet of dry filter paper (3MM) to blot dry the membrane. Rinse briefly in 20x SSC.
7. Wrap the filter in plastic wrap while still damp. Fix RNA to the filter by irradiating the filter RNA side down on a UV transilluminator for 3-5 minutes. The filter is then ready for hybridization.

Note: For storage, nylon filters should be sealed in plastic bags and kept at 4°C.

Subprotocol 5
Preparation of Northern Blots

For Northern blotting, RNA is first size-separated by denaturing agarose gel electrophoresis, blotted to nitrocellulose or, more commonly, to a nylon filter membrane, and hybridized. By analogy to Southern blotting which involves blotting of DNA, this procedure has been named Northern blotting. Usually, Northern blots are used to detect transcription of specific mRNA molecules in different cells, tissues, or organs, and to study regulation of transcription in response to various stimuli. Furthermore, Northern hybridization may be used to determine the length of RNA transcripts. Alternative splicing, for example, may be detected by variable transcript lengths. Densitometry and comparison with controls permits quantification of RNA molecules in order to characterize the level of expression.

Changes in size or level of expression of mRNA may be the cause for altered amount of protein product or altered biological activity.

Materials

Use RNase-free reagents or DEPC-treated reagents only. Wear gloves when working with RNA.

- Formaldehyde/agarose gel containing size separated RNA
- Formaldehyde, 37% solution (handle in a fume hood)
- Formamide (handle in a fume hood)
- 10x MOPS (0.2 M 3-[N-morpholino] propanesulfonic acid, 0.5 M sodium acetate pH 7.0, 0.01 M EDTA)
- Loading buffer (50% glycerol, 0.1 mg/ml bromophenol blue, 0.1 mg/ml ethidium bromide [optional, if direct UV-visualization of RNA is desired], in 1x MOPS buffer)
- Water bath (50°C)
- 20x SSC (3 M NaCl, 0.3 M sodium citrate)
- Membrane filter. The use of nylon membranes is recommended due to their superior handling properties.

Procedure

In advance

1. Run a formaldehyde/agarose gel according to the following protocol with 10-20 μg total RNA per lane or 0.5-1 μg poly(A)$^+$ RNA, and photograph the gel on a UV-transilluminator alongside a ruler (see Chap. 4).

a. Incubate total RNA (10-20 μg) for 5 minutes at 65°C in the following solution:

Total RNA (10-20 μg)	6.0 μl
Formamide	12.5 μl
10x MOPS buffer	2.5 μl
Formaldehyde (37%)	4.0 μl

b. Quickly chill on ice, add 2.5 µl loading buffer, and load on to a formaldehyde/agarose gel prepared as follows:

Agarose	1.0 g
10x MOPS buffer	10.0 ml
Water	73.0 ml

c. Heat to 100°C to dissolve the agarose, then cool to 50°C in a water bath. Add 17 ml formaldehyde, mix and pour the gel immediately. Work under a fume hood and wear gloves. Use an electrophoresis chamber that is reserved for RNA electrophoresis.

Northern blots

1. While the gel is still running, cut a sheet of filter membrane to the size of the gel. Soak in 20x SSC.
2. Set up the transfer apparatus as shown in Figure 1. Fill a tray with blotting buffer (20x SSC). Make a platform above the buffer level, e.g. by inverting a gel tray and cover it with a wick made from 3 sheets of filter paper (Whatman 3MM) saturated with 20x SSC. The wick should hang into the buffer on both sides of the platform. Use a 10 ml glass pipet to roll out air bubbles and to flatten the paper wick.
3. Pour several ml of 20x SSC on top of the paper wick, then place the gel without further treatment upside down on the wick. Carefully remove any air bubbles trapped between gel and paper wick. Surround the gel with cling film or plastic wrap to prevent buffer flow around the gel.

Note: The use of a glass pipet may be helpful to roll out air bubbles. Take care not to distort or squeeze the gel.

4. Cover the gel with several ml of 20x SSC. Place the filter membrane on top of the gel. Make sure that no air bubbles are trapped between the gel and the filter. Mark the side of the filter membrane that faces the gel with a soft pencil.
5. Place 3 sheets of 3MM paper cut to size and pre-wetted with 20x SSC on top of the filter membrane.
6. Place a stack of dry absorbent paper (blotting paper, or paper towels) on top of the 3MM paper (approximately 5-8 cm high).
7. Place a glass plate on top of the stack of paper, and put a 0.75-1 kg weight on top.

8. Allow transfer to proceed for 12-16 hours. Make sure that there is enough 20x SSC in the reservoir (approximately 3 l).
9. After blotting, carefully disassemble the blotting apparatus. Transfer the membrane together with the gel to a dry filter paper, gel side up. Mark the position of the gel and the wells on the filter membrane by using a soft pencil. Peel off the gel.
10. Wash the filter membrane briefly in 20x SSC to remove any traces of agarose.

Note: It is important to wash the filter in 20x SSC. Washing with lower ionic strength buffer may remove RNA from the filter.

11. Place the filter membrane briefly on a dry sheet of 3MM paper to blot dry.
12. Fix the RNA to the filter membrane. Expose the nylon filter RNA side down to UV light for 3-5 minutes as described above (see Chap. 10).
13. The filter is now ready for hybridization or can be stored at 4°C. Nylon filters should be sealed in plastic bags.

Subprotocol 6
Hybridization of RNA Blots

Usually, RNA blots are hybridized at 42°C in hybridization buffers containing 50% formamide. Higher temperatures increase the rate of RNA degradation and are therefore not recommended. The following is a standard protocol.

Materials

- DNA or RNA probe labeled to high specific activity (approximately 10^9 cpm/μg) with e.g. random primed labeling (DNA), or in vitro transcription (RNA). For protocols see Chap. 9.
- RNA containing Northern filter
- 20x SSPE (3.6 M NaCl, 0.2 M NaH_2PO_4, 0.02 M EDTA, adjust the pH to 7.5 with 10 N NaOH [approximately 6 ml/l])

- 100x Denhardt's solution (2 g/100 ml bovine serum albumine, 2 g/100 ml Ficoll 4000 [e.g. Pharmacia], 2 g/100 ml polyvinylpyrrolidine [PVP])
- Denatured, fragmented salmon sperm DNA (10 mg/ml). After complete dissolution pass 10 times through a 17-gauge hypodermic needle. Alternatively, sonicate at 70 W for 2 minutes on ice, then denature by boiling for 10 minutes. Chill quickly on ice.
- 10% SDS in water
- 20x SSC
- Dextran sulfate
- Ultrapure water
- Prehybridization buffer. To make 10 ml, add

20x SSPE	2.5 ml
100x Denhardt's	0.5 ml
Salmon sperm DNA	0.1 ml
Formamide	5.0 ml
Water	0.9 ml
10% SDS	1.0 ml

The solution should be prepared freshly from stock solutions before use. Add SDS as last component.

- Hybridization buffer is identical to prehybridization buffer (see above). 10% (w/v) dextran sulfate may be added to the hybridization solution if sensitivity is a problem. However, increased background problems sometimes occur with dextran sulfate. If adding dextran sulfate, reduce the amount of water appropriately. Heat the buffer to 42°C to dissolve dextrane sulfate. Mix gently.
- Wash buffer 1 (2x SSC, 0.1% SDS)
- Wash buffer 2 (0.25x SSC, 0.1% SDS)
- Hybridization bag
- Heat sealer (industrial standard is preferable over household plastic bag sealer)
- Water bath (42°C) or hybridization oven (42°C)

- Heating block (95°C)
- X-ray cassette with intensifying screen
- X-ray film (e.g. Kodak or Dupont)

Procedure

Hybridization

1. Wet the RNA filter in 5x SSPE.
2. Transfer the wet RNA filter into a plastic bag. Seal the bag at four sides. Cut off one corner.
3. Preheat prehybridization solution to 42°C. Add 0.1 ml/cm^2 to the bag through the open corner. Try to avoid bubbles by careful pipetting. Squeeze out all air bubbles from the bag. Seal the bag completely.
4. Incubate the bag in a shaking water bath or between two glass plates in an oven for 2-4 hours at 42°C. Make sure that there are no air-bubbles on the filter surface.
5. Denature the probe for 10 minutes at 95°C. Immediately chill on ice for 5 minutes.
6. Cut off a corner of the hybridization bag. Empty the prehybridization buffer into a 15 or 50 ml Falcon tube. Add probe at approximately 10-20 ng/ml (random primed labeling). In general, 10^6-10^7 cpm/ml should suffice. Mix gently. Add the hybridization solution to the filter by using a disposable plastic pipet.

Note: If it is necessary to improve sensitivity (see above), prepare hybridization solution with dextran sulfate (10% w/v) freshly, preheat to 42°C, then add denatured probe. Mix gently, and add the solution to the filter. Do not add radioactive probe directly into the bag.

7. Squeeze out all air bubbles through the open corner of the bag. Use a paper towel to remove traces of hybridization buffer that come out. Seal the bag carefully. Avoid leakage. If necessary, seal the bag in a second bag to prevent contamination.

Note: Wear gloves when working with radioactive probe. Work according to your laboratory's safety rules. Dispose of radioactive waste properly.

8. Incubate for 12-16 hours at 42°C in a shaking waterbath or between two glass plates in an oven.

9. After hybridization, open the bag at one corner and squeeze out the hybridization mix into a 15 or 50 ml Falcon tube. Take care not to spill radioactive solution.

Note: The hybridization solution may be reused once.

10. Open the bag completely. Use blunt forceps to transfer the filter to a tray for the subsequent washes. Immediately rinse the filter with wash buffer 1.

Note: Take care that the filter does not dry.

11. Wash three times for 5 minutes with wash buffer 1 at room temperature.
12. Wash two times for 5 minutes with wash buffer 2 at room temperature.
13. Wash one to three times for 15 minutes with preheated wash buffer 2 at 45°C. Between the buffer changes, check the radioactivity remaining in the center of the blot, where specific signal is expected, in comparison with the margins of the blot, when background signal is expected.

Note: The washes at 42°C containing decreased cation concentration represent stringency washes. If higher stringency is required, increased time (30 minutes) or decreased salt concentration (0.1x SSC) may be used. The washes can be carried out in glass trays kept in shaking water baths. Alternatively, washes can be carried out in plastic boxes on a rocker.

14. After sufficient washing, **briefly** blot the filter on a sheet of 3MM paper. Seal the damp filter in a plastic bag or plastic wrap.

Note: The filter should not be dried completely if it is to be reused. Probe cannot be stripped from the filter if the filter had been dried.

15. Autoradiography of the filter: place the blot (wrapped in plastic seal, RNA side up) on the bottom of the X-ray cassette, then lay the film on top. For convenience, the intensifying screen can be affixed to the cover of the cassette. Close the cassette. Expose at –70°C for one day up to two weeks.

Note: is convenient to place two films on top of the blot. The film on top can be developed earlier to check the autoradiograph while the other film is still exposed to the blot. One film should be taped to the blot so that it cannot be displaced.

16. When removing the cassette from the freezer, allow sufficient time (30 minutes to one hour) for temperature equilibration. In a dark room, remove the film from the cassette and develop in an automatic X-ray processor.

Subprotocol 7
Membrane Stripping

Procedure

Probes can be removed from hybridized membranes when rehybridization is desired. Membranes must never be allowed to dry during or after hybridization and washing. Many protocols have been described for membrane stripping. A simple one is the following:

Membrane stripping

1. Boil a solution of 0.1 % SDS.
2. Place the membrane in a glass tray. Pour the solution on the membrane. Allow to cool to room temperature.
3. Check the membrane with a Geiger counter to monitor a decrease in activity.

Note: Filters that have been stored for a long time (5 half-lives) of ^{32}P have less than 5% of their initial activity. They may be used without stripping after the absence of signal has been documented by exposure for an appropriate time.

References

Hames BD, Higgins SJ (ed.) (1985) Nucleic acid hybridisation: a practical approach. IRL Press, Oxford New York Tokyo

Southern, EM (1975) Detection of specific sequences among DNA fragments separated by agarose gel electrophoresis. J Mol Biol 98:503-517

Specimen Preparation Techniques for Cytogenetic Analysis by Fluorescence in situ Hybridization

STEFAN JOOS, MARTIN BENTZ, ANTON H. N. HOPMAN AND PETER LICHTER

Introduction

Analysis of chromosomal DNA by **fluorescence in situ hybridization (FISH)** is used in many areas of biological research as well as in clinical cytogenetics. A major advantage of this method for the clinical applications is that chromosomal changes can be diagnosed not only on **metaphase chromosomes** but also in **interphase nuclei** (see Figure 1 and Chap. 12). Thus, the analysis is not dependent on the preparation of metaphase chromosomes, which are often difficult or impossible to obtain. Furthermore, preparation of metaphase chromosomes often requires stimulation of cells by mitogens. Such mitogenic stimulation frequently results in selective growth of a subset of cells in the sample. This may lead to the diagnosis of chromosomal aberrations that do not reflect the composition of the tumor in vivo.

Interphase analysis permits the investigation of a wide range of different tissues by means of molecular cytogenetics. The following descriptions provide techniques for preparation of clinical materials commonly used for analysis by FISH. For further information on this topic the reader is referred to the literature (see for example Manuelidis 1985; Burns et al. 1986; Hopman et al. 1988; Emmerich et al. 1989a,b; Walt et al. 1989; Hopman et al. 1990; Van Dekken et al. 1990; Hopman et al. 1991; Arnoldus et al. 1991; Hopman et al. 1992; Scherthan and Cremer 1993; Lichter et al. 1996). Recent technical developments of high diagnostic relevance include comparative

Stefan Joos, Deutsches Krebsforschungszentrum, Im Neuenheimer Feld 280, Heidelberg, 69120, Germany (*phone*), Martin Bentz, Universität Heidelberg, Medizinische Klinik and Polyklinik V, Heidelberg, 69120, Germany, Anton H. N. Hopman, University of Limburg, Dept. Molecular Cell Biology & Genetics, Postbus 616, Maastricht, 6200, The Netherlands, *Correspondence to* Peter Lichter, Deutsches Krebsforschungszentrum, Abt. Organisation komplexer Genome, Im Neuenheimer Feld 280, Heidelberg, 69120, Germany (*phone* +49-6221-424-609; *fax* +49-6221-424-639; *e-mail* P.Lichter @dkfz-heidelberg.de)

genomic hybridization (CGH) (Kallioniemi et al. 1992; du Manoir et al.1993; Joos et al. 1993; Joos and Lichter 1997) and multicolor FISH (Speicher et al. 1996; Schröck et al. 1996; Lichter 1997).

Preparation of chromosomes from peripheral blood cells

Because of their easy accessability peripheral blood lymphocytes are most commonly used for cytogenetic analyses of constitutional chromosomal aberrations. In general, heparinized blood is cultured in the presence of mitogens (e.g. phytohemagglutinin, pokeweed mitogen or concanavalin A) to re-stimulate non-cycling lymphocytes. Cells are arrested in metaphase by application of colcemid, then treated with a hypotonic medium to increase the nuclear volume. Finally, cells are fixed in methanol/acetic acid and spread on glass slides. The protocol given below can be used for primary cells from normal or leukemic individuals as well as for lymphoblastoid cell lines. In the latter case, no stimulation by mitogens is needed. It should be noted that the yield of metaphase chromosomes is often much lower when clinical samples are investigated.

Subprotocol 1
Cultivation of Blood Lymphocytes

Materials

- NH_4-heparinized blood.
- Culture medium: RPMI 1640 supplemetned with fetal calf serum (10%), L-glutamine (1%), penicillin (1%), streptomycin (1%), phytohemagglutinin (PHA) (1.5% in water).

Procedure

Cultivation of lymphocytes

1. Add 0.5-1 ml heparinized blood to 10 ml culture medium.
2. Incubate at 37°C (optional: in an atmosphere of 5% CO_2) for 71 h. Carefully resuspend settled cells once per day.
3. Proceed as described in Subprotocol 2.

Subprotocol 2
Preparation of Metaphase Chromosomes and Interphase Nuclei from Blood Lymphocytes

Materials

- Colcemid solution: prepare stock solution of 10 μg/ml in water and store at -20°C.
- Hypotonic solution (per l): 3g KCl; 0.2 g EGTA; 4.8 g HEPES; adjust pH to 7 with 1 M KOH.
- Fixative solution: 3 parts methanol (abs.), 1 part acetic acid (glacial).

Procedure

In advance

Pre-cleaned microscopic slides are washed in ethanol for several hours, dipped several times in water and dried immediatedly before cells are dropped onto them.

Preparation of chromosomes and nuclei

1. Following incubation of blood cells in culture medium for 71 h add colcemid stock solution to a final concentration of 0.1 μg/ml.
2. Incubate for 10-60 min at 37°C (incubation time should be optimized for different cell sources).
3. Aliquot into two 15 ml tubes (polystyrene) and centrifuge at 200 g for 10 min in a clinical centrifuge.
4. Remove supernatant, resuspend cell pellet in remaining medium, and add a few drops of hypotonic solution.
5. Add 5 ml of hypotonic solution to each tube, resuspend and incubate for 15 min at 37°C.
6. Spin at 200 g for 10 min.
7. Remove supernatant, resuspend the pellet, add a few drops of ice-cold fixative solution.
8. Slowly (!) add 5 ml of cold fixative solution to each tube while constantly whirling the suspension.
9. Centrifuge at 200 g in a cooled (4°C) centrifuge.

10. Repeat steps 7-8.
11. Incubate the tube on ice for 30 to 60 min.
12. Spin at 200 g in a cooled (4°C) centrifuge and repeat steps 7-9 at least 5 more times (keep tubes on ice).
13. Resuspend cells in a small volume (e.g. 0.5 - 1 ml) fixative solution.
14. Drop the cells from a distance of about 10-50 cm on pretreated slides (see above). Slides should be kept in a humidified enviroment while the cells are dropped. Check concentration of cells by microscopy. If the concentration of cells is too high, add fixative solution, if it is too low, spin again and resuspend in smaller volume.
15. Keep the slides in a humidified chamber for another 5 min.
16. Air dry slides.
17. Dehydrate slides in a series of 70%, 90%, 100% ethanol for 5 min each.
18. Air dry again and keep slides at room temperature for one day.
19. Slides can either be used directly for in situ hybridization or stored at -70°C for several month. Try to prevent water condensation when thawing the slides before FISH.
20. If no pepsin digestion step is included (see Subprotocol 3) proceed with fluorescence in situ hybridization protocol (see Chap. 12).

The penetration of probe molecules can be significantly facilitated by mild digestion with pepsin prior to FISH. This may be a critical step for obtaining strong hybridization signals in clinical samples. Therefore, we describe a brief method for pepsin digestion of chromosome spreads (Raap et al. 1992). The optimal amount of pepsin may vary for different sources of cells.

Subprotocol 3
Pepsin Digestion of Blood Lymphocytes

Materials

- Phosphate-buffered saline (PBS).
- 0.01 M HCl.

- Pepsin (e.g. from porcine stomach mucosa, 500 - 3000 units/mg).
- Digestion solution: 1-2 mg pepsin/100 ml in 0.01 M HCl.
- 0.5% Paraformaldehyde-PBS-solution (freshly prepared).

Procedure

In advance

0.5% Paraformaldehyde-PBS-solution: In a fume hood (!) dissolve paraformaldehyde in PBS/5mM $MgCl_2$, heat at 80°C, adjust to pH 7 using 1M NaOH. Under these conditions, most of the paraformaldehyde solution is dissolved; the remainder is removed by paper filtration.

Pepsin digestion

1. Wash slides containing fixed chromosomes in PBS for 5 min at 37°C.
2. Incubate in digestion solution for 10 min at 37°C.
3. Wash slides in PBS for 5 min at room temperature.
4. Immerse slides in 0.5% paraformaldehyde-PBS-solution for 5 min at room temperature.
5. Wash slides 3 times in PBS for 5 min each.
6. Dehydrate slides in a series of 70%, 90%, 100% ethanol 5 min each.
7. Proceed with fluorescence in situ hybridization protocol (see Chap. 12).

Subprotocol 4
Preparation of Fibroblasts for FISH (Adherently Growing Cells)

Subprotocol 2 can also be applied for the pretreatment of adherently growing cells. There are some minor modifications, however, that are described in the following for the preparation of human fibroblasts.

Materials

- Colcemid solution: prepare stock solution of 10 µg/ml in water and store at -20°C.
- Trypsin-EDTA solution: 2.5 g trypsin and 0.38 g EDTA dissolved in 1000 ml PBS (-Mg^{2+}, -Ca^{2+}).
- Fetal calf serum (FCS).

Procedure

Preparation of fibroblasts

1. Add colcemid solution to adherently growing cells (final concentration of 0.1 μg/ml). Incubate for 2 h at 37°C.
2. Suspend mitotic cells by strong shaking (flasks can be knocked on bench); monitor by microscopy; when mitotic cells are in suspension, proceed with step 4.
3. If cells remain adherent to the surface, discard culture medium and add trypsin-EDTA-solution to cover the cells. Incubate for about 10-20 min, then stop the digestion by adding medium (5-10 times the volume of the trypsin-EDTA-solution) or FCS (same volume as the trypsin-EDTA-solution).
4. Transfer cell suspension into a 50 ml tube. Spin for 15 min at 150g.
5. Remove supernatant and proceed as described in Subprotocol 2 (step 5).

Subprotocol 5
Treatment of Blood Smears

Interphase cytogenetics using FISH can be performed on routine blood smears that have been previously Wright's stained for hematological analysis (see Figure 1). Thus, cell morphology can be related to specific chromosomal aberrations, allowing the detection of cell lineages involved in tumorigenesis (see Anastasi et al. 1992; Bentz et al. 1993b; Stilgenbauer et al. 1992). The protocol presented below has been successfully used on previously stained blood smears that have been stored for up to one year at room temperature in the clinical hematology laboratory.

Materials

- Phosphate-buffered saline (PBS).
- TE: 10mM Tris-HCl, pH 7.5; 1 mM EDTA
- Methanol/acetic acid (19:1)
- Methanol/acetic acid (3:1)
- 70%, 90%, 100% ethanol.

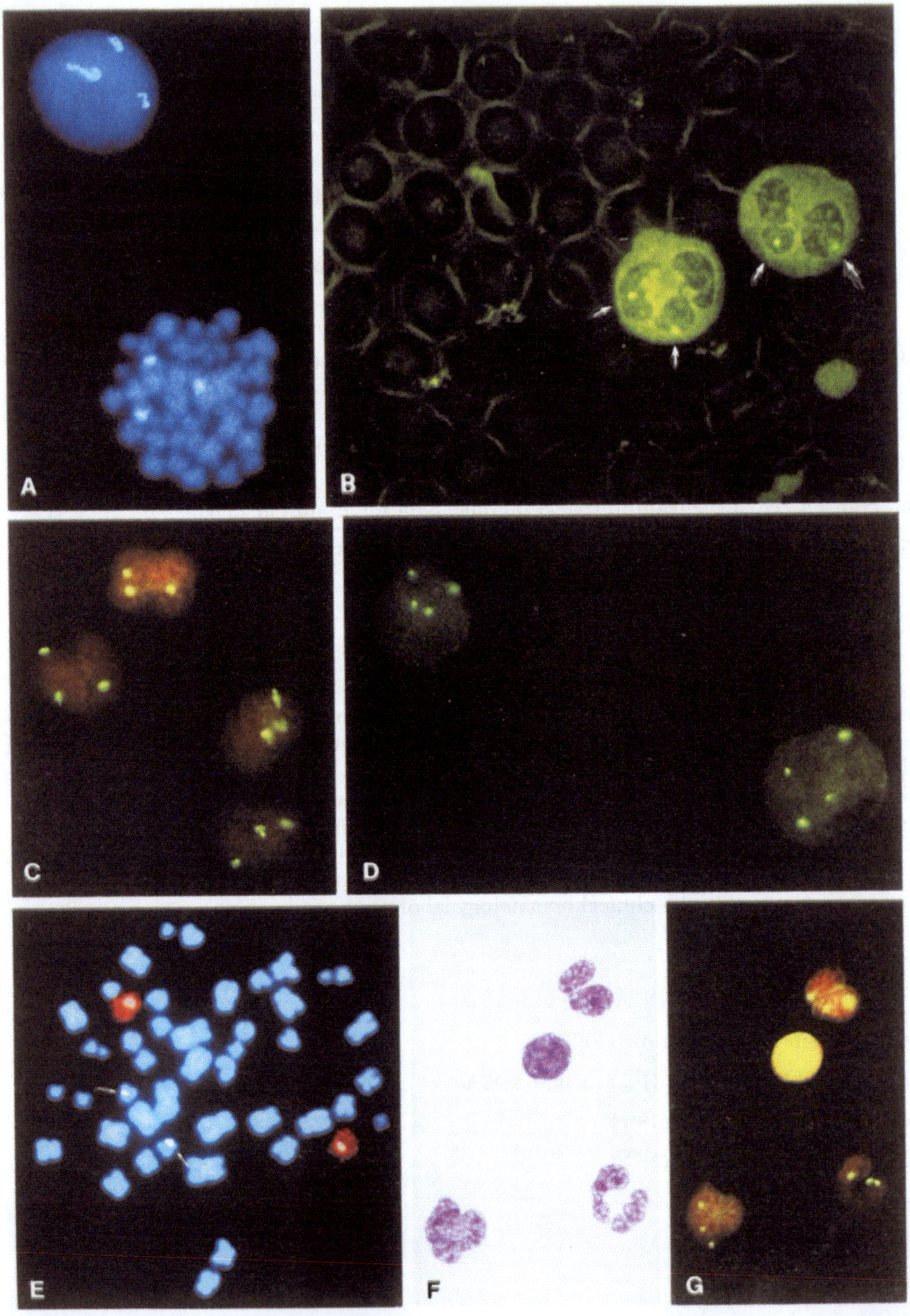
A
B
C
D
E
F
G

Blood smears

With fresh blood smears (up to two weeks) start at step 3. For older blood smears, extensive rehydration (steps 1 and 2) is required.

Procedure

1. Incubate stained blood smear in PBS overnight at 37°C.
2. Repeat step 1 using TE pH 7.5.
3. Incubate in methanol/acetic acid (19:1) for 10 min at room temperature (first fixation step).
4. Repeat with methanol/acetic acid (3:1) for 10 min at room temperature (second fixation step).
5. Dehydrate in a series of 70%, 90%, 100% ethanol, 5 min each, at room temperature.
6. Proceed with fluorescence in situ hybridization protocol (see Chap. 12).

Fig. 1. Fluorescence in situ hybridization on blood smears. Photomicrographs of hybridization signals obtained with an alphoid probe specific for chromosome 8 (A-D, F, G); E dual colour hybridization. **A** Hybridization of a chromosome 8 specific alphoid probe to a metaphase spread of a leukemia patient with trisomy 8: three signals are seen both in the metaphase and in the interphase nucleus; **B** Hybridization to a blood smear of a proband: note the well preserved morphological features of the segmented cells. The hybridization signals are indicated by arrows; **C** Hybridization to a blood smear of a leukemia patient with trisomy 8: three signals are seen in all nuclei; **D** Nuclei on a blood smear of a patient with acute myeloid leukemia with trisomy 8 and a subclone with tetrasomy 8 after in situ hybridization: all cells belong to the subclone with tetrasomy 8; **E** Metaphase of a patient with myeloid leukemia, who had a small percentage of trisomy 8 by interphase cytogenetics and a marker chromosome. This experiment was performed to address the question, if the marker chromosome contains material of chromosome 8. Hybridization was done with a chromosome 8 specific plasmid library (detected via rhodamine, red signals) and a chromesome 17 specific alphoid probe (detected with FITC). Only the two #8 homologs are stained by the DNA library; **F, G** Consecutive morphological (F) and cytogenetic (G) assessment of a blood smear of a patient with myeloid leukemia and trisomy 8: in all myeloid cells, three signals can be detected. In the lymphocyte, only two hybridization signals were seen, which are in a different focal plain. (Reprinted from Benz et al. 1993, Macmillan Press Ltd.)

Subprotocol 6
Bone Marrow Smears

Successful in situ hybridization to bone marrow smears was described by several groups (Lee et al. 1993; Anastasi et al. 1991; Bentz et al. 1993a). The following protocol is suitable for bone marrow smears stored for several months (Lee et al. 1993). Depending on the probe used for hybridization, a proteolytic digestion step may be needed to obtain sufficient signal intensities. However, this may significantly impair the morphological evaluation of the samples.

Materials

- Digestion solution: 1 μg/ml proteinase K dissolved in phosphate-buffered saline (PBS) (see note below).
- Phosphate-buffered saline (PBS).
- Methanol/acetic acid (3:1).
- 70% acetic acid.
- RNase stock solution: 10 mg/ml RNase dissolved in water. In order to inactivate DNase, heat at 100°C for 15 min, and keep at room temperature for > 15 min.
- Hybridization mix: 50% formamide, 2x SSC, 10% dextran sulfate.
- 70%, 90%, 100% ethanol.

Procedure

Bone marrow smears

1. Wash Wright's-stained bone marrow smears for 5 min in PBS.
2. Incubate in digestion solution for 7-8 min at room temperature (see note below).
3. Wash 3 times in PBS, then 3 times in water.
4. Incubate in methanol/acetic acid (3:1) for 45 sec at room temperature (first fixation step).
5. Incubate in 70% acetic acid for 40 sec at room temperature (second fixation step).

6. Incubate slides in RNAse solution (0.1 mg/ml) for 1 h at 37°C.
7. Dehydrate in 70%, 90%, 100% ethanol for 5 min each.
8. Air dry and proceed as described in Chap. 12.

Note: As an alternative to proteinase K treatment, digestion can be performed using pepsin in a concentration of 50-200 μg/ml in 0.01 M HCl

Subprotocol 7
Clinical Material Obtained from Solid Tumors: Pretreatment of Slides with Poly-L-Lysine

The evaluation of interphase nuclei by FISH is especially useful for the cytogenetic analysis of solid tumor material, since in many cases metaphase chromosomes cannot be prepared. In principle, there are two strategies for specimen preparation:

- isolation of single cells from solid tumors which are spotted on slides (see Subprotocol 8), and then subjected to in situ hybridization (Hopman et al. 1988; Van Dekken et al. 1990; Hopman et al. 1991; Arnoldus et al. 1991; Hopman et al. 1992) or
- direct hybridization to sections of frozen or paraffin-embedded tissues (Burns et al. 1986; Emmerich et al. 1989b; Walt et al. 1989; Hopman et al. 1991; Arnoldus et al. 1991; Hopman et al. 1992; Burns et al 1985; Pringle et al. 1987; Naoumov et al. 1988).

Only the second strategy provides information about the location of cytogenetically aberrant cells in a histological context. Frozen and paraffin-embedded tissue sections require special pre-treatments that are described in Subprotocols 9 and 10, respectively.

In general, successful hybridization to material derived from solid tumors depends on efficient permeabilization procedures, which facilitate probe penetration into cells. For this purpose, proteolytic digestion is most commonly used. It should be noted that proteolytic digestion must be optimized for each sample source. In paraffin-embedded tissue sections the effect of pepsin digestion can be enhanced by additional treatment with sodium thiocyanate (Hopman et al. 1992; Hopman et al. 1991). In many applications tissue sections are mounted on glass slides prior to treatment. Alternatively, the treatment (as well as the hybridization procedure) can be performed in suspension, followed by mounting. The adhesion of mounted tissue sections and spotted cells is significantly facilitated by using poly-L-lysine coated glass slides (described in Subprotocol 7).

For variations of these protocols, that may be important for specific applications the reader is referred to the literature (see for example example Manuelidis 1985; Scherthan and Cremer 1993; Rawlins et al. 1991; Chumakov et al. 1992; Klinger et al. 1992; Stock et al. 1993).

Materials

Poly-L-Lysine stock solution: 0,1 mg/ml poly-L-lysine (Mol. Wt. 70.000-150.000) dissolved in water. Keep aliquots frozen at -20°C.

Procedure

Pretreatment of slides

1. Immerse glass slide in stock solution of poly-L-lysine and incubate for 5 min. Take care that the solution does not dry!
2. Wash 5 times with distilled water.
3. Air dry and store at room temperature. The slides can be stored for several weeks.

Subprotocol 8
Preparation of Single Cell Suspensions from Solid Tumor Material

Materials

- 70% ethanol (-20°C).
- Poly-L-lysine coated slides
- Phosphate-buffered saline (PBS)
- Digestion solution: Pepsin from porcine stomach mucosa (2500 - 3000 units/mg). Dissolve 50 - 400 μg pepsin/ml in 0.01 M HCl
- 1% paraformaldehyde, dissolved in 0.1 M phosphate buffer (pH 7.2-7.4) containing 5 mM $MgCl_2$
- 70%, 90% and 100% ethanol

- Denaturation solution: 70% deionized formamide/2x SSC/50 mM sodium phosphate, adjusted to pH 7
- Hybridization solution: 50% deionized formamide/2x SSC/10% dextran sulfate
- Rubber cement

Procedure

Preparation of single cell suspension

1. Place sample of solid tumor (0.5 - 1g, if available) in an ice-cold petri dish and mince with a scalpel.
2. Suspend cells in 70% ethanol (-20°C). If necessary, further homogenize tissue in a Dounc homogenizer. The material can be stored in 70% ethanol at -20°C.
3. Suspend tumor cells to a concentration of approximately 5×10^5 cells/ml. Place 10 µl of cell suspension on poly-L-lysine coated slides. Mark area containing cells using a diamond pen on the opposite side of the slide.
4. Air dry for 15 min.
5. Add 200 µl of digestion solution and incubate for about 15 min (see note 1 below). It may be necessary to modify the incubation time empirically.
6. Briefly wash slides 5 times in water, then 5 times in PBS.
7. Fix the cells in 1% formaldehyde, dissolved in 0.1 M phosphate buffer for 10 min at room temperature.
8. Briefly wash slides 5 times in water and 5 times in PBS.
9. Dehydrate slides in 70%, 90% and 100% ethanol for 5 min each and air dry.
10. Denature slides in denaturation solution at 70°C for 2 min in a Coplin jar.
11. Immediatedly dehydrate in ice-cold 70%, 90% and 100% ethanol (5 min each) and air dry.
12. Add hybridization solution containing denaturated probe (see note 2 below) and hybridize overnight at 37°C in a humidified chamber. For detection of hybridized probes see Chap. 12).

Note: Small tumor samples (cytological specimens) can be placed directly in a tube containing digestion solution including pepsin **before** the cells are put on slides.

Note: Denaturation of the probe as well as the target DNA can be performed in a single step: After adding the probe to the specimen, a cover slip is applied and sealed with rubber cement. The slide is heated for about 3 min at 70°C and then transferred to 37°C.

Subprotocol 9
Preparation of Frozen Sections

Materials

- Poly-L-lysine coated slides
- Digestion solution: Pepsin from porcine stomach mucosa (2500 - 3000 units/mg). Dissolve 50-400 μg pepsin/ml in 0.01 M HCl
- 1% paraformaldehyde, dissolved in 0.1 M phosphate buffer (pH 7.2-7.4) containing 5 mM $MgCl_2$.
- Methanol-acetone (1:1)
- 70%, 90% and 100% ethanol series.
- 70% ethanol (-20°C)
- Phosphate-buffered saline (PBS)
- Tween-20

Procedure

Preparation of frozen sections

1. Mount frozen tissue section on poly-L-lysine coated slides and air dry at room temperature.
2. Fix sections in methanol-acetone (1:1) for 20 min at -20°C.
3. Wash slides twice in PBS containing 0.5% Tween-20 for 5 min each.
4. Incubate slides for 10 min in 200 μl of digestion solution at 37°C. It may be necessary to modify the incubation time empirically.

5. Briefly wash slides 5 times in water then 5 times in PBS.
6. Fix the cells in 1% paraformaldehyde (dissolved in 0.1 M phosphate buffer) for 10 min at room temperature.
7. Briefly wash slides 5 times in water then 5 times in PBS.
8. Dehydrate slides in 70%, 90% and 100% ethanol for 5 min each, then air dry.
9. Proceed as described in Subprotocol 8 (step 10-12).

Subprotocol 10
Paraffin-Embedded Tissue Sections

Materials

- 0.2 M HCl
- 1% H_2O_2 dissolved in methanol
- Xylene
- Methanol
- Phosphate-buffered saline (PBS)
- 1 M sodium thiocyanate (NaSCN)
- Digestion solution: Pepsin from porcine stomach mucosa (2500 - 3000 units/mg). Dissolve about 4 mg pepsin/ml in 0.2 M HCl
- Denaturation solution: 70% deionized formamide/2x SSC/50mM sodium phosphate, pH 7
- 70%, 90% and 100% ethanol
- Hybridization solution: 50% deionized formamide/2x SSC/10% dextran sulfate

Procedure

In advance

Glutaraldehyde-activated, poly-L-lysine coated slides: Coat the slides as described in Subprotocol 7 and activate the poly-L-lysine coated slides with glutaraldehyde: incubate the coated slides in 2.5% glutaraldehyde dissolved in phosphate-buffered saline (PBS) for 1 h at room temperature then wash in PBS and water. (As an alternative one can also use alkylamine coated slides).

Paraffin-embedded tissue sections

1. Paraffin sections (about 4-6 μm thick) in water (40°C) are mounted on glutaraldehyde-activated-poly-L-lysine coated slides.
2. Air dry and bake slides at 56°C overnight.
3. Deparaffinate 3 times in 100% xylene for 10 min and twice in absolute methanol for 5 min. each.
4. Immerse slides in 1% H_2O_2 dissolved in absolute methanol for 30 min.
5. Wash twice in 100% methanol for 5 min then air dry.
6. Immerse slides in 1M sodium thiocyanate at 80°C for 10 min (this step is optional, see above).
7. Wash twice in water for 5 min.
8. Incubate slides 5 - 60 min at 37°C in digestion solution. If sections were incubated with sodium thiocyanate (step 6) reduce incubation time to 5-15 min.
9. Wash 5 times in water, then 5 times in PBS briefly.
10. Dehydrate the sections in 70%, 90% and 100% ethanol for 3 min, then air dry.
11. Denature slides in denaturation solution at 80°C for 4-10 min in a Coplin jar (see note below).
12. Immediately dehydrate in ice-cold 70%, 90% and 100% ethanol (5 min each) and add hybridization solution containing denaturated probe.
13. Hybridize overnight at 37°C in a humidified chamber, then detect signals as described in Chap. 12.

Note: Denaturation of the probe as well as the target DNA can be performed in a single step. After adding the probe solved in 10-15 μl of hybridization solution to the tumor cells the slide is heated on a warm plate for about 4-10 min at 80°C.

Comments

The quality of the hybridization results depends on several parameters such as the efficiency of the permeabilization procedure (see above). For practical reasons, it is important to consider the potential of various types of probes on the different materials described in this chapter. The intensity of hybridization signals clearly correlates with the size of the targeted sequences (except when the target sequences are dispersed, as in the case of chromosome painting probes). For most applications, especially with regard to diagnostic settings, > 80% of the targets need to be labeled. With metaphase chromosome spreads, suspended cells, and blood cells or bone marrow smears such efficiencies can be achieved with probes of >10-50 kb (e.g. cosmid clones). However, in tissue sections, in particular in paraffin-embedded material, the chromosomal target region has to be larger. Thus, alphoid or satellite probes recognizing tandemly repeated sequences and, therefore, resulting in strong hybridization signals are useful tools for this application. However, although such probes are very powerful for the diagnosis of numerical chromosomal changes, their ability to detect structural aberrations, such as translocations, deletions, inversions and duplications, is very limited. For this purpose local probes which result in strong hybridization signals are needed. Such probes may be provided by contigs of cosmids and/or yeast artificial chromosomes (YACs) which will become increasingly available in the course of the Human Genome Project.

References

Anastasi J, Le Beau MM, Vardiman JW, Fernald AA, Larson RA, Rowley JD (1992) Detection of trisomy 12 in chronic lymphocytic leukemia by flourescence in situ hybridization to interphase cells: A simple and sensitive method. Blood 79: 1796-1801

Anastasi J, Vardiman JW, Rudinsky R, Patel M, Nachman J, Rubin CM, LeBeau MM (1991) Direct correlation of cytogenetic findings with cell morphology using in situ hybridization: an analysis of suspicious cells in bone marrow specimens of two patients completing therapy for acute lymphoblastic leukemia. Blood 77: 2456-2462

Arnoldus EPJ, Noordermeer IA, Peters ACB, Raap AK, van der Ploeg M (1991) Interphase cytogenetics reveals somatic pairing of chromosome 17 centromeres in normal human brain tissue, but no trisomy 7 or sex chromosome loss. Cytogenet Cell Genet 56: 214-216

Bentz M, Döhner H, Schröder M, Pohl S, Lichter P (1993a) Detection of chromosomal abnormalities on previously stained peripheral blood and bone marrow smears. In: Büchner T, Hiddemann W, Wörmann B (ed) Acute leukemias IV: Prognostic factors and treatment strategies. Springer, Heidelberg

Bentz M, Schröder M, Herz M, Stilgenbauer S, Lichter P, Döhner H (1993b) Detection of trisomy 8 on blood smears using fluorescence in situ hybridization. Leukemia 7: 752-757

Burns J, Chan VTW, Jonasson JH, Fleming KA, Taylor S, McGee JO (1985) Sensitive system for visualizing biotinylated DNA probes hybridized in situ. J Clin Pathol 38: 1085-1092

Burns J, Redfern DRM, Esiri MM, McGee JOD (1986) Human and viral gene detection in routine paraffine embedded tissue by in situ hybridization with biotinylated probes: viral localization in herpes encephalitis. J Clin Pathol 39: 1066-1073

Chumakov IM, Le Gall I, Billaut A, Ougen P, Soularue P, Guillou S, Rigault P, Bui H, De Tant, Barillot E, Abderrahim H, Cherif D, Berger R, Le Paslier D, Cohen D (1992) Isolation of chromosome 21-specific yeast artificial chromosomes from a total human genome library. Nature Genet 1: 222-225

Du Manoir S, Speicher MR, Joos S, Schröck E, Popp S, Döhner H, Kovacs G, Robert-Nicoud M, Lichter P, Cremer T (1993) Detection of complete and partial chromosome gains and losses by comparative genomic hybridization. Hum. Genet. 90: 590-610

Emmerich P, Jauch A, Hofmann M-C, Cremer T, Walt H (1989) Interphase cytogenetics in paraffin embedded sections from human testicular germ cell tumor xenografts and in corresponding cultured cells. Lab Investigation 61: 235-242

Emmerich P, Loos P, Jauch A, Hopman AHN, Wiegant J, Higgins M, White BN, van der Ploeg M, Cremer C, Cremer T (1989) Double in situ hybridization in combination with digitized image analysis: a new approach to study interphase chromosome topography. Exp Cell Res 181: 126-140

Hopman AHN, Moesker O, Smeets WGB, Pauwels RPE, Vooijs GP, Remakaers FCS (1991) Numerical chromosome 1, 7, 9, and 11 aberrations in bladder cancer detected by in situ hybridization. Cancer Res 51: 644-651

Hopman AHN, Poddighe P, Moesker O, Ramaekers FCS (1992) Interphase cytogenetics: an approach to the detection of genetic aberrations in tumors. In: Herrington M (ed) Diagnostic Molecular Pathology. A practical approach. IRL Press Inc., Oxford New York Tokyo

Hopman AHN, Ramaekers FCS, Raap AK, Beck JLM, Devilee P, van der Ploeg M, Vooijs GP (1988) In situ hybridization as a tool to study numerical chromosome aberrations in solid bladder tumors. Histochemistry 89: 307-316

Hopman AHN, Remaekers FCS, Vooijs GP (1990) Interphase cytogenetics in solid tumors. In: Polak JM, McGee JO (ed) In situ hybridization, principles and practice. Oxford University Press, Oxford

Hopman AHN, van Hooren E, van de Kaa CA, Vooijs PGP, Ramaekers FCS (1991) Detection of numerical chromosome aberrations using in situ hybridization in paraffin sections of routinely processed bladder cancers. methods Pathol 4: 503 - 513

Joos S, Lichter P (1997) Comparative genomic hybridization. In Spector D (ed) Cells - A Cold Spring Habor Laboratory Manual. Cold Spring Habor Laboratory Press, New York

Joos S, Scherthan H, Speicher MR, Schlegel J, Cremer T, Lichter P (1993) Detection of amplified genomic sequences by reverse chromosome painting using genomic tumor DNA as probe. Hum. Genet. 90: 584-589

Kallioniemi A, Kallioniemi OP, Sudar D, Rutovitz D, Gray F, Waldman F, Pinkel D (1992) Comparative genomic hybridization for molecular cytogenetic analysis of solid tumors. Science 258: 818-821

Klinger K, Landes G, Shook D, Harvey R, Lopez L, Locke P, Lerner T, Osathanondh R, Leverone B, Houseal T, et al (1992) Rapid detection of chromosome aneuploidies in uncultured amniocytes by using fluorescence in situ hybridization (FISH). Am J Hum Genet 51: 55-65

Lee W, Han K, Drut RM, Harris CP, Meisner LF (1993) Use of fluorescence in situ hybridization for retrospective detection of aneuploidy in multiple myeloma. Genes, Chromosomes & Cancer 7: 137-143

Lichter P (1997) Multicolor FISHing: what's the catch? Trends Genet 13: 475-479

Lichter P, Fischer K, Joos S, Fink T, Baudis RK, Potkul S, Ohl S, Solinas-Toldo S, Weber R, Stilgenbauer S, Bentz M, Döhner H (1996) Efficacy of current molecular cytogenetic protocols for the diagnosis of chromosome aberrations in tumor specimens. Cytokines Mol. Ther. 2: 163-170

Manuelidis L (1985) In situ detection of DNA sequences using biotinylated probes. Focus 7: 4-8

Naoumov NV, Alexander GJM, Eddleston ALF, Williams R (1988) In situ hybridization in formalin fixed, paraffin wax embedded liver specimens; method for detecting human and viral DNA using biotinylated probes. J Clin Pathol 41: 793-798

Pringle JH, Homer CE, Warford A, Kendall CH, Lauder I (1987) In situ hybridization: alkaline phosphatase visualization of biotinylated probes in cryostat and paraffin sections. Histochem. J. 19: 488-496

Raap AK, Wiegant J, Lichter P (1992) Multiple fluorescence in situ hybridization for molecular cytogenetics. In: Kessler C (ed) Techniques and methods in Molecular Biology: Non-radioactive labeling and detection of biomolecules. Springer Verlag, Berlin, Heidelberg, New York

Rawlins DJ, Highett MI, Shaw PJ (1991) Localization of telomeres in plant interphase nuclei by in situ hybridization and 3D confocal microscopy. Chromosoma 100: 424-431

Scherthan H, Cremer T (1993) Methology of non-isotopic in situ hybridization in paraffin-embedded tissue sections. In: Adolph KW (ed) methods in Molecular Genetics. Academic Press

Schröck E, du Manoir S, Veldman T, Schoell B, Wienberg J, Ferguson-Smith MA, Ning Y, Ledbetter DH, Bar-Am I, Soenksen D, Garini Y, Ried T (1996) Multicolor spectral karyotyping of human chromosomes. Science 273: 494-497

Speicher M, Ballard SG, Ward DC (1996) Karyotyping human chromosomes by combinatorial multi-fluor FISH. Nature Genet. 12: 368-375

Stilgenbauer S, Döhner H, Bulgay-Mörschel M, Weitz S, Bentz M, Lichter P (1992) High frequency of monoallelic retinoblastoma gene deletion in B-Cell chronic lymphoid leukemia shown by interphase cytogenetics. Blood 81: 2118-2124

Stock C, Ambros IM, Mann G, Gadner H, Amann G, Ambros PF (1993) Detection of 1p36 deletions in paraffin sections of neuroblastoma tissues. Genes, Chromosomes & Cancer 6: 1-9

Van Dekken H, Pizzolo JG, Kelsen DP, Melamed MR (1990) Targeted cytogenetic analysis of gastric tumors by in situ hybridization with a set of chromosome-specific DNA probes. Cancer 66: 491-497

Walt H, Emmerich P, Cremer T, Hofmann M-C, Bannwart F (1989) Supernumerary chromosome 1 in interphase nuclei of atypical germ cells in paraffin-embedded human seminiferous tubules. Lab Investigation 61: 527-531

Chromosome Analysis by Fluorescence in situ Hybridization

MARTIN BENTZ, STEFAN JOOS AND PETER LICHTER

Introduction

In 1969 visualization of nucleic acid sequences in situ was first described using isotopically labeled rRNA (Gall and Pardue 1969). Since then, in situ hybridization employing radioactively labeled probes has been widely used. With the advent of nonisotopic procedures for labelling nucleic acids in the early 1980s, fluorescence in situ hybridization (FISH) was developed and has become increasingly popular. Compared with radioactive in situ hybridization, FISH has a number of specific advantages including increases in speed, sensitivity and resolution (for review see Raap et al. 1990; Lichter and Ward 1990; Trask 1991).

In situ hybridization permits the detection of single-copy nucleic acid sequences in biological and clinical specimens. Using a suitable probe, entire chromosomes or chromosomal subregions may be visualized permitting detection of alterations in copy number or structure (translocations, etc.). This can be achieved using metaphase chromosome spreads or interphase nuclei ("interphase cytogenetics"), examples for which are shown in Figure 1 and 2 (Joseph et al. 1984; Cremer et al. 1986). FISH is based on the specific base pairing of sequences of the DNA probe and complementary sequences in cells of the specimen. The DNA probes are chemically modified by the introduction of **reporter molecules** (e.g. biotin or digoxigenin) and are detected using fluorochromes conjugated to specific **reporter-binding molecules** (as avidin or antibodies directed against the reporter molecules). Alternatively, the probes can be labeled directly with fluorochromes.

Martin Bentz, Univ. Heidelberg, Medizinische Klinik und Poliklinik V, Hospitalstr. 3, Heidelberg, 69115, Germany, Stefan Joos, Deutsches Krebsforschungszentrum, Abt. Organisation komplexer Genome, Im Neuenheimer Feld 280, Heidelberg, 69120, Germany, *Correspondence to* Peter Lichter, Deutsches Krebsforschungszentrum, Abt. Organisation komplexer Genome, Im Neuenheimer Feld 280, Heidelberg, 69120, Germany (*phone* +49-6221-424609; *fax* +49-6221-424639; *e-mail* p.lichter@dkfz-heidelberg.de)

The combined use of several probes modified with different reporter systems in a single experiment allows the simultaneous visualization of multiple nucleic acid targets within the same preparation and significantly increases the utility of FISH.

Many fields of biomedical research have been greatly influenced by applications of FISH including cytogenetics, gene mapping and cell biology. The implications of these applications are reviewed in detail elsewhere (Joos 1994; Bentz et al.1994; LeBeau 1993). In this chapter, we describe a basic protocol for FISH to chromosomal DNA.

Components of a FISH experiment

- **Specimens:** For most applications, conventional methanol/acetic acid fixed metaphase spreads are used. However, FISH is applicable to a variety of specimen preparations important in clinical settings and cell biol-

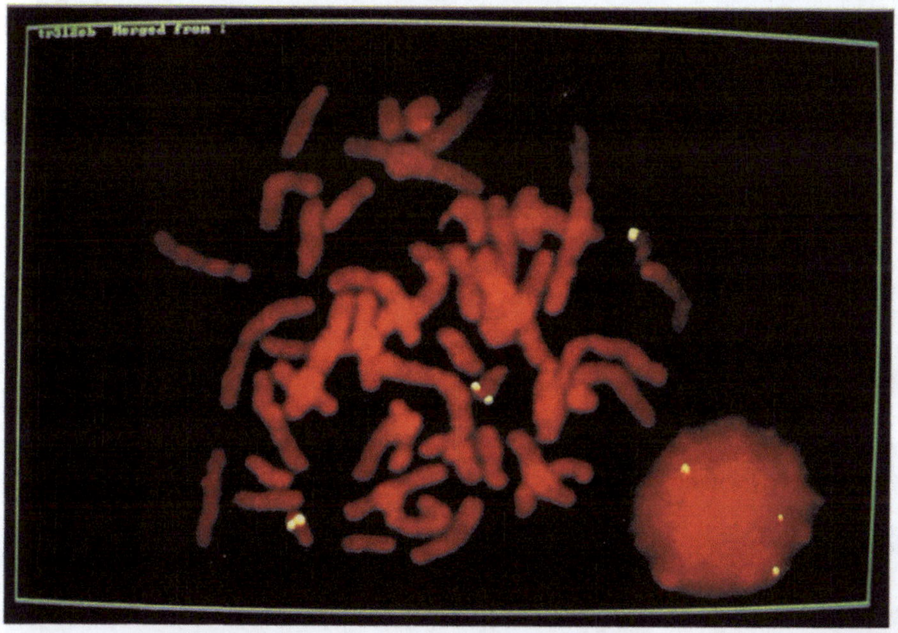

Fig. 1. In situ hybridization with the cosmid DNA probe c512 to a metaphase spread and a nucleus of lymphocytes from a Down's syndrome patient. The biotin labelled probe was detected with avidin-FITC; chromosomes and the interphase nucleus are counterstained with propidiumiodide. Note the highly specific labelling allowing the rapid diagnosis of trisomy 21: on the metaphase, the signals are visible on both chromatids (all signals are doublets) of all three chromosomes 21. Corresponding three signals are seen in the nucleus. Reprinted from Lichter and Cremer 1992, Oxford University Press

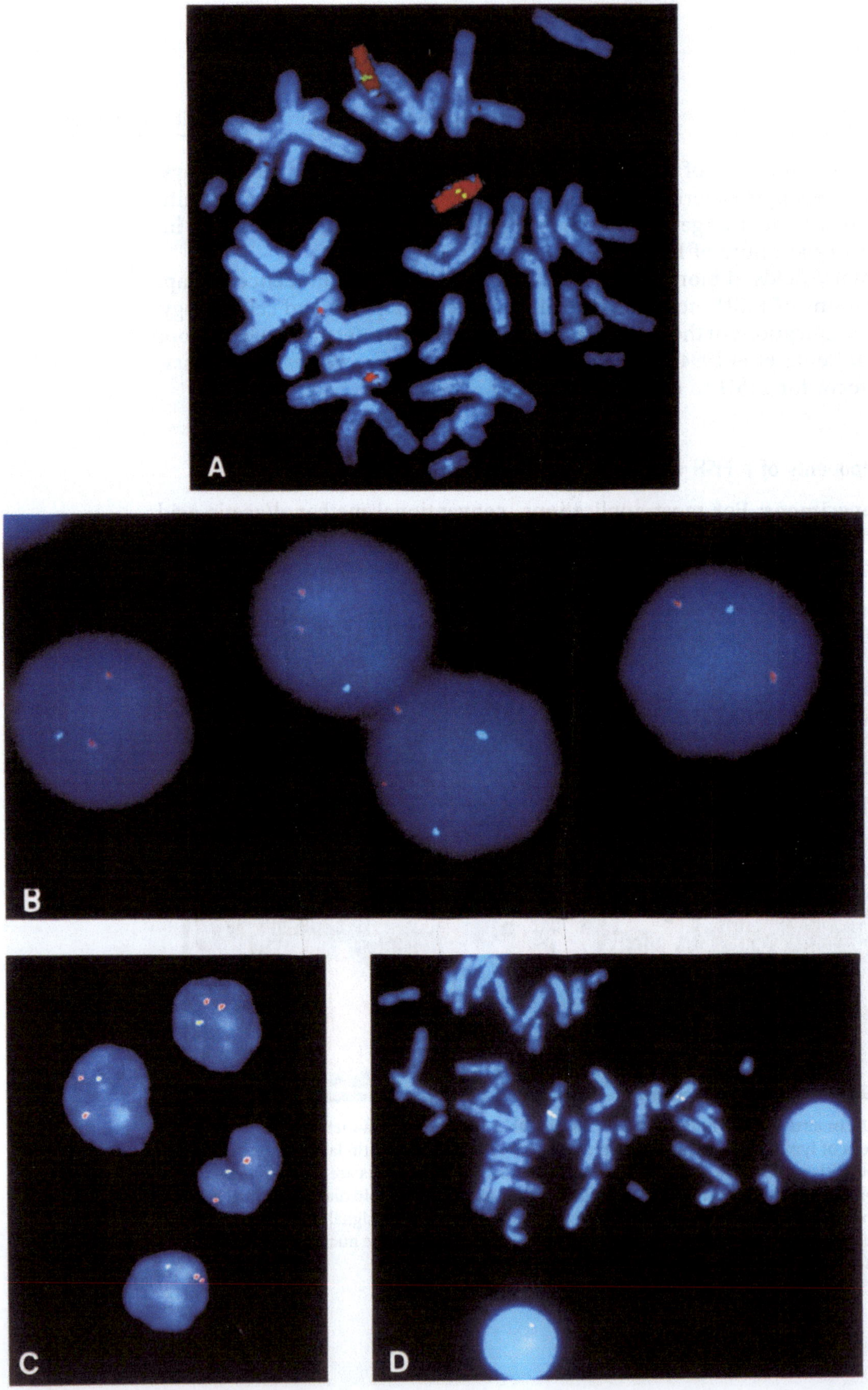
A
B
C
D

ogy: Hybridization can be performed on routine clinical material (Manuelidis 1985; Hopman et al. 1988; Anastasi et al. 1991; Bentz et al. 1993) as well as preparations with well-preserved three-dimensional morphology (Zirbel et al. 1993).

- **Probes:** A variety of probe types that can be categorized according to the complexitiy of their target sequences are used for FISH. Clusters of chromosome-specific repeated sequences are detected by alphoid or satellite probes. Composite probes allow specific "painting" (Cremer et al. 1988; Lichter et al. 1988; Pinkel et al 1988) of chromosomes or chromosomal regions. Other probes such as most genomic DNA fragments cloned in plasmid-, phage-, cosmid-, P1 or yeast artificial chromosome vectors detect unique loci in the genome. However, for the selection of probes from this category, the size of the targeted DNA sequence must be taken into consideration, as this correlates fairly well with the hybridization efficiency. With cosmids (25 to 40 kb of cloned insert) signals of both chromosome homologs are usually visualized in about 90% of cells. In contrast, probes targeting less than 5 kb of single copy sequences detect a maximum of 50% of the complementary sequences (Joos et al. 1994). These estimates refer to hybridizations performed on routine cytogenetic preparations.
- **Probe labelling:** The nucleic acid probes are labeled by the introduction of specific reporter groups (e.g. biotin, digoxigenin or FITC). In general,

Fig. 2. Analysis of deletions of the retinoblastoma gene (Rb-a) using phage clones spanning the whole 200 kb of the retinoblastoma gene: **A** Metaphase spread of a healthy proband counterstained with DAPI showing specific Rb-1 hybridization signals detected via FITC (yellow). Painting of chromosomes 13 was achieved by cohybridization of chromosome-specific library DNA detected via rhodamine (red). Cross-hybridizations of library DNA occurs on the short arms of other D- and G- group chromosomes. **B** Lymphocyte interphase nuclei of a patient with chronic lymphocytic leukemia (CLL) after hybridization with Rb-1 (green) and c518 that was used as a control for hybridization efficiency (red), and counterstain with DAPI. Three cells show a Rb-1 deletion (only one green signal), whereas one cell has two copies of Rb-1 (two green signals). Note the two c518 signals (red) in each nucleus indicating a high hybridization efficiency. **C** Blood smear of the same patient after hybridization with Rb-1 and c518. Three round nuclei and one lobulated nucleus likely to represent three lymphocytes and one granulocyte, respectively, are visualized by DAPI counterstain. The granulocyte has two Rb-1 copies (two green signals), whereas the other cells show an Rb-1 deletion (only one green signal). The high efficiency is shown by two c518 signasl (red) per cell. **D** Metaphase spread and interphase nuclei of the same patient counterstained with DAPI. While the metaphase shows Rb-1 signals on each chromosome 13 homolog, both interphase nuclei exhibit a deletion of one Rb-1 copy (only one signal per cell). Reprinted from Stilgenbauer et al. 1993, Saunders Company

this is achieved by enzymatic incorporation of modified nucleotides, for example during a nick translation reaction (Langer et al. 1981).

- **Fluorochromes:** Directly labeled probes contain fluorochromes as reporter groups (e.g. FITC or rhodamine). Alternatively, labeled probes can be detected via fluorochromes conjugated to reporter-binding molecules (e.g. FITC conjugated to avidin or an antibody). Similar to other histochemical detection procedures, different combinations of primary and secondary reporter-binding molecules can be used for indirect immunofluorescence. It should be noted that when multiple fluorochromes are used they must not exhibit highly overlapping emission ranges. Useful combinations of fluorochromes for probe detection and chromosomal counterstain are: rhodamine or texas red with propidium iodide, and AMCA with DAPI or Hoechst 3325. The use of multicolor hybridizations is described below.
- Visualization of the probes by **epifluorescence microscopy:** For many applications of FISH, a standard epifluorescence microscope equipped with appropriate filter sets is sufficient. However, data management and documentation are greatly facilitated by digital imaging microscopy. This requires specialized equipment, such as a sensitive CCD (charge-coupled device) camera connected to the standard microscope. Digital imaging microscopy is mandatory for advanced applications such as recording of weak signals, quantitative measurement of signal intensities or three dimensional analysis of fluorescently labeled specimens.

Subprotocol 1
Specimen Preparation

Procedure

For most applications methanol/acetic acid-fixed metaphase preparations are used. Such metaphase preparations are prepared as described in Chap. 11, then air-dried and treated as follows:

1. Dehydrate in a series of 70%, 90% and 100% ethanol for 5 min at room temperature.
2. Air-dry slides.

3. For short term use, slides can be stored at room temperature for up to a few days. For long term storage slides should be "aged" overnight at room temperature. then placed in containers. Containers are sealed in plastic bags containing Drierite and stored at -70°C. In our experience, for slides used within 6 months storage at -20°C is sufficient. The slides are thawed immediately prior to the hybridization experiment. The re-freezing of slides is not recommended.

Subprotocol 2
Probe Labelling by Nick Translation

Nick translation is the most frequently applied method for labelling DNA probes for use in fluorescence in situ hybridization experiments. Although the labelling efficiencies of nick translation and primer extension (Feinberg and Vogelstein 1984) are comparable, we prefer nick translation. Nick translation permits easy adjustment of the probe size, which for optimum hybridization conditions should be 100 - 500 nucleotides, by varying the DNase concentration. The protocol is based on Langer et al. (Langer et al. 1981). It should be noted that the conditions of primer extension can also be adjusted to optimize probe length. Subprotocol 2 describes labelling with biotin and digoxigenin. These reporter molecules are most popular due to their sensitivity and commercial availability.

Materials

Reagents

- Solution of template DNA (purified from RNA)
- 10x reaction buffer containing 0.5 M Tris-HCl (pH 8.0), 50 mM $MgCl_2$, 0.5 mg/ml BSA
- 0.1 M ß-mercaptoethanol prepared by adding 0.1 ml of ß-mercaptoethanol to 14.4 ml of double-distilled water
- 10x nucleotide stock containing 0.5 mM dATP, 0.5 mM dCTP, 0.5 mM dGTP, 0.5 mM biotin-16-dUTP and 0.12 mM dTTP (alternatively, other biotinylated dUTP derivatives can be used)
- Escherichia coli DNA polymerase I
- DNase I solution: 3 mg DNAse I in 1 ml 0.15 M NaCl, 50% glycerol
- Column buffer: 10mM Tris-HCl (pH 8.0), 1 mM EDTA, 0.1% SDS

- For centrifuge columns: Sephadex G50 (medium), 1 ml syringes, glass wool
- For minigels: agarose, TBE loading buffer (0.089 M Tris-borate, 0.089 M boric acid, 0.002 M EDTA) suitable (1kb) size standard

Supplies

- Water bath, 15°C
- Horizontal agarose gel electrophoresis apparatus

Procedure

Biotinylation of DNA probes by Nick translation

This method describes the labelling of 2μg probe DNA in a 100 μl reaction volume.

Biotinylation

1. Combine: 2μg of probe DNA, 10μl of 10x reaction buffer, 10μl of ß-mercaptoethanol, 10μl of nucleotide stock, 20 units of DNA polymerase I and the tested volume of a 1:1000 dilution of DNase I, adjust to 100 μl with double distilled water (enzymes should be added last).
2. Incubate for 2 hours at 15°C.
3. Place the reaction mixture on ice. It should be kept on ice until the actual size of the reaction product has been determined.
4. Check the length of the probe molecules by gel electrophoresis: Remove 10 μl of the reaction mixture and add gel loading buffer. Denature by boiling in a waterbath for 2-3 minutes. After another 3 minutes on ice, the sample is loaded on a standard 1-2 % agarose minigel with a suitable size marker and run at 15 V/cm for 30 min. For visualization of DNA, stain the gel in 0.5 μg/ml ethidium bromide and photograph under UV-illumination.
5. For optimum hybridization conditions, the probe (which is visible as a smear) should be 100 - 500 nt in length. Depending on the result of the electrophoresis proceed as follows:
   a. If the probe size is within the desired range proceed to step 6.
   b. If the probe size is too large, add more DNase I, incubate at 15°C (usually higher amounts of DNase I are added for another 30 minutes), repeat step 4.

c. If the probe is incompletely digested, purify probe and repeat the nick translation.
d. If part of the probe is too short, repeat reaction using less DNase I.

6. In order to inactivate the DNase, 2 µl of 0.5M EDTA (final concentration 15 mM) and 1 µl of 10% SDS (final concentration 0.1%) are added, and the reaction is heated at 68°C for 10 minutes.

7. Unincorporated nucleotides are separated from the labeled probe by gel filtration using a spin column:
a. Pack a 1 ml syringe with sialinized glass wool up to the 0.2 ml mark. Add buffer-equilibrated Sephadex G50 to the 1ml mark. Place the column in a 15 ml tube and centrifuge at 3000 rpm for 6 min at room temperature.
b. Remove flow-through, fill buffer and repeat centrifugations until the column is tightly packed to the 1 ml mark. Add 100 µl of column buffer and centrifuge again at 3000 rpm for 6 min. Repeat this washing step three times. After the last washing step, make sure that the volume of the flowthrough equals the volume of the loaded buffer, i.e. 100 µl.
c. Prior to centrifugation of the probe solution, put a 1.5 ml reaction tube in the 15 ml tube beneath the syringe. Load the probe solution onto the column and centrifuge as before. The flow-through is collected in the reaction tube and now contains the labeled probe at a concentration of 20 ng/µl. The probe is ready for use in an in situ hybridization experiment or can be stored at -20°C.

Digoxigenin labelling by Nick translation

Labelling of DNA probes with Digoxigenin is carried out in exactly the same manner as biotinylation by nick translation with one exception: 10x nucleotide stock contains:

- 0.5 mM dATP, 0.5 mM dCTP, 0.5 mM dGTP, 0.125 mM digoxigenin-11-dUTP, 0.375 mM dTTP.

When new reagents are used or when no hybridization signals are obtained, it is advisable to check the incorporation of the biotin or digoxigenin labeled nucleotides. This can be achieved using a single, colorimetric dot blot assay (Subprotocol 3).

Subprotocol 3
Dot-Blot Assay to Test Labelling

Materials

Reagents

- DNA dilution buffer: 0.1 mg/ml sheared salmon sperm DNA, 6x SSC (1x SSC: 0.15M NaCl, 0.015 sodium citrate, pH 7.0)
- Series of dilutions of biotinylated standard DNA (commercially available standard or previously used, well-labeled probe) in DNA dilution buffer, e.g. 1, 3, 10, 20 pg/μl standard DNA
- AP 7.5 buffer: 0.1 M Tris-HCl pH 7.5, 0.1 M NaCl and 2 mM $MgCl_2$
- AP 9.5 buffer: 0.1 M Tris-HCl pH 9.5, 0.1 M NaCl and 50 mM $MgCl_2$
- Blocking buffer: 3% BSA in AP 7.5 buffer
- TE buffer: 10 mM Tris-HCl pH 7.5 and 1 mM EDTA
- 75 mg/ml nitroblue tetrazolium (NBT) in 70% dimethylformamide and 50 mg/ml 5-bromo-4-chloro-3-indolyl phosphate (BCIP) in dimethylformamide

Supplies

- Nitrocellulose filter

Procedure

Dot-blot for testing labeling

1. Aliquot 1μl of the dilutions of standard DNA and, in parallel, 1 μl of identical dilutions of test DNA on a nitrocellulose filter.
2. Bake the filter at 80°C for 1 hr in a vacuum oven.
3. Wash filter with AP 7.5 buffer for 1 min at room temperature.
4. Seal filter in a plastic bag containing 10 ml of blocking solution. Incubate at 37°C for 30 min.
5. Open the bag at one end, remove the blocking solution and add a freshly prepared solution of streptavidin-conjugated alkaline phosphatase (diluted to a concentration of 1μg/ml in AP 7.5). Re-seal the bag and incubate at 37°C for 30 min.
6. Remove the filter from the plastic bag and wash with AP 7.5 buffer (twice for 5 minutes at room temperature) followed by AP 9.5 buffer (10 minutes at room temperature).

7. Reseal the filter in a plastic bag containing the developing solution (33 µl of NBT added to 10 ml AP 9.5 buffer). After careful mixing (do not vortex!) 25 µl of BCIP are added. The resulting solution is again mixed gently. Incubate at 37°C until color development is appropriate, usually 15 to 60 min.
8. Remove the filter from the plastic bag and wash with TE buffer to stop the color reaction.
9. After air-drying, the assay results can be evaluated: the color intensities of the test and control DNA should be comparable. For optimal hybridization results, the signal of the lowest dilution should be visible.

When the dot-blot assay is carried out to check probes labeled with digoxigenin, an anti-digoxigenin labeled alkaline phosphatase is used instead of Streptavidin-conjugated alkaline phosphatase.

Subprotocol 4
Probe Mixture and Denaturation for Fluorescence in situ Hybridization

This section describes a standard in situ hybridization which is used for probes in which all sequences contribute to the hybridization signal (such as chromosome specific repetitive DNA probes, unique cDNAs or genomic DNA probes lacking interspersed repetitive sequences [ISR]) (Subprotocols 4, 5 and 6). Subprotocol 7 describes the modifications required for chromosomal in situ suppression (CISS) hybridization for use with genomic DNA fragments or cDNAs containing IRS segments as probes. Protocols are suitable for a 10 µl hybridization volume which is sufficient to cover an area of 18x18 mm on a glass slide. If hybridization of larger areas is desired, the volumes of the solutions should be scaled-up accordingly.

Materials

- 3 M sodium acetate, pH 5.2
- Deionized formamide (molecular biology grade). The conductivity of the solution should be below 100 µSiemens.
- Hybridization buffer: 4x SSC, 20% dextran sulphate. The solution is best prepared as follows: Prepare 20x SSC and 50% dextran sulphate solu-

tions. After careful dissolving, autoclave the dextran sulphate solution or filter through a nitrocellulose filter. Combine 200 µl of 20x SSC, 400 µl of 50% dextran sulphate and 400 µl of double distilled water. Store at 4°C until use.

Procedure

Probe mixture and denaturation

1. Combine 20 to 60 ng of single-copy DNA and 3-5 µg of sheared salmon sperm DNA (used as carrier). If the resulting volume is below 10µl, lyophilize. For larger volumes, precipitate DNA by adding 1/20 volume of 3 M sodium acetate and 2 volumes of 100% ethanol. Mix well and incubate at -70°C for 30 minutes. Centrifuge in an Eppendorf centrifuge at 12,000 rpm for 10 minutes at 4°C. Discard the supernatant, wash the pellet by adding 500 µl of 70% ethanol and centrifuge again (12,000 rpm, 10 min, 4°C). Discard the supernatant, lyophilize.
2. Resuspend in 5 µl deionized formamide by vortexing for several minutes at room temperature.
3. Add 5 µl of hybridization buffer and vortex again for 5 to 10 minutes.
4. Denature probe DNA at 75°C for 5 minutes followed by incubation on ice for another 5 minutes. After this step, the probe is ready for hybridization.

Note: Certain probes, such as chromosome-specific sequences of the alphoid family of DNA repeats, share considerable homologies with related sequences on different chromosomes. Therefore, specific signals are only obtained when highly stringent hybridization conditions are applied. This can be achieved, for example, by increasing the formamide concentration in the hybridization medium to 60 % or more.

Subprotocol 5
Denaturation of the DNA on the Slides

Materials

- Denaturation solution: 70% deionized formamide, 2x SSC, 50 mM sodium phosphate, pH 7.0
- Ice-cold ethanol: 70%, 90% and 100%

Procedure

Denaturation

1. Select an appropriate area on the slide for hybridization and mark the opposite side with a diamond pen.
2. Incubate slides in an oven at 60°C prior to denaturation. This will prevent lowering of the temperature of the denaturation solution when added to the slides.
3. Fill denaturation solution into a Coplin jar and put it into a water bath heated to 70°C. Check temperature of the denaturation solution with a thermometer inside the jar. (This is a critical step!). For good results, the temperature should be 70°C.
4. Transfer the prewarmed slides (not more than three at a time) into coplin jar containing the denaturation solution for **exactly** 2 minutes.
5. Immediately transfer the slides into the Coplin jars with 70%, 90% and 100% ethanol (on ice) for 5 minutes each.
6. After air drying, the slides are ready for hybridization.

When the denaturation is not sufficient, no efficient hybridization can be performed. Over-denatured chromosomes usually appear fuzzy when counterstained with DAPI or may even lose their morphological features.

Hybridization

1. Add 10 µl of hybridization mixture containing the denatured probe onto the denatured target DNA on the slides.
2. Place an 18x18 mm coverslip on top of the hybridization droplet. Take care not to trap air bubbles.
3. Seal the edges of the coverslip with rubber cement, and place the slides in a humidified chamber. Incubate overnight at 37°C.

After hybridization, detection of the probe sequences on the slides is performed using the following protocol.

Subprotocol 6
Detection

Materials

- Wash solution A: 50% formamide (inexpensive grade), 2x SSC, pH 7.0.
- Wash solution B: 1x SSC - 0.1x SSC, p 7.0. The stringency should be adjusted empirically for each probe: e.g. if a probe also hybridizes weakly to a second region, stringency should be increased. If a probe of a different species is used, stringency conditions should be decreased.
- Wash solution C: 4x SSC, 0.1% Tween20; pH 7.
- Wash solution D: 2x SSC, 0.05% Tween 20.
- Blocking solution: 3% BSA, 4x SSC, 0.1% Tween 20.
- Detection buffer: 1% BSA, 4x SSC, 0.1% Tween 20.
- Antifade-mounting solution. Such solutions considerably decrease the bleaching of the illuminated slides. Several commercial reagents are available; of these Vecta Shield (Vector Laboratories) proved to be very efficient. Alternatively, DABCO antifading medium can be prepared as follows: 0.233g of DABCO (1,4-Diazabicyclo[2.2.2]octane), 20 mM Tris-HCl, pH 8.0, 90% glycerol. Store this solution in the dark at 4°C.

Procedure

Detection

Note: During the entire procedure, the slides must not be allowed to dry completely.

1. Prewarm washing solutions A and B in a 42°C and 60°C water bath, respectively.
2. After taking the slides out of the humidified chamber, carefully remove the rubber cement using a forceps.
3. Place slides in a Coplin jar containing wash solution A prewarmed to 42°C, agitate in a shaking water bath for 10 minutes until coverslips fall off. Transfer slides to another jar containing wash solution A, and agitate for 5 minutes. Change the wash solution A twice more. Shake for 5 minutes each.

4. Transfer the slides to a Coplin jar containing prewarmed (60°C) wash solution B, and wash for 5 minutes. Change solution twice, each time washing for 5 minutes.
5. Remove the slides from the jar, drain, and apply 200 µl of blocking solution. Cover with a 22x40mm coverslip, place in a humidified chamber and incubate at 37°C for at least 30 minutes.
6. Remove the coverslip from each slide, drain excess fluid, and add 200 µl of detection solution containing the fluorochrome-conjugated reporter-detecting reagent (e.g. 5µg/ml fluorescein-conjugated avidin or 6 µg/ml rhodamine conjugated anti-digoxigenin). Incubate in a humidified chamber at 37°C for 30 minutes. All subsequent steps should be carried out in light-protected Coplin jars (e.g. wrapped with aluminium foil).
7. Remove the coverslip, transfer the slides into wash solution C, and wash 3 times for 5 minutes at 42°C (shaking water bath).
8. Place slides in a Coplin jar containing counterstaining solution (e.g. 2x SSC, 200ng/ml DAPI) and agitate at room temperature for 20 minutes.
9. Transfer slides into a jar containing wash solution D. Incubate for 1 to 2 minutes at room temperature.
10. Remove each slide from the Coplin jar, add 20 to 30 µl of antifade solution and cover with a 22x40 mm coverslip. Place slides in suitable boxes, which should be kept at 4°C for long-term storage.

Subprotocol 7
Chromosomal in situ Suppression (CISS) Hybridization

This protocol is used when the probes contain IRS sequences, as is the case for almost all genomic DNA probes. For the suppression of non-specific signals arising from these repetitive segments, the DNA probe is pre-mixed with an excess of unlabeled competitor DNA and an additional pre-annealing step is performed after probe denaturation (Lichter et al. 1988; Pinkel et al 1988). Again, a protocol is provided for a 10 µl hybridization volume.

Probes as small as a few kb as well as probes targeting many megabases can be used for CISS hybridization. Therefore the concentration of the individual probe in the hybridization volume must be adjusted according to the complexitiy of the target region. The following list provides the quantities of probes required for a 10 µl reaction:

- Single cosmids, phages, plasmids: 40-80 ng
- Isolated inserts derived from individual sorted human chromosomes: 100-300 ng
- Gel-purified or PCR amplified single YACs: 50-150 ng

Materials

- Sheared salmon sperm DNA
- Cot1-DNA fraction of human genomic DNA at a size of approximately 500nt
- Solutions and labeled probe as described in Subprotocols 4 and 6

Procedure

Hybridization

1. Combine the appropriate amount of labeled probe DNA (see above), 2-4 μg of human competitor DNA (Cot1-fraction) and add salmon sperm DNA to a total of 10 μg DNA.
2. Precipitate DNA by adding 1/20 volume of sodium acetate, pH 5.2 and 2 volumes of ethanol. Incubate at -70°C for 30 minutes. After centrifugation at 12,000 rpm for 10 minutes, in most cases a pellet will be visible. Discard supernatant and wash with 70% ethanol. Re-centrifuge and discard supernatant.
3. Lyophilize sample and resuspend the pellet in 5 μl of deionized formamide (see Subprotocol 4).
4. Add 5 μl of denaturation buffer.
5. Denature the DNA by incubation at 75°C for 5 minutes.
6. Quickly transfer the microcentrifuge tube containing the probe solution to 37°C. Incubate for 5-20 minutes (or even longer) to allow partial reannealing.
7. Apply pre-anneled probe mixture to the denatured DNA on the slides.
8. Continue as described in Subprotocols 4, 5 and 6.

Subprotocol 8
Signal Amplification

Hybridization signals may be weak especially when small probes are used. In these cases, amplification of the signal can facilitate the evalution. Whereas the signal intensity is increased considerably, amplification also results in a marked decrease in the signal-to-noise ratio. Therefore amplification is a useful procedure only in slides with low background signals. The following technique using avidin mediated detection was first described by Pinkel et al. (Pinkel et al. 1986). The method is based on the binding of antibodies directed against the reporter binding molecule (e.g. avidin). These antibodies carry reporter groups which can be detected by fluorochrome coupled secondary molecules. Subprotocol 8 is designed for the amplification of biotin-labeled probes.

Materials

- Wash solution C, wash solution D, detection buffer, antifade mounting solution as in Subprotocol 6.
- Biotinylated anti-avidin antibodies.

Procedure

Signal amplification

1. Carefully remove the coverslip from the slide.
2. Transfer slides into wash solution C and incubate 3 times for 10 minutes at 42°C.
3. Remove slides from the Coplin jar, drain excess fluid, add 200 µl of detection buffer containing 1-5 µg/ml biotinylated anti-avidin antibody, cover with a 22x40 mm coverslip and incubate in a humidified chamber at 37°C for 30 minutes.
4. Wash the slides 3 times for 5 minutes in a Coplin jar containing wash solution C at 42°C.
5. Remove slides from the Coplin jar, drain and add 200 µl of detection buffer containing 5 µg/ml avidin conjugated fluorochrome.
6. Proceed with washing steps and chromosome counterstaining as described in Subprotocol 6.

Subprotocol 9
Multicolor Fluorescence in situ Hybridization

FISH allows the simultaneous detection of multiple target sequences. The targets are marked with differentially labeled DNA probes which are visualized via different fluorochromes. For multicolor hybridization experiments, only two steps of the standard protocols must be modified.

Procedure

Multicolor fluorescence

1. Instead of only one probe, several probes coupled with different reporter molecules are combined prior to DNA precipitation (Subprotocol 4.).
2. The appropriate reporter binding molecules must be added to the detection buffer (Subprotocol 6). Each reporter binding molecule must be conjugated with a fluorochrome that can be spectrally distinguished from other fluorochromes used in the same experiment.

The number of targets that can be simultaneously visualized by multicolor FISH can be increased when the probes are modified by different reporter groups and detected by spectrally distinct fluorochromes. Such combinatorial labelling was demonstrated by several groups (Ried et al. 1992; Wiegant et al. 1991; Dauwerse et al. 1992; Nederlof et al. 1992). In addition, the combined use of two fluorochromes at different ratios further increases the number of simultaneously detected sequences (Ried et al. 1992; Nederlof et al. 1992). Using the combinatorial labeling approach with five different fluorescent dyes, the simultaneous visualization of each human chromosome in a specific color has become possible (so called "Multiplex-FISH": Speicher et al., 1996; or "Spectral Karyotyping" Schröck et al., 1996). For the first time, these techniques provide a general screening test for chromosomal aberrations in a single experiment (Lichter 1997).

Comments

Microscopy

Signals from painted whole chromosomes, chromosomal subregions and single probes as small as 5 -10 kb are visible with a conventional fluorescence microscope. Special attention should be paid to the selection of appropriate filter sets. For certain experiments, the use of filter sets selective for a particular fluorochrome is preferred. Other applications require the simultaneous visualization of different fluorochromes which can be achieved by the use of multi-band pass filters. These may be advantageous for the evaluation of multi-color experiments. They also provide a solution to the so-called registration problem: when several images of one object are taken (by using single bandpass filters) and overlayed to obtain a combined image, more or less pronounced shifts are detected, which are based on small differences in the optical pathway due to variations in filter and mirror positions. Advanced applications of FISH rely on the quantitative analysis of fluorescence intensities: multicolor hybridization by **ratio labelling** (see above) and, in particular, the powerful approach of **comparative genomic hybridization** (CGH; Kallioniemi et al. 1992). For the description of the CGH approach, we refer the interested reader to the literature (Kallioniemi et al. 1992; Du Manoir et al. 1993; Joos et al. 1993; Forozan et al. 1997). For such applications optical detection systems are needed which record signal intensities in a linear mode over a wide range of intensities. Cooled **CCD** (charge-coupled device, (Hiraoka et al. 1987) **cameras** provide powerful tools for this application due to their wide dynamic range. They are currently the most sensitive systems and, thus, are also very useful for the analysis of weak signals, e.g. obtained using very small probes (smaller than 2-5kb). Although **confocal laser scanning microscopy** is also capable of sensitive digital imaging, this procedure is primarily designed for three-dimensional analyses of fluorescently labeled specimens. Moreover, the use of fluorochromes is limited to those which are excited by the lasers available in the instrument. In addition, confocal laser microscopes are much more expensive than the most sensitive camera systems. Therefore, for two-dimensional analyses, cooled CCD cameras are preferred. When purchasing equipment for digital imaging microscopy, attention has to be paid to both the memory and storage capacity of the hardware as well as to the software that is available for special applications.

References

Anastasi J, Vardiman JW, Rudinsky R, Patel M, Nachman J, Rubin CM, LeBeau MM (1991) Direct correlation of cytogenetic findings with cell morphology using in situ hybridization: an analysis of suspicious cells in bone marrow specimens of two patients completing therapy for acute lymphoblastic leukemia. Blood 77: 2456-2462

Bentz M, Döhner H, Cabot G, Lichter P (1994) Fluorescence in situ hybridization in leukemias: The FISH are spawning! Leukemia 8: 1447-1452

Bentz M, Schröder M, Herz M, Stilgenbauer S, Lichter P, Döhner H (1993) Detection of trisomy 8 on blood smears using fluorescence in situ hybridization. Leukemia 7: 752-757

Cremer T, Landegent J, Brückner A, Scholl HP, Schardin M, Hager HD, Devilee P, Pearson P, van der Ploeg M (1986) Detection of chromosome aberrations in the human interphase nucleus by visualization of specific target DNAs with radioactive and non-radioactive in situ hybridization techniques: diagnosis of trisomy 18 with probe L1.84. Hum Genet 74: 346-352

Cremer T, Lichter P, Borden J, Ward DC, Manuelidis L (1988) Detection of chromosome aberrations in metaphase and interphase tumor cells by in situ hybridization using chromosome specific library probes. Hum Genet 80: 235-246

Dauwerse JG, Wiegant J, Raap AK, Breuning MH, van Ommen GJB (1992) Multiple colors by fluorescence in situ hybridization using radio-labeled DNA probes create a molecular karyotype. Hum Mol Genet 1: 593-598

Du Manoir S, Speicher MR, Joos S, Schröck E, Popp S, Döhner H, Kovacs G, Robert-Nicoud M, Lichter P, Cremer T (1993) Detection of complete and partial chromosome gains and losses by comparative genomic in situ hybridization. Hum Genet 90: 590-610

Feinberg AP, Vogelstein B (1984) A technique for radiolabelling DNA restriction fragments to high specific activity (Addendum). Anal Biochem 137: 266-267

Forozan F, Karhu R, Kononen J, Kallioniemi A, Kallioniemi OP (1997) Genome screening by comparative genomic hybridization. Trends Genet 13: 405-409

Gall JG, Pardue ML (1969) Formation and detection of RNA-DNA hybrid molecules in cytological preparations. Proc Natl Acad Sci USA 63: 378-383

Hiraoka Y, Sedat JW, Agard DA (1987) The use of a charge-coupled device for quantitative optical microscopy of biological structures. Science 238: 36-41

Hopman AHN, Ramaekers FCS, Raap AK, Beck JLM, Devilee P, van der Ploeg M, Vooijs GP (1988) In situ hybridization as a tool to study numerical chromosome aberrations in solid bladder tumors. Histochemistry 89: 307-316

Joos S, Fink TM, Rätsch A, Lichter P (1994) Mapping and chromosome analysis: the potential of fluorescence in situ hybridization. J Biotechnol 35: 135-153

Joos S, Scherthan H, Speicher MR, Schlegel J, Cremer T, Lichter P (1993) Detection of amplified genomic sequences by reverse chromosome painting using genomic tumor DNA as probe. Hum Genet 90: 584-589

Joseph AM, Gosden JR, Chandley AC (1984) Estimation of aneuploidy levels in human spermatozoa using chromosome specific probes and in situ hybridization. Hum Genet 66: 234-238

Kallioniemi A, Kallioniemi O-P, Sudar D, Rutovitz D, Gray JW, Waldman F, Pinkel D (1992) Comparative genomic hybridization for molecular cytogenetic analysis of solid tumors. Science 258: 818-821

Langer PR, Waldrop AA, Ward DC (1981) Enzymatic synthesis of biotin-labeled polynucleotides: Novel nucleic acid affinity probes. Proc Natl Acad Sci USA 78: 6633-6637

LeBeau MM (1993) Detecting genetic changes in human tumor cells. Have scientists “gone fishing“ Blood 81: 1979-1983

Lichter P (1997) Multicolor FISHing: what’s the catch? Trends Genet 13:475.479

Lichter P, Cremer T (1992) Chromosome analysis by non-isotopic in situ hybridization. In: Rooney DE, Czepulkowski (Eds) Human cytogenetics - A practical approach, Vol I. Oxford University Press, Oxford, pp 157-192

Lichter P, Cremer T, Borden J, Manuelidis L, Ward DC (1988) Delineation of individual human chromosomes in metaphase and interphase cells by in situ suppression hybridization using recombinant DNA libraries. Hum Genet 80: 224-234

Lichter P, Ward DC (1990) Is non-isotopic in situ hybridization finally coming of age? Nature 345: 93-95

Manuelidis L (1985) In situ detection of DNA sequences using biotinylated probes. Focus 7: 4-8

Nederlof PM, Flier van der S, Vrolijk J, Tanke HJ, Raap AK (1992) Fluorescence ratio measurements of double-labeled probes for multiple in situ hybridization by digital imaging microscopy. Cytometry 13: 839-845

Pinkel D, Gray JW, Trask B, van den Engh G, Fuscoe J, van Dekken H (1986) Cytogenetic analysis by in situ hybridization with fluorescently labeled nucleic acid probes. Cold Spring Harbor Symp Quant Biol 51: 151-157

Pinkel D, Landegent J, Collins C, Fuscoe J, Segraves R, Lucas J, Gray JW (1988) Fluorescence in situ hybridization with human chromosome-specific libraries: detection of trisomy 21 and translocations of chromosome 4. Proc Natl Acad Sci USA 85: 9138-9142

Raap AK, Dirks RW, Jiwa NM, Nederlof PM, van der Ploeg M (1990) In situ hybridization with hapten-modified DNA probes. In: Racz P, Haase AT, Gluckman JC (eds) Modern Pathology of AIDS and Other Retroviral Infections. S. Karger, Basel

Ried T, Landes G, Dackowski W, Klinger K, Ward D (1992) Multicolor fluorescence in situ hybridization for the simultaneous detection of probe sets for chromosome 13, 18, 21, X and Y in uncultured amniotic fluid cells. Hum Molec Genet 1: 307-313

Rooney DE, Czepulkowski (1992) Human Cytogenetics - A practical approach, Vol I Oxford University, Oxford 157-192

Schröck E, Du Manoir S, Veldman T, Schoell B, Wienberg H, Ferguson-Smith, MA, Ning Y, Ledbetter DH; Bar-Am I Soenksen D, Garini Y, Ried T (1996) Multicolor spectral caryotyping of human chromosomes. Science 273: 494-498

Speicher MR, Ballard SG, Ward DC (1996) Karyotyping human chromosomes by combinatorial multi-fluor FISH. Nature Genet 12: 368-375

Trask B (1991) Fluorescence in situ hybridization: applications in cytogenetics and gene mapping. TIG 7: 149-154

Stilgenbauer S, Döhner H, Bulgay-Mörschel M, Weitz S, Bentz M, Licher P (1993) High frequency of monoallelic retinoblastoma gene deletion in B-Cell chronic lymphoid leukemia shown by interphase cytogenetics. Blood 81: 2118-2124

Wiegant J, Ried T, Nederlof P, van der Ploeg M, Tanke HJ, Raap AK (1991) In situ hybridization with fluoresceinated DNA. Nucleic Acids Res 19: 3237-3241

Zirbel RM, Mathieu UR, Kurz A, Cremer T, Lichter P (1993) Evidence for a nuclear compartment of transcription and splicing located at chromosome domain boundaries. Chromosome Research 1:93-106

Chapter 13

Sequencing

KLAUS DEICHMANN

Introduction

Principles of sequencing

Knowledge of the DNA primary sequence leads to a closer understanding of the structural organization of DNA, the deduction of an amino acid sequence and allows for the analysis of (clinically relevant) mutations. The first sequencing methods were quite labor-intensive and time-consuming (Wu and Taylor 1971; Robertson et al. 1973; Sanger et al. 1973; Ziff et al. 1973). However, two rapid and universally applicable methods have emerged: Maxam and Gilbert used chemical agents producing defined fragments of DNA (Maxam and Gilbert 1977). These fragments are made visible by polyacrylamide gel electrophoresis using radioactive markers. The original DNA sequence can be deduced from the length of fragments formed. The second technique has been developed by Sanger in 1977 (Sanger et al. 1977). This method, the chain termination method, is most widely used nowadays, and is based on enzymatic steps (Davis et al. 1986; Sambrook et al. 1989; Ausubel et al. 1990).

The chain termination method

Usually, the DNA fragment to be sequenced, the insert, is cloned into a vector, a DNA molecule of known structure and sequence. Sequencing is performed on this construct in a single-stranded state. Single-strand DNA can be obtained, either by denaturation of double-stranded DNA or directly by the use of single-stranded vectors. An oligonucleotide primer, that is com-

Klaus Deichmann, Universitäts-Kinderklinik, Mathildenstrasse 1, Freiburg, 79106, Germany (*phone* +49-7612704300; *fax* +49-7612706372; *e-mail* deichman@kkl200.ukl.uni-freiburg.de)

plementary to the 3'-end of the known vector sequence adjacent to the insert, is synthesized. This primer anneals to the denatured template, thereby initiating in vitro synthesis of a complementary DNA strand by the use of a DNA polymerase. Sanger's method is based on the use of small amounts of 2',3'-dideoxynucleotide 5'-triphosphates (ddNTPs). These lack at the 3'-position of deoxyribose a hydroxyl residue necessary for DNA elongation (see Figure 1 and 2). Once a ddNTP is incorporated in the nascent DNA strand, chain elongation will be terminated. DNA polymerization is performed using all four dNTPs in addition to a small amount of one of the ddNTPs. A certain fraction of the forming chains will be terminated at each site where the complementary ddNTP is incorporated in the chain. Figure 3 shows the reaction products of a sequencing reaction performed on a vector insert construct using a small amount of ddATP.

Looking carefully at the results shown in Figure 3, it becomes evident that the question of the position of a nucleotide in the template has been converted into the question of the length of the terminated chains that will form by the use of the complementary ddNTP. Including a radioactively labeled nucleotide in the synthesis makes the chains visible by autoradiography when separated on a polyacrylamide gel according to their size (Figure 4 and 5).

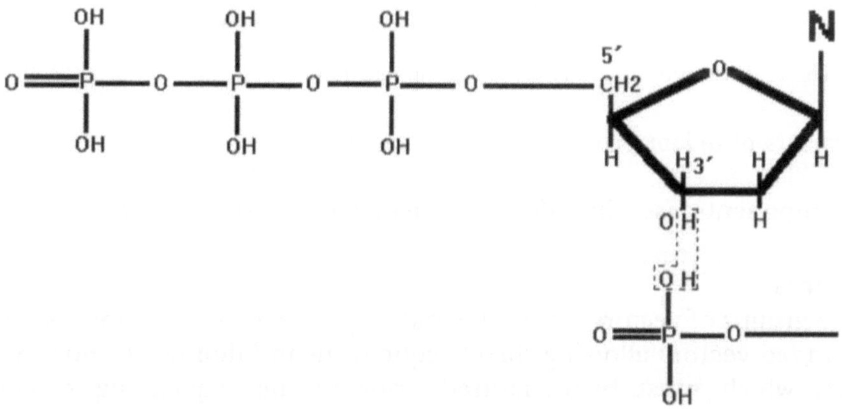

Fig. 1. A 2'-deoxy nucleotide (dNTP) forming a phosphodiester bond at the 3'-end after cleavage of the pyrophosphate of the next nucleotide.

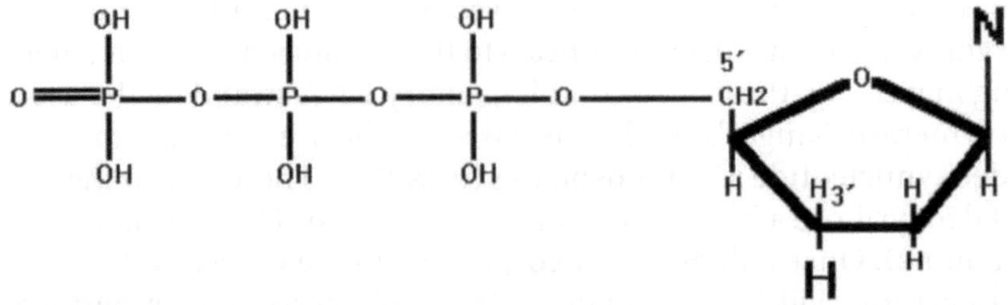

Fig. 2. A 2'-,3'-dideoxy nucleotide is lacking at its 3'-end the OH-group, which is essential for chain elongation.

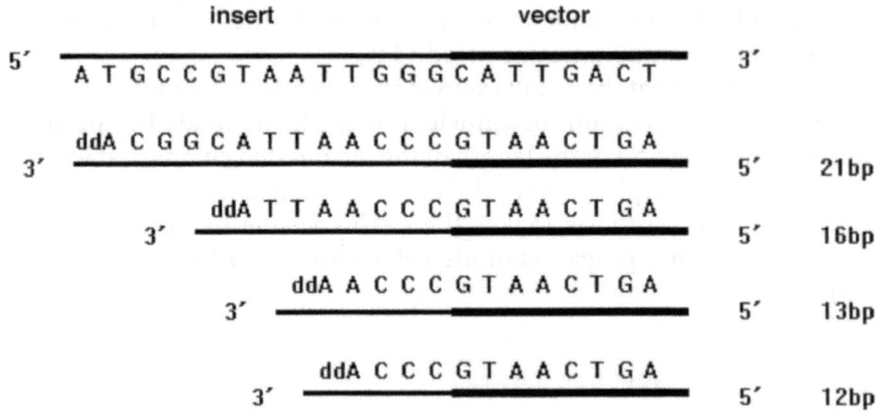

Fig. 3. The reaction products after a sequencing reaction

Components of dideoxy chain termination sequencing

The components used in dideoxy sequencing are listed below:

- **Vectors**
  Two groups of vectors are used in dideoxy sequencing reactions: single-stranded vectors allowing direct sequencing and double-stranded vectors, which must be denatured prior to the sequencing reaction (Mead and Kemper 1986).

- **M13 derivatives**
  M13 is a filamentous, male-specific coliphage (Messing et al. 1977; Wu et al. 1983; Sambrook et al. 1989) existing in two forms: a double-stranded replicative form (RF), present during infection of suitable *E. coli* host cells and a single-stranded form, which is packed into a protein coat and extruded out of the cell. The modern sequencing vectors derived

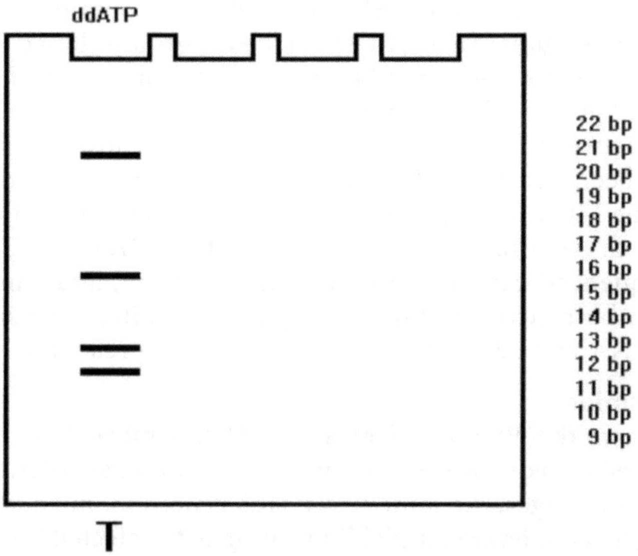

Fig. 4. The sequencing ladder resulting from the ddATP reaction of Figure 3 after separation on a polyacrylamide gel.

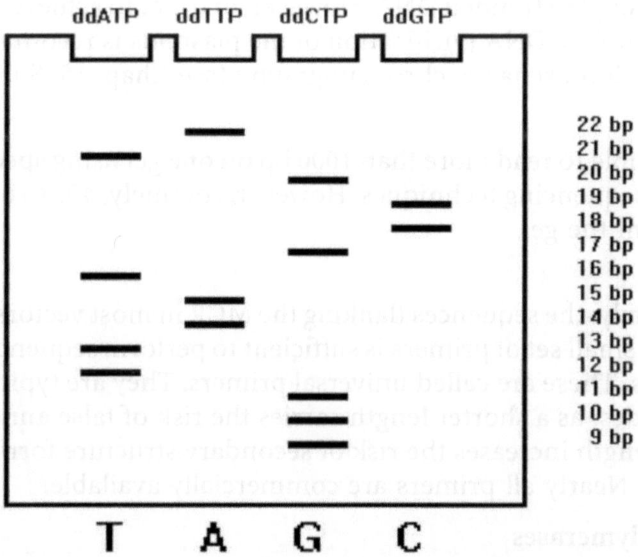

Fig. 5. The sequencing ladders resulting from all 4 ddNTP reactions of Figure 3 giving the complete sequence.

from M13 have been extensively modified. They contain a so-called multiple cloning region (MCR), in which the inserts are cloned into defined, unique restriction sites. In addition they contain a part of the beta-galactosidase gene of *E. coli* (lacZ gene), allowing by color selection a destinction between vectors bearing an insert from those lacking an insert. All have been constructed in pairs differing by the order of restriction sites in the MCR (e.g. M13mp18/M13mp19). This allows direct sequencing of both the complementary strands of the DNA insert. The fragment to be sequenced can be ligated into the RF form, and single-stranded DNA can be recovered from the supernatant after extrusion. The extruded DNA can be used directly in sequencing reactions.

- **pUC plasmids**
  These plasmids (Vieira and Messing 1982; Sambrook et al. 1989) are widely used in sequencing reactions. They contain an MCR as described for the M13 phages, the same *E. coli* lacZ gene, and, in addition, a β lactamase gene, confering ampicillin resistance to infected host cells. Once the host cells have been transformed with the insert-containing plasmid, the recombinant plasmid will replicate to a high copy number. Prior to the sequencing reactions, the plasmids are denatured by heat or alkali, yielding single-stranded DNA. However, in order to achieve good quality of the template DNA purification of the plasmids is recommended, e. g. by use of ion exchange chromatography (see Chap. 15, Subprotocol 5).

- **Inserts**
  It is possible to read more than 1000 bp on one gel using special gel-casting and sequencing techniques. However, routinely, 400 to 600 bp can be read from the gel.

- **Primers**
  Fortunately, the sequences flanking the MCR in most vectors are similiar so that a small set of primers is sufficient to perform sequencing on most plasmids. These are called universal primers. They are typically 15 to 29 nucleotides, as a shorter length carries the risk of false annealing and a longer length increases the risk of secondary structure formation of the primers. Nearly all primers are commercially available.

- **DNA polymerases**
  Four different polymerases are used for sequencing:
  - **Klenow fragment of *E. coli* DNA Polymerase I** (Sanger et al. 1977), the "classical" sequencing enzyme. The Klenow fragment lacks the 5'- to -3' exonuclease activity of native DNA polymerase I and provides faithful synthesis of a complementary strand from a single-stranded template.

- **Modified T7 DNA Polymerase (Sequenase™)** (Tabor and Richardson 1987), shows a higher processivity, i.e. less premature termination of chain elongation due to fall-off of the enzyme, enabling reading of longer sequences with less background. Further advantages are reduced 3'- to - 5' exonuclease activity and more efficient incorporation of nucleotide analogs.
- **Taq Polymerase** (Innis et al. 1988) (see also Chap. 14) is used at a temperature of 60°C. This allows a higher stringency of primer hybridization and a decrease in secondary structure formation of the template that may otherwise lead to artifacts on the gel. The high rate of polymerisation adds to these advantages. The major practical advantage lies in the thermal cycling of the sequencing reaction. By heating the reaction to 95°C without inactivation of the enzyme, the hybrid of template and new chain is denatured. This allows the template to be used again during the next sequencing cycle which leads to a linear amplification of the formed chains and decreases the amount of template required. This process also increases the signal intensity and decreases the sensitivity to reaction conditions.
- **Avian Myeloblastosis Virus Reverse Transcriptase** (Karanthanasis 1982; Graham et al. 1986; Mierendorf and Pfeffer 1987) reads better through GC-rich stretches than does Klenow fragment.

- **Radiolabeled deoxynucleotide triphosphates (dNTPs)**
  $\alpha[^{32}P]$dNTP or $\alpha[^{35}S]$dNTP may be used. ^{35}S has the advantage of high resolution and operator safety (Biggin et al. 1983). At -21° C it remains stable for 4 to 6 weeks. There is, however, the disadvantage that gels have to be dried prior to autoradiography.

- **Reaction buffer**
  Most enzymes work only under appropriate electrolyte and pH conditions.

- **Dideoxynucleotide and deoxynucleotide triphosphates (ddNTPs and dNTPs)**
  These reagents are available commercially in high concentrations. This permits the necessary adjustment of the concentration for each sequencing reaction. Secondary structures formed by DNA stretches with dyad symmetry, which lead to compression bands on a sequencing gel, can usually be resolved by the use of **nucleotide analogs** such as **dITP** (inosine in place of guanosine) and **7-deaza-dGTP** (Barnes et al. 1983; Gough and Murray 1983; Mizusawa et al 1986; Innis et al. 1988). These are ideal substrates for Taq polymerase and modified T7 polymerase.

- **Stop Solution**
  The stop solution inactivates the enzyme and denatures the hybrids which have formed.

- **Polyacrylamide Gels**
  Good quality is essential for sharp bands and a uniform pattern. Always use fresh acrylamide/bisacrylamide solution and electrophoresis grade materials. Long range sequencing can be achieved by long gels (80 to 100 cm), gradient gels differing in acrylamide concentration from top to bottom, or by using wedged spacers.

Mechanism of dideoxy chain termination sequencing

The sequencing protocol using the chain termination method as described by Sanger differs somewhat according to the polymerase used. The following refers to the use of modified T7 DNA polymerase. After preparation of the recombinant vectors, sequencing is carried out in 4 steps.

1. First **annealing** has to be performed using approximately equimolar amounts of template DNA and primer. Heating to 65°C prevents formation of secondary structures of the template DNA. Cooling slowly over several minutes allows annealing of the primer.
2. For the **labeling reaction,** equimolar amounts of dNTPs are added, one of which is radio-labeled (usually $\alpha[^{35}S]$dATP). The polymerase is added and polymerization performed for a few minutes (this step must be performed under the right buffer conditions). The result is radioactively labeled complementary strands of heterogenous size ranging from a few base pairs to about 200 bp.
3. For the **termination reaction** the labeled mixture is divided into 4 tubes and equimolar amounts of all 4 dNTPs are added as well as a small amount of ddNTP (a different one in each of the 4 tubes). The enzyme will still be active. Incubation is carried out for several minutes, and chain termination occurs in each position where a dideoxy nucleotide is incorporated (see Figure 3).
4. By adding the **stop solution** to the reactions, the enzyme is inactivated and the template-terminated chain hybrids are denatured, prior to loading them on a sequencing gel (usually a denaturing polyacrylamide gel).

Automated fluorescent sequencing

This method is based on the use of dye-labeled dideoxynucleotides as terminators of the DNA elongation (Prober et al. 1987; Wilson et al. 1988). Replacing normal ddNTPs and radioactive labeled dNTPs with the dye-labeled terminators provides 3 advantages: First, radioactive work can be avoided, facilitating sequencing. Second, all four termination reactions can be performed in a single tube provided that four different dyes are used, one for each of the ddNTPs. The method reduces tube handling and pipetting by a factor of four. Third, using different dyes, all reactions may be loaded into one lane for gel electrophoresis. The bands are then recognized by their ability to absorb light of different wavelengths when passing a scanner. Modern sequencing machines combine the advantages of Taq polymerase sequencing and the use of dye-labeled ddNTPs - the basis for modern **automated sequencing.**

The protocol given below describes reactions using the same dye. The reaction products are suited for use in many of the modern sequencing machines.

Multiplexing

Multiplexing represents a method to speed up sequencing by running several different sequencing reactions in the same 4 lanes of a gel at one time (Church and Kieffer-Higgins 1988). The different reactions are not radiolabeled, but initiated by the use of different primers. The sequencing gel is blotted onto a membrane after electrophoresis has been performed, and the blot is hybridized with a radioactive probe complementary to one of the primers, thereby giving signals in all four lanes corresponding to one template. After reading the film, the probe is washed off and hybridization performed using the next primer-probe. Multiplexing is a useful tool to save time and material when sequencing large amounts of DNA (e.g. the Human Genome Project).

Laboratory protocols

The following two sequencing protocols will provide standard reaction conditions using modified T7 DNA polymerase (Sequenase^{TM}) or Taq polymerase. They should be optimized for a particular application. Additional information may be found in the troubleshooting section. Most commercial suppliers offer sequencing kits with standard protocols. It is always advisable to start with the standard protocols, optimizing reaction conditions as experience accumulates.

Subprotocol 1
Standard protocol for T7 DNA polymerase (Sequenase™)

Materials

Reagents

- DNA template (single-stranded or double-stranded DNA)
- universal oligonucleotide primer(-s)
- modified T7 DNA polymerase (e.g. Sequenase™, USB)
- dNTPs (dATP, dCTP, dGTP, dTTP)
- ddNTPs (ddATP, ddCTP, ddGTP, ddTTP)
- 5 ×reaction annealing buffer (Tris-HCl 200 mM (pH7.5), $MgCl_2$ 100 mM, NaCl 250 mM)
- stop solution (95% Formamide, 20 mM EDTA, 0.05% Bromophenol Blue, 0.05% Xylene Cyanol FF)
- $\alpha[^{35}S]$dATP (1000 Ci/mmol)

Solutions

- Ultrapure H_2O
- Dithiothreitol (DTT) 100 mM
- 2 M NaOH
- 3 M sodium acetate, pH 4.5
- Ethanol 70% and 100%
- 10 ×TBE buffer (121 g Tris base, 7.4 g EDTA, 53.4 g Boric acid in 1000 ml A.D., pH adjusted to 8.3)
- P/CIA = salt saturated phenol:chloroform:isoamyl alcohol (25:24:1 v/v)

Supplies

- water bath at 37°C
- 1.7 ml microcentrifuge tubes
- microcentrifuge
- Ethanol / dry ice bath

Procedure

1. In advance, prepare the following on ice:

In advance

a. **dNTPs (labeling mix, 5 x concentrated):** Prepare a stock solution (7.5 μM each dNTP):

dCTP (10 mM)	3.8 μl
dGTP (10 mM)	3.8 μl
dTTP (10 mM)	3.8 μl
ultrapure water	5 ml
total volume	5 ml
store in aliquots at -20°C.	

b. **"A, T, G, C" termination mixes:** Prepare 4 stock solutions (80 μM each dNTP, 8 μM each ddNTP):

dCTP (10 mM)	8 μl
dGTP (10 mM)	8 μl
dTTP (10 mM)	8 μl
dATP (10 mM)	8 μl
ddNTP (10 mM)	0.8 μl
NaCl (50 mM)	967 μl
total volume	1 ml
store in aliquots at -20°C.	

c. **primers:** Dilute to a final primer:template molar ratio of about 1:1; start with 3 μg/ml using a universal primer of 17 nucleotides in length.

d. **DNA template:** M13 single-stranded template: adjust concentration to 1 μg / 7 μl, use directly in the sequencing reaction.

Plasmid double-stranded template must be denatured prior to use as follows:

1. Add 2 μl 2 M NaOH to 3 μg plasmid in 8 μl water.
2. Mix, incubate for 10 min at room temperature.
3. Add 3 μl 3 M NaAc and 7 μl water.
4. Add 60 μl (=3 vol.) ice-cold ethanol (100%), mix, place on dry ice / ethanol for 10 min.

5. Centrifuge for 10 min at 5°C (Eppendorf microcentrifuge).
6. Discard supernatant, wash pellet once with ethanol (70%) at -20°C.
7. Dry pellet briefly under vacuum, resuspend in 7 µl water.

Annealing

2. In a 1.7 ml microcentrifuge tube, combine
- 7 µl template solution
- 1 µl primer solution
- 2 µl annealing buffer

3. Using denatured plasmid DNA as template, incubate at 37°C for at least 15 min; using M13 DNA, heat to 65°C for 2 min.

4. Cool tube to room temperature over at least 30 min.

5. Meanwhile:
- Dilute the 5 ×labeling mix 5-fold with distilled water.
- Dilute **Sequenase**™ to a concentration of 1.5 U/µl using the appropriate dilution buffer supplied with the enzyme, keep on ice!.
- Thaw label, dNTP-solutions (labeling-mix) and termination-mixes and place on ice.
- Label 4 microcentrifuge tubes "A", "C", "G", "T".
- Pipette 2.5 µl of the corresponding termination mixes into each of the tubes; close lids and incubate at 37°C in a water bath.

Labeling

6. To the annealed template-primer-mix add:
- 1 µl DTT 100 mM
- 2 µl diluted labeling-mix
- 0.5 µl $\alpha[^{35}S]$dATP
- 2 µl diluted **Sequenase**™

giving a total volume of 15.5 µl. Mix carefully and incubate at room temperature for 5 min.

Termination

7. Transfer 3.5 µl of the labeling reaction into each of the tubes containing the termination mixes (use fresh tips and avoid contamination), spin 1 sec to mix and incubate for exactly 5 min at 37°C.

8. Add 4 µl stop solution to each of the tubes and spin for 1 sec to mix.

9. Pipette 3 µl of each of the reaction products into separate microcentrifuge tubes. Close lids firmly and heat to 75-80°C. Spin to collect the mixture at the bottom of the tube, and keep on ice until ready to load the gel. (Reactions can be stored for later use at -20°C).

Subprotocol 2
Standard Protocol for Sequencing with Taq Polymerase

In the following, a standard protocol for sequencing with the chain termination method using Taq polymerase (AmpliTaq™, Perkin-Elmer Cetus) and dye-terminators is given. Taq polymerase allows linear amplification of extension products using a thermal cycler. Furthermore, its heat-resistance allows the direct use of double-stranded DNA without prior denaturation. Annealing, labeling and termination are performed in one tube simply by altering the temperature.

Materials

Reagents

- DNA template (single-stranded or double-stranded DNA)
- Universal oligonucleotide primer(-s)
- Thermus aquaticus polymerase ("Taq") (e.g. AmpliTaq™, Perkin-Elmer Cetus)
- dNTPs (dATP, dCTP, dGTP, dTTP)
- Dye-ddNTPs (e.g. DyeDeoxy™ Terminators, Perkin-Elmer Cetus)
- Stop solution (5 vol. deionized formamide plus one vol. 50 mM EDTA (pH 8.0))

Solutions

- Ultrapure H_2O
- Mineral oil
- Phenol:H_2O:Chloroform (68:18:14 v/v)
- 2 M sodium acetate, pH 4.5
- Ethanol 70% and 100%
- 5 × TACs buffer (400 mM Tris-HCl, 10 mM $MgCl_2$, 100 mM $(NH_4)_2SO_4$, pH adjusted to 9.0)

Supplies

- Thermal cycler (e.g. Perkin Elmer Cetus Model 480, Biomed Thermal Cycler 60)
- 0.6 ml microcentrifuge tubes
- 1.7 ml microcentrifuge tubes
- Eppendorf microcentrifuge

Procedure

In advance

1. In advance, prepare the following on ice:
   a. **dNTPs:** Prepare a stock solution (150 μM each dNTP):

dCTP (10 mM)	15 μl
dGTP (10 mM)	15 μl
dTTP (10 mM)	15 μl
dATP (10 mM)	15 μl
ultrapure water	940 μl
total volume	1000 μl
store in aliquots at -20°C.	

Note: Sequencing with Taq polymerase is suited for use of **dITP** or **7-deaza-dGTP** to avoid compression bands. In this case 75 μl (750 μmol) dITP (10 mM) is substituted for dGTP.

   b. **Reaction mix:** Prepare a stock solution. The total volume should be n+1 times the volume for one reaction (38 μl):

5 × TACS buffer	8 vol.
dNTP mix	2 vol.
DyeDeoxy™ A-Terminator	2 vol.
DyeDeoxy™ C-Terminator	2 vol.
DyeDeoxy™ G-Terminator	2 vol.
DyeDeoxy™ T-Terminator	2 vol.
AmpliTaq™ (8 U/μl)	1 vol.
can be kept at 4°C for up to a month.	

   c. **Primers:** Dilute to a final molar ratio of about 3:1 to 4:1; start with 5 μg/ml using a universal primer of 17 nucleotides in length.
   d. **DNA template:** Adjust concentration:
   - M13 single-strand template: 500 ng/μl
   - Plasmid double-strand template: 1 μg/μl

Standard reaction

2. In a 0.6 ml microcentrifuge tube, combine:
   - 1 µl template solution
   - 1 µl primer solution
   - 9.5 µl reaction mix
   - 8.5 µl water
3. Overlay with a drop of mineral oil.
4. Preheat the thermal cycler to 96°C.
5. Insert the tubes and start immediately:
   - 96°C for 30 sec (**denaturation** of template and extension products)
   - rapid thermal ramp to 50°C
   - 50°C for 15 sec (**annealing** of primer and template)
   - rapid thermal ramp to 60°C
   - 60°C for 4 min (**extension**, thereby **termination** and **labeling**)

Repeat 25 times (25 cycles).
Finally, rapid thermal ramp to 4°C.

Note: You may have to adjust plateau and ramping times depending on the thermal cycler used. For more information, refer to the manual of your machine.

DNA extraction

6. Add 80 µl of water to the reaction mixture (giving a total volume of 100 µl).
7. Pipet the **under**lying aqueous layer (containing the extension products) into a fresh 1.7 ml Eppendorf tube.
8. Add 100 µl P/CIA, vortex briefly, spin using an Eppendorf centrifuge to separate the phases, transfer the **over**lying aqueous layer into a new tube.
9. Repeat the last step once.
10. Precipitate the DNA by adding 15 µl 2 M sodium acetate and 300 µl of 100% ethanol.
11. Spin for 15 min, wash the pellet with 70% ethanol and dry.
12. Redissolve the pellet in 4 µl of the loading solution.
13. Heat the sample to 90°C for 2 min, transfer immediately on to ice.
14. Load the sample on to the automated sequencer according to manufacturer's instructions.

Subprotocol 3
Denaturing Polyacrylamide Sequencing Gels

The following protocol describes casting and running a standard polyacrylamide sequencing gel. Quality and running conditions are as important for the outcome of sequencing as the sequencing reaction itself. No protocol will be appropriate for all purposes, so the length of the gel as well as the polyacrylamide concentrations will differ with the particular application.

Materials

Reagents

- Acrylamide
- N,N'- methylenebisacrylamide
- Urea (ultrapure)
- TEMED
- 10% Ammonium Persulfate
- Dichlorodimethylsilane (e.g. Methyl-Silane™, LKB)
- Amberlite MB1

Solutions

- Ultrapure H_2O
- 10% acetic acid/10% methanol in water
- 10% methanol in water
- 10 × TBE buffer (121 g Tris base, 7.4 g EDTA, 53.4 g Boric acid in 1000 ml A.D., pH adjusted to 8.3)

Supplies

- Sequencing gel apparatus
- vacuum gel-drying apparatus
- glass plates (34 ×40 and 34 ×38 cm for conventional sequencing gel)
- 2 spacers, 0.4 mm thick
- shark's tooth combs
- tray (40 ×50 cm)
- filter paper (e.g. Whatman No.1)

- plastic wrap
- syringes
- clamps
- tape
- 0.45 μm disp. filter

Procedure

In advance

1. Prepare a stock solution of 40% acrylamide/bisacrylamide:

Acrylamide	190 g
Bisacrylamide	10 g
Ultrapure water	ad 1 l

   Add 20 g Amberlite MB1 and stir for 30 min to deionize, filter through a 0.45 μm disp. filter, keep at 4°C in the dark (aluminium wrapped) for up to 1 week. Always use fresh solutions for gels.

2. Clean the glass plates using detergent, and rinse thoroughly with water. Dry plates, and rinse again with ethanol.
3. Treat one side of each plate with Methyl-Silane™ (wearing gloves and working in a fume hood). Rinse the plates thereafter with distilled water. This step is essential to prevent the gel from sticking to the plates later.
4. Place 2 spacers on each side of the large plate and place the small plate on top, both treated sides facing inward, thus forming a casting frame.
5. Tape the plates on both sides and the bottom thoroughly, taking special care on the corners, and clamp on both sides using several clamps.
6. Set the plates at a 30° angle with the top of the plates elevated.

Gel casting

7. In a beaker, mix:

40% acrylamide/bisacrylamide	15 ml
Urea	42 g
10 × TBE buffer	10 ml
Ultrapure water	ad 100 ml

Filter using a 0.45 μm disp. filter and degas with vacuum for 5 min. Urea is responsible for the denaturing conditions of the gel.

8. Add 60 μl TEMED and 600 μl ammonium persulfate (10%) and swirl to mix.
9. Pour the solution carefully between the plates, using a 20 ml pipette. Start on one side and take care to avoid introducing air bubbles; carefully insert the **flat** side of the shark's tooth combs from the top, about 5 mm deep.
10. Allow the gel to polymerize for at least 2 hours.
11. Mount the plates in the sequencing gel apparatus after removing the tape; clamp so that the smaller plate faces the upper buffer chamber.
12. Fill upper and lower chambers with 1 x TBE. After this step the TBE in the upper chamber should be in contact with the top of the gel, and the TBE in the lower chamber should be in contact with the bottom. The gel thus forms a conductive bridge between the 2 chambers.
13. Invert shark's tooth combs so that the teeth indent about 0.5 mm into the top of the gel.
14. Using a syringe with needle, thoroughly rinse the spaces between the teeth with 1 ×TBE buffer to remove unpolymerized acrylamide and urea diffusing out of the gel.
15. Pre-run gel at 40 W/ca. 1500 V for about 1 hour. The plates should be hot after the pre-run.
16. Load half of the first set of denatured sequencing reactions into 4 adjacent wells after washing the wells again with 1 x TBE buffer. Always start from the left side. Leave 4 empty slots before loading the next 4 samples (see step 18.)
17. Run the gel at 1200-1500 V (about 30 mA, 43 W). Continue electrophoresis until the xylene cyanol front is just about to reach the bottom of the gel.
18. Load the second half of the same denatured reactions in the next 4, adjacent wells and continue electrophoresis as before.
19. Stop electrophoresis when the bromophenol blue of the samples loaded last reaches the bottom of the gel. This procedure will give you a long run and a short run, i.e. the first 200 bp are read from the set of samples loaded last on the gel, the following 200-400 bp are read from the set of samples loaded first, always reading from the bottom to the top.

20. Take the glass plates out of the chamber and remove the small plate carefully so that the gel remains attached to the other plate; cut off a small part of the left upper corner of the gel.

21. Place the gel on the plate in a suitable tray and fill with a solution containing 10% acetic acid/10% methanol. Incubate for 15 min. This step will wash out the urea from the gel.

22. Carefully remove the solution by aspiration, and rinse again with 10% methanol for an additional 5 min.

23. Layer a piece of Whatman filter paper cut to size on to the gel. Avoid entrapping air bubbles between the gel and the paper. The gel will stick to the paper. Invert the assembly, and position it on the edge of the lab bench. Peel the paper downward with gel attached, while carefully lifting up the plate. Cut off one corner corresponding to the cut corner of the gel. Cover the free gel side with plastic wrap and transfer to the vacuum gel drying apparatus.

24. Dry for 60 min at 80°C.

25. Remove the plastic wrap. In the darkroom place the gel in a cassette and place a sheet of X-ray film in direct contact with the gel. Bend one corner of the film corresponding to the cut-off corner of the gel. This ensures the correct alignment of autoradiogram and gel.

26. Expose overnight at room temperature.

Troubleshooting

There are many problems that may arise when performing sequencing reactions. Some are due to improper reagents or DNA, others are due to formation of secondary DNA structure during DNA polymerization as well as during electrophoresis. The following chapter will list problems and their possible underlying causes and will recommend changes of the protocol. It may also help to adapt the technique to your specific needs.

- **Weak signals or even blank films**
  - Did you follow all the steps or did you omit anything when performing the sequencing reactions?
  - Bad template DNA preparation? Problem should not occur using commercial control DNA.

- Loss of activity of the polymerase? Always store the enzyme at -20°C, dilute just as much as you need and keep the dilution on ice until use. Use the diluted enzyme as quickly as possible. Beware of so-called frost-free freezers which warm up above 0°C daily!
- Labeled nucleotides may be too old. ^{32}P should be less than 2 weeks old, ^{35}S label can be kept at -20°C for up to 6 weeks.

- **Diffuse or fuzzy bands throughout the lanes**
  - Bad template DNA, contaminated with protein?
  - Polyacrylamide gel of poor quality? Always use electrophoresis grade reagents and prepare fresh solutions and buffers in all critical experiments.
  - Gel run too hot/too cold? Adapt the voltage. Gels should be run at a temperature of about 50-55°C.
  - Samples incompletely denatured? Always heat the samples to at least 75°C and place on ice prior to loading the gel.

- **Bands through all 4 lanes at the same position; above this site faint bands**
  - This is most probably due to strong secondary structures of the template. The enzyme will pause at these structures. The underlying sequences most probably contain dG and dC rich residues and/or dyad symmetries where hairpin loops are formed. This means that all extensions in the termination reaction will stop at the same position, and the enzyme will fall off the forming strand, no matter which of the 4 ddNTPs has been used. Try to sequence the template on both strands, lower the dNTP concentration in the termination reaction, reduce the reaction time to 1-2 min and raise the amount of template DNA to keep the enzyme from reaching the structure. You can add *E. coli* single strand binding protein (USB) to the reaction. This may prevent the DNA from forming secondary structure. However, the protein has to be removed by proteinase K prior to loading the samples on the gel. It is best to use *Taq polymerase* sequencing, as high temperature during the extension will prevent the template from forming secondary structures.

- **Bands at the same position in 2 or 3 lanes at specific regions**
  - This is most probably due to compression artifacts. These occur, when secondary DNA structures of the polymerized strands, especially those containing dG and dC rich residues, do not remain fully denatured under the conditions of electrophoresis. Try to substitute dGTP with dITP or 7-deaza-dGTP during the sequencing reactions. How-

ever, you should be aware, that dITP destabilizes DNA lowering the yield. Formamide, added to the polyacrylamide solution, should also prevent fragments from forming secondary structures during electrophoresis.

- **Bands at the same position in 2 or 3 lanes throughout the gel**
  - Spontaneous deletions may occur during phage growth leading to 2 different template DNAs. Limit the time allowed for growth to 6 hours.
- **Faint bands at the bottom of the gel**
  - If you are primarily interested in the first 50 bp adjacent to the primer, dilute the dNTP-concentration during the labeling reaction by a third, increase the ddNTP-concentration by 3-fold, increase the amount of template and restrict the time allowed for labeling and termination to 1-2 min. Use a short electrophoresis run (1-2 hours).
- **Faint bands at the top of the gel**
  - Increase the concentration of dNTPs by 2-fold and reduce the concentration of ddNTPs by a half during the termination reaction.
- **Faint bands all over the gel**
  - Increase the amount of template and reduce the time for termination to 1 min. If too long a time is chosen, extraneous nuclease activity may cause degradation of the newly synthesized DNA strand.

References

Ausubel R, Brent R, Kingston RE, Moore DD, Seidman JG, Smith JA, Struhl K (eds) (1990) Current protocols in molecular biology, Volumes 1 and 2. Greene Publishing Associates and Wiley-Interscience

Barnes WM, Bevan M, Son PH (1983) Kilo-sequencing: creation of an ordered nest of asymmetric deletions across a large target sequence carried on phage M13. Methods Enzymol 101:98

Biggin MD, Gibson TJ, Hong GF (1983) Buffer gradient gels and ^{35}S label as an aid to rapid DNA sequence determination. Proc Natl Acad Sci U.S.A. 80:3963

Church GM, Kieffer-Higgins S (1988) Multiplex DNA Sequencing. Science 240:185

Davis LG, Dibner MD, Battey JF (eds) (1986) Basic methods in molecular biology. Elsevier Science Publishing Co., Inc., New York

Gough JA, Murray NE (1983) Sequence diversity among related genes for recognition of specific targets in DNA molecules. J Mol Biol 166:1

Graham A, Steven J, McKechnie D, Harris WJ (1986) Direct DNA sequencing using avian myeloblastosis virus and Moloney murine leukemia virus reverse transcriptase. Bethesda Res Lab Focus 8(2):4

Innis MA, Myambo KB, Gelfand DH, Brow MAD (1988) DNA sequencing with Thermus aquaticus DNA polymerase and direct sequencing of polymerase chain reaction-amplified DNA. Proc Natl Acad Sci U.S.A. 85:9436

Karanthanasis S (1982) M13 DNA sequencing using reverse transcriptase. Bethesda Res Lab Focus 4(3):6

Maxam AM, Gilbert W (1977) A new method for sequencing DNA. Proc Natl Acad Sci U.S.A. 74:560

Mead DA, Kemper B (1986) in Vectors: A Survey of molecular cloning vectors and their uses. Butterworth Publishers, Massachusetts USA

Messing J, Gronenborn B, Muller-Hill B, Hofschneider PH (1977) Filamentous coliphage M13 as a cloning vehicle. Insertion of a HincII fragment of the lac regulatory region in M13 replicative form in vitro. Proc Natl Acad Sci U.S.A. 74:3642-3646

Mierendorf RC, Pfeffer D (1987) Direct sequencing of denatured plasmid DNA. Methods Enzymol 152:556

Mizusawa S, Nishimura S, Seela F (1986) Improvement of the dideoxy chain termination method of DNA sequencing by use of deoxy-7-deazaguanosine triphosphate in place of dGTP. Nucleic Acids Res 14:1319

Prober JJ, Trainor GL, Dam RJ, Hobbs FW, Robertson CW, Zagursky RJ, Cocuzza AJ, Jensen MA, Baumeister K (1987) A system for rapid DNA sequencing with fluorescent chain-terminating dideoxynucleotides. Science 238:336

Robertson HD, Barrel BG, Weith HL, Donelson JE (1973) Isolation and sequence analysis of a ribosome-protected fragment from bacteriophage phiX 174 DNA. Nature New Biol 241:38

Sambrook J, Fritsch EF, Maniatis T (eds) (1989) Molecular Cloning, a laboratory manual. 2nd edition. Cold Spring Harbor Laboratory Press

Sanger F, Donelson JE, Coulson AR, Kössel H, Fischer D (1973) Use of DNA polymerase I primed by a synthetic oligonucleotide to determine a nucleotide sequence in phage f1 DNA. Proc Natl Acad Sci U.S.A. 70:1209

Sanger F, Nicklen S, Coulson AR (1977) DNA sequencing with chain-terminating inhibitiors. Proc Natl Acad Sci U.S.A. 74:5463

Tabor S, Richardson CC (1987) DNA sequence analysis with a modified bacteriophage T7 DNA polymerase. Proc Natl Acad Sci U.S.A. 84:4767

Vieira J, Messing J (1982) The pUC plasmid, an M13mp7 derived system for insertion mutagenesis and sequencing with synthetic universal primers. Gene 19:259-268

Wilson RK, Yuen AS, Clark SM, Spence C, Arakelian P, Hood LE (1988) Automation of dideoxynucleotide DNA sequencing reactions using a robotic workstation. BioTechniques 6(8):776

Wu R, Grossmann L, Maldave K (eds), Messing J (1983) Methods in Enzymology 101 (part C): Recombinant DNA. Academic Press, New York

Wu R, Taylor E (1971) Nucleotide sequence analysis of DNA. II. Complete nucleotide sequence of the cohesive ends of bacteriophage lambda DNA. J Mol Biol 57:491

Ziff EB, Sedat JW, Galibert F (1973) Determination of the nucleotide sequence of a fragment of bacteriophage phiX 174 DNA. Nature New Biol 241:34

Chapter 14

Polymerase Chain Reaction

FRIEDHELM HILDEBRANDT AND IVA SINGH-SAWHNEY

Introduction

The polymerase chain reaction (PCR) (1) is a rapid technique for in vitro amplification of a specific DNA fragment by use of two short single-stranded primers flanking this fragment. Through repeated cycles of heat denaturation of the double-stranded DNA template, primer annealing, and primer extension using a heat-stable polymerase, the fragment of interest is amplified exponentially up to a million fold. Starting from very small amounts of DNA such as even that contained in a single cell, μg amounts of PCR product may be produced. The PCR product can be further modified by cloning into a vector, labeling for use as a probe, or directly sequenced. Starting with very small amounts of template the technique gives a high yield and can be modified in many ways. In this chapter the components and mechanisms of a PCR are described, followed by a standard PCR proctocol. Subsequently, modifications of this protocol for specific experimental purposes are discussed.

Components of a PCR

The components required for a PCR are shown in Figure 1A.

- The **DNA template** (Figure 1Aa) is shown here as a double stranded DNA molecule, in which lines symbolize the hydrogen bonds between the complementary bases. The template can be of any length. It contains within it a sequence of a well-defined length which is the **region of interest** to be amplified.

Correspondence to Friedhelm Hildebrandt, Universitäts-Kinderklinik, Mathildenstrasse 1, Freiburg, 79106, Germany (*phone* +49-761-270-4301; *fax* +49-761-270-4533; *e-mail* hildebra@kkl200.ukl.uni-freiburg.de), Iva Singh-Sawhney

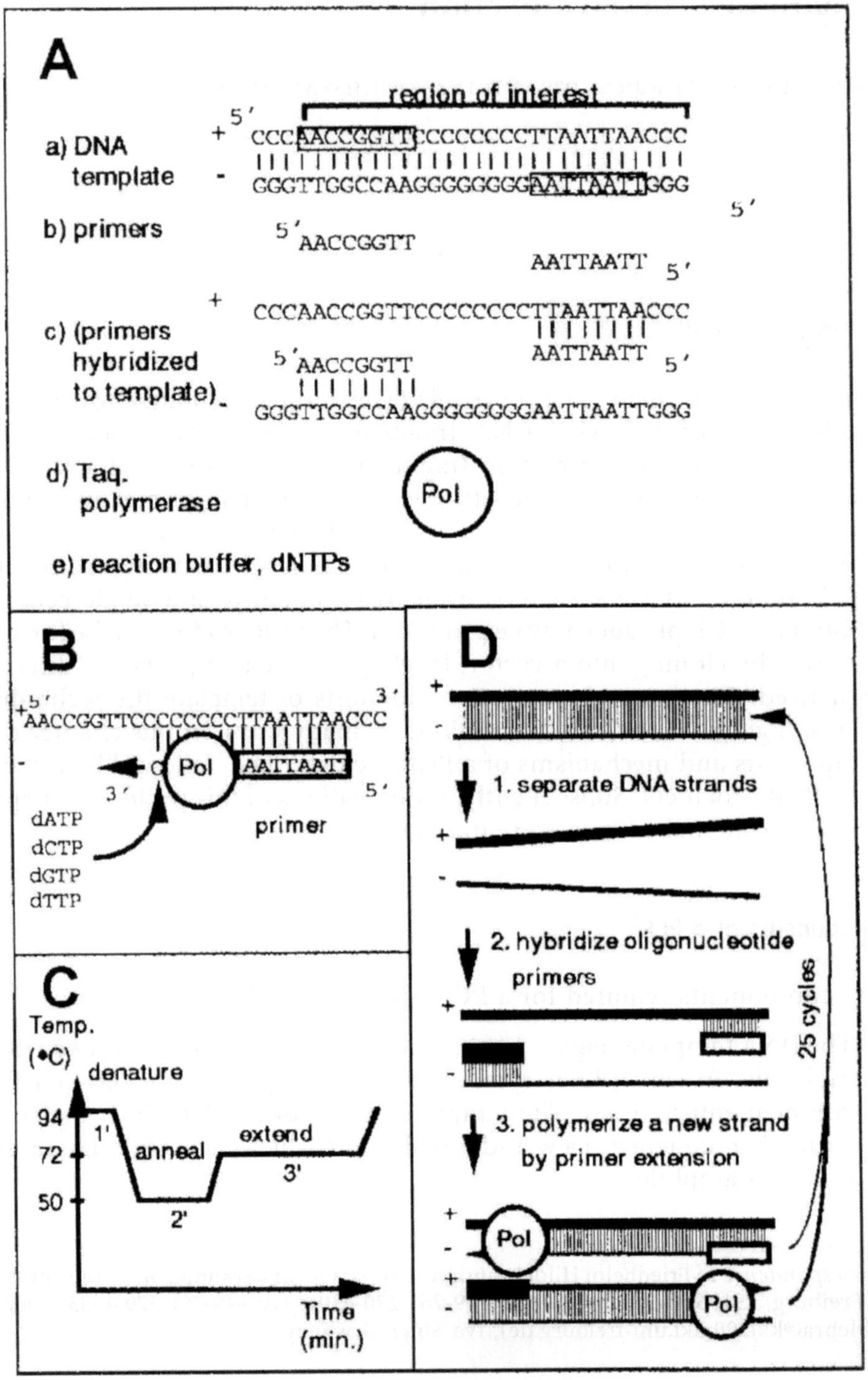

Fig. 1. Mechanisms of a polymerase chain reaction (PCR). A Components of a PCR; B Polymerase reaction; C Heat cycle for a PCR reaction; D A PCR cycle has 3 steps

- A **PCR primer pair** (Figure 1Ab) consists of two single-stranded oligonucleotides of about 18-28 nucleotides length. The sequence of one primer is identical to one template strand at one end of the region of interest, while the sequence of the other primer is identical to the opposite strand at the opposite end of the region of interest (Figure 1a). Therefore, after heat denaturation of the two template strands, the two primers are able to hybridize to their respective complementary strands at both ends of the region of interest (Figure 1Ac). It is important to choose the template strand from which the primer sequence is derived such that the 3' ends of the primers point towards the region of interest.
- The fact that some **sequence information** (Figure 1Aa) at the extreme ends of the region of interest is necessary for the synthesis of the PCR primers constitutes the most serious constraint to using the otherwise very versatile method of PCR for experimental purposes. There are various ways of overcoming the need for sequence information.
- **Taq polymerase** (Figure 1Ad) is a heat-stable DNA-polymerase. It recognizes a primer-template hybrid as a substrate (Figure 1B), binds to the hybrid, reads the template sequence and uses the complementary deoxynucleotide triphosphates present in the reaction mixture to extend the primer by formation of new phosphodiester bonds, thereby synthesizing a complementary DNA strand.
- A **reaction buffer** (Figure 1Ae) for the Taq polymerase contains Tris, KCl, $MgCl_2$, and gelatin.
- All four **deoxynucleotide triphosphates** (dNTPS) (Figure 1Ae) are required for synthesis of the complementary DNA strand (Figure 1B).

PCR mechanism

In every PCR cycle the reaction mixture is incubated sequentially at three different temperatures (Figure 1C) to allow for template denaturation, primer annealing, and primer extension. This process can be automated using a programable temperature cycling machine. The events that occur during the three steps of one PCR cycle are shown in Figure 1D.

- In the **denaturation step** separation of the two template strands by heat denaturation occurs at 94°C.
- In the **annealing step** the primers are hybridized to the template strands at a temperature around 50°C.

- In the **primer extension step** the polymerase at its optimum temperature of 72°C extends the primer sequence by synthesizing a new DNA strand that is complementary to each template molecule, thereby doubling the copy number of the target region. The PCR cycle is then repeated 25 to 40 times, thus leading to exponential accumulation of the PCR product.

The size of the PCR product is defined by the distance between the 5' ends of the two primer annealing sites (Figure 1Aa). Therefore, the PCR product can be visualized as a distinct band on agarose gel electrophoresis.

No single PCR protocol will be appropriate for all experimental purposes. Therefore, in the following a standard PCR protocol is given with subsequent discussion of the optimization of individual parameters for specific applications. Additional information on PCR procedures can be found in books on PCR methodology (1, 2).

Materials

Reagents

- DNA template
- Oligonucleotide primer pair synthesized on a oligonucleotide synthesizer
- Taq polymerase (e.g. Perkin Elmer Cetus)
- dNTPs (dATP, dCTP, dGTP, dTTP)
- 10 × reaction buffer for Taq polymerase (e.g. Perkin Elmer Cetus: 1 × buffer is Tris-HCl 50 mM (pH8.3), $MgCl_2$ 1.5 mM, KCl 25 mM, Tween 20 0.05%, autoclaved gelatin or nuclease free bovine serum albumin 100 µg/ml)

Solutions

- Ultrapure H_2O
- Mineral oil

Supplies

- Automated temperature cycling machine
- 500 µl Eppendorf tubes or heat resistant microtiter plate, fitting to the machine
- Gel electrophoresis apparatus to analyse the PCR product

Procedure

In advance

1. In advance prepare the following on ice:
   a. dNTPs (Perkin Elmer Cetus): prepare a stock solution (1mM each dNTP):

dATP (10 mM)	75 µl
dCTP (10 mM)	75 µl
dGTP (10 mM)	75 µl
dTTP (10 mM)	75 µl
ultrapure water	450 µl
total volume	750 µl

   b. **primers**: dilute to 20 µM (= 20 pmol/µl)
   c. **DNA template**: dilute to 10^5-10^6 template molecules/5 µl (3 x 10^5 template molecules is equivalent to 1 µg of human genomic DNA, 10 ng of yeast DNA, 1 ng of *E. coli* DNA, and 0.3% of an M13 phage plaque)

Standard PCR protocol

2. This protocol provides for a 100 µl reaction, but can easily be scaled down to a total volume of 10 µl. For all PCR reactions performed with the same primer pair, a master mix is generated containing every component except the DNA template. For *n* reactions a master mix for *n+1* reaction volumes should be prepared to have sufficient material available.
   In a 1.7 ml Eppendorf prepare a master mix on ice by pipetting *n+1* times the volume for one reaction combining the following components:

Component	Volume for one reaction	Final concentration	Stock concentration	Amount
a. PCR buffer	10 µl	1×	10 ×	
b. dNTPs	5 µl	50 µM each dNTP	1 mM each dNTP	5 nmol each dNTP
c. Primer pair	2 × 1µl	0.2 µM each	20 µM each	20 pmol each
d. Taq. polymerase	0.5 µl	0.025 U/µl	5 U/µl	2.5 U
e. Ultrapure H_2O	77.5 µl[a]			

[a] The later addition of 5 µl of DNA template will make a total volume of 100 µl per reaction (see below).

3. Aliquot 1 reaction volume of master mix into each PCR tube.
4. Add 5 µl of DNA template to each PCR sample (10^5-10^6 molecules, i.e. approximately 1 fmol), mix.
5. Overlay with 50 µl of mineral oil.

Note: For multiplex PCR (i.e. coamplification using multiple primer pairs) adjust the volume by reducing the amount of water added.

6. After an initial denaturation at 95°C for 4 min perform 25 to 35 PCR cycles in an automated temperature cycling machine using the following temperatures and times:

	Temperature	Time
Denaturation	95°C	30 sec
Annealing	55°C	30 sec
Extension	72°C	1.5 min

Note: For optimal temperature and time settings see Comments, Optimizing reaction conditions.

7. After the last cycle, final extension is performed for 5 min to fill in all single-stranded ends.
8. Analyse the sample by performing agarose or polyacrylamide gel electrophoresis using 1/5 to 1/2 of the reaction volume.

Note: When removing the PCR reaction under the layer of mineral oil, immerse the pipet tip under the oil layer, eject a little air, touch the bottom of the tube and retract the tip by half a millimeter. Otherwise, mineral oil will be pulled down along the wall of the tubes. Traces of mineral oil usually do not interfere with electrophoresis.

Troubleshooting

- **Missing PCR product**
  If no PCR product is generated, the following measures can be taken:
  - Run a standard postive control with a PCR product similar in size to the expected product (e.g. β-globin primers of Perkin Elmer Cetus).
  - Alter the template amount used. Sometimes the yield can be low because of too much template. For genomic DNA try 1 µg to 20 ng of template.

- Prepare a Southern blot from the agarose gel of the PCR product and hybridize with a labeled internal oligonucleotide, to determine whether the expected product is present in small amounts.
- Try nested PCR on an aliquot of the first PCR product using 2 primers derived from a sequence internal to the primers used in the first PCR.

- **False PCR products**
  If PCR products of a size other than the expected size are seen on electrophoresis, the following measures can be taken:
  - Determine the optimal extension time using a control primer pair giving rise to a similar size product.
  - Try different annealing temperatures starting from 50°C, increasing the temperature by 5°C in each experiment.
  - Use a 3:1 mixture of 7-deaza-dGTP:dGTP to reduce secondary structure formation of the template.
  - Optimize other components of the PCR reaction (see above).

- **Primer dimers**
  When designing a PCR primer pair it is important to avoid complementarity for the 3' ends of both primers and self-complementarity for the 3' end of each primer alone. Otherwise primers can hybridize to primers and the artifact of primer dimers or primer multimers will result. CG-rich 3'ends and palindromic sequences within a primer can give rise to the same problem.

- **Misincorporation**
  To overcome artifacts resulting from misincorporation by the Taq polymerase, the authentic sequence of a target region can be found by sequencing several independent PCR products and deriving a consensus sequence.

- **Avoiding template contamination**
  Use pipets, dedicated exclusively for use in PCR. Positive displacement pipets are preferable. For preparation of the PCR reaction, use a room (or at least a laminar flow work bench) that is kept separate from the area for product analysis. For the reactions with identical primer pairs use a master mix containing all components except the template. Always run a negative control containing no template using this master mix. Use different pipets when preparing the master mix and when adding template. Do not cross-contaminate your samples when adding mineral oil. Personalize all stock solutions for PCR.
  If a second PCR is to be performed with nested primers, the first PCR can be performed substituting dUTP for dTTP. Any contaminating template

from the first PCR can be destroyed with the help of uracil N-glycosylase before the second PCR is performed (3).

Comments

Optimizing reaction conditions

- **Specificity**
  Specificity of PCR is on the order of 10^7, since a specific fragment of 300bp can be isolated from a complex template such as human genomic DNA (3×10^9bp). Specificity (or "stringency") is increased at higher annealing temperatures and lower $MgCl_2$ concentrations.
  All PCR products, specific or non-specific, will accumulate exponentially once the second primer anneals to an extension product of the first primer. Therefore, the **first few cycles** are the most critical, when trying to avoid the generation of non-specific products. It may be advisable to start with an **initial denaturation** of 4 min at 95°C to assure complete strand separation before the first cycle, thus avoiding artifacts due to incomplete denaturation, secondary structure formation of the template, and formation of so-called "shuffle clones" which are derived from in vitro recombination or template strand switching.
  The PCR can be performed on crude preparations of blood or directly on unpurified cells. In these instances Taq polymerase can be added as the last PCR component after initial incubation at 95°C has been started, to denature any proteases that may be present in the template. This "hot start" approach will also reduce the generation of non-specific products, by avoiding unwanted primer annealing and extension at low temperatures (which may occur when heating up the reactions from ice to 95°C). Rather than using mineral oil, the PCR can be overlaid with a wax (Stratagene) on top of which the Taq polymerase is pipetted. When a temperature above 90°C is reached, the polymerase will melt through the wax and gain contact with the reaction mixture ("hot start"). *E. coli* ssDNA binding protein and anti-Tag-antibodies have also been used to increase specific amplification (4).
  If too many cycles are performed, the exponential accumulation of the specific product will plateau and non-specific products may "catch up", thus impairing specificity as well as yield.

- **Yield**
  For amplification from 3×10^5 templates and from 10^3 templates, a sufficient cycle number will be 25-30 and 40 cycles, respectively. Whereas dNTPs and

primers are used in large excess, yield is mostly determined by amount of template and activity of Taq polymerase (see below).

- **Polymerase processivity**
The processivity of the polymerase is defined as the rate of nucleotide incorporation at 72°C is approximately 35 to 100 nt/sec (5). Whereas the record for PCR product length currently lies around 30 kb, fragments of up to 3 kb be amplified routinely. Processivity can be improved by adding accessory proteins (such as bovine serum albumin at 100 µg/ml), lowering the salt concentration, or by using novel DNA polymerases.

- **Polymerase fidelity**
The misincorporation rate for Taq polymerase is thought to be on the order of 10^{-4}-10^{-5}/cycle. Misincorporation is reduced by decreasing $MgCl_2$ concentration, decreasing dNTP concentration, raising annealing temperatures, and decreasing extension times. Fidelity will, however, be impaired at dNTP concentrations <10 µM or if concentrations of the four dNTPs differ. The authentic sequence of a target region can be determined despite some misincorporation by sequencing several independent PCR products and generating a consensus sequence.

- **Taq polymerase**
The concentration of Taq polymerase can be optimized for a template using 0.5 to 5 U/100µl and monitoring the PCR product yield by agarose gel electrophoresis. If insufficient enzyme is used, low yield will result. Excessive enzyme will give rise to non-specific PCR products. The activity of the enzyme varies by 2 orders of magnitude between 20 and 85°C. The half-life of Taq polymerase activity is estimated to be >2 h, 40 min, and 5 min at 92.5°C, 95°C, and 97.5°C, respectively.

- **dNTPs**
dNTP stocks (10 mM) are stored at -20°C. Concentration of each dNTP can be altered between 10 µM and 50 µM (up to 200 µM) for optimal results. Decreasing the amount of dNTPs should result in improved specificity and fidelity (6). However, fidelity will be less at dNTP concentrations <10 µM or if concentrations of the four dNTPs differ.

- **Magnesium**
Magnesium concentration influences product specificity, enzyme processivity and fidelty, and specificity of primer annealing. The optimal magnesium concentration may vary with different templates and primer pairs. It can be altered from 0.5 to 10 mM, the standard being 1.5 mM.

- **Reaction buffer**
  Gelatin and bovine serum albumin (100 μg/ml) as well as nonionic detergents (e.g. Tween 20, 0.05%) can be used in the reaction buffer to help stabilize the polymerase. DMSO may be helpful in multiplex PCR.
- **Annealing temperatures**
  The optimal annealing temperature for the primer template hybrid must be determined empirically (7) and depends on the length of the primers, their GC content, and the percentage of mismatch match between primer and template sequence. As a rule one starts at 50°C, then increases the temperature by 5°C increments in each subsequent experiment until a product of the correct size is generated. The melting temperature (T_m) of a primer-template hybrid is defined as the temperature where 50% of the primer-template hybrids are formed (Figure 2). One estimate for T_m can be calculated from the following formula:

$$T_m[°C] = 4\ (C+G) + 2\ (A+T) + 8 \qquad \text{(for 17–25mers)}$$

(C, G, A, T is the number of cytosines, guanine, adenines, and thymines, respectively, in the oligonucleotide.)

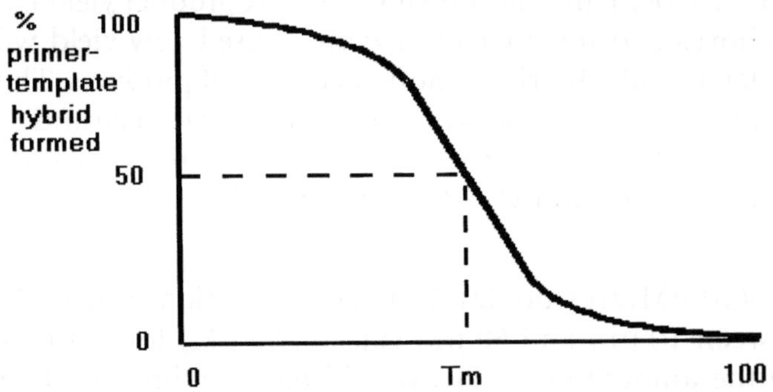

Fig. 2. Definition of the melting temperature (Tm) for a primer template hybrid.

The optimal PCR annealing temperature will be about 5°C below T_m and will reside between 55 °C and 72°C. It seems to be important to design the pairs of primers such that the difference in their respective T_m is within 5°C. Empirically, it also seems advisable two maintain the GC content below 55%. Primer annealing usually requires only a few seconds. High stringency (high annealing temperatures) are especially important during the first PCR cycles to increase specificity (see above).

- **Extension times**
  An extension time of 1 min will be sufficient for a PCR product length of up to 2 kb. For longer products the extension time may have to be prolonged to 5 or even 10 min.
- **Primer synthesis**
  PCR primers are synthesized using an oligonucleotide synthesizer. Typically, a length of 18-28 nt (10-34 nt) is used. CG content should not exceed 50-60%. The difference in the calculated melting points between the two primers (see above) should be ≤ 5°C.
- **Primer concentration**
  An optimal primer concentration is in the range of 0.1 to 0.5 μM. Above this concentration mispriming and lower yield of the specific product are to be expected.
- **Primers for specific experimental purposes**
  Examples of primers for specific experimental purposes are described briefly below together with the respective references to the literature. Mismatched primers are often used for site-directed mutagenesis and to add a 5'-extension to primers which may contain additional sequence such as restriction enzyme recognition sites, promoters, etc. When using PCR primers that contain mismatches it is important to note that the PCR product will always contain the sequence of the primers rather than the template at its ends (Figure 3).

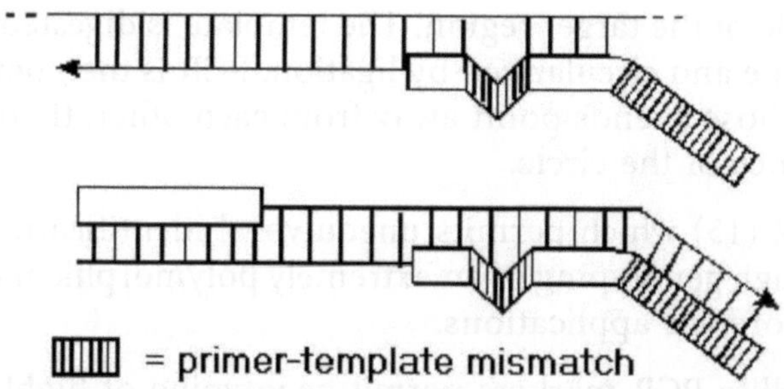

Fig. 3. In regions of primer-template mismatch, primer sequence is incorporated in the PCR product.

Also frequently used are degenerate primer mixtures for the isolation of sequence-related novel genes. To help with the design of an optimal primer pair for a certain sequence, computer programs such as “TM-Oligo“ are available.

Applications

Modifications and applications of PCR for specific purposes

PCR is a very versatile method. A multitude of modifications of the PCR have been successfully applied for experimental purposes. In this section such special applications of PCR are listed, classified according to emphasis on the use of a specific template, primer, cycling protocol, product detection method, quantitation of PCR, or polymerase used. Many of these methods represent approaches to partially overcome the most severe limitation to PCR, the requirement for specific sequence information flanking the target region.

Genomic template

- PCR is used in prenatal diagnosis on chorionic villus biopsy specimens, amniocytes, in vitro fertilized eggs, prefertilization polar bodies, and on maternal peripheral blood and preimplantation embryos (8) for sex-determination of the fetus.
- PCR is used on degraded genomic DNA ("jumping PCR" (9)), on fossil DNA (10), and on paraffin embedded samples (11).
- PCR is used in forensic applications on single hair, sperm etc..
- Inverse PCR (12, 13, 14) is used if sequence information is available only on one side of the target region. The template is digested with a restriction enzyme and circularized by ligation. PCR is then performed with 2 primers whose 3' ends point away from each other, thereby amplifying the sequence of the circle.
- MVR-PCR (15) which permits unequivocal identification of an individual through genotyping of an extremely polymorphic marker is mainly used for forensic applications.
- Microsatellite-PCR markers permit genotyping of highly polymorphic dinucleotide-repeat markers for use in linkage analysis (16).
- Inter-Alu repeat PCR employs primer complementary to repetitive sequences (IRSs) of genomic DNA to characterize overlapping YAC clones (17).
- Transposon-related PCR strategies (18, 19) use transposons to provide sequence information after integration of the transposon into large cloned fragments of genomic DNA.

- Arbitrarily primed PCR (APPCR) or random amplified polymorphic DNA (RAPD) use short random primer mixes to characterize cloned fragments of genomic DNA (20, 21).
- Sequence tagged sites (STSs) are genomic landmarks at 100kb distance (22) as defined by the human genome projects.
- PCR across gene rearrangements is widely used in viral diagnostics, break point determination in oncogenes (84), and determination of rearrangents in immunoglobulin genes.
- "Footprinting" is used to identify binding regions for DNA-binding proteins (23).
- Microdissection by PCR (24) permits the generation of probes for fluorescence in-situ hybridization of chromosomal segments.
- PCR on single cells (25).

RNA template

PCR can be applied to the study of gene expression. Starting from mRNA and using reverse transcriptase the first cDNA strand is synthesized, which then serves as the template for the PCR.

- Anchored PCR (26) or "RACE: rapid amplification of cDNA ends" (27) is a two-step approach to clone a full-length cDNA (see also "PCR-cloning").
- Multiple oligonucleotide primed amplification of cDNA (MOPAC) (28) uses degenerate primer mixtures to clone novel genes or cDNAs under the hypothesis of sequence similarity between gene families and genes from different species (29).
- mRNA phenotyping is a method for studying the level of expression of specific transcripts from specific cell types within a heterogeneous cell population (30).
- In-situ PCR represents a PCR-based approach to in situ hybridization (31).
- PCR permits the creation of specific libraries from a sub-population of cells.

Mitochondrial template

- The mitochondrial genome (32) can easily be studied by PCR, due to its low complexity. The origins of human evolution have extensively been studied by this approach.

Infectious agents

- The presence of DNA from viral, bacterial, fungal, and parasitic pathogens (33) can be studied in the background of host DNA. For the human immunodeficiency virus quantitation of the relative amounts of different strains can be studied by automated sequencing following PCR (34).

Specific primers

- Degenerate primer mixtures. Designing degenerate primers is one approach for partly overcoming the main limitation of PCR, namely the requirement for knowledge of some specific sequence of the template for primer design. Amino acid sequence can be used to design mixtures of primers that represent best guesses on the codon usage in the target template (35).
- Mismatched primers can be used for the introduction of novel sequence into a template for site-directed mutagenesis (see Figure 3).
- Linker primers contain additional sequence at their 5' ends which have no correlate in the template (see Figure 3). These sequences can consist of restriction enzyme recognition sites for directed subcloning of the PCR product (36, 37), promoters for the in vitro transcription of RNA from the PCR product, translation initiation sites for in vitro translation from the PCR product (38), sequence encoding epitopes of monoclonal antibodies (39), or CG-clamps for denaturing gradient gel electrophoresis (see Chapter 6).
- Nested primers are used for re-amplification with a primer pair internal to the 2 outer primer annealing sites. A labeled nested primer can be used in Southern blotting of PCR products. "Drop-in, drop-out nested primers" are employed to enhance specific amplification in highly sensitive single-template PCR (40).
- One-sided PCR uses one specific and one non-specific primer (e.g. insert to vector).

- Labeled primers can be employed for many purposes. Biotinylated primers can be used for purification of PCR products using streptavidin coated magnetic beads. Fluorescence-labeled primers can be used for automated detection of PCR products.
- Multiplex PCR is the coamplification of several PCR products using several primer pairs in the same reaction mixture.

Specific cycling protocols

- Asymmetric PCR describes one approach to overcome the problem of shadow bands resulting from the opposite PCR primer in direct sequencing of a PCR product (41).
- "Hot PCR" is used to incorporate a radioactive label during PCR (e.g. a-[^{32}P]-dCTP) during PCR).

Allele detection and sequencing of PCR products

- Allele-specific oligonucleotides (ASOs) describes the use of labeled oligonucleotides for differential dot blot hybridization to PCR products that exhibit point mutations (42, 43, 44, 45).
- Reverse dot blot method is the reverse of the ASO method: the labeled PCR product is hybridized to several dot blotted ASOs (46).
- Denaturing gradient gel electrophoresis (DGGE) (47, 48) and temperature gradient gel electrophoresis (TGGE) are used to discriminate heteroduplexes and homoduplexes of PCR products from templates containing point mutations (see Chapter 6).
- Single-strand conformational polymorphism (SSCP) allows for discrimination of point mutations by differential migration of single strands in electrophoresis (49) (see Chapter 6).
- Oligomer ligation assay (OLA) is a technique for allelic discrimination by use of a heat-stable ligase (50).
- Restriction enzyme digestion of PCR products can be used for allele discrimination of loss or addition of a restriction enzyme recognition site.
- Direct sequencing of PCR products has been described in many protocols (51) all aimed at avoiding "shadow sequence" derived from the second PCR primer.

Quantitative PCR

The PCR cycle in which extension from both primers occurs for the first time is the cycle in which exponential accumulation of the PCR product begins. This event is of stochastic nature. If in 2 different experiments the difference for this first event is 2 cycles, the product amount will differ by a factor of 4. There- fore, when using PCR for quantitation of template amounts, the inherent problem of the exponential nature of product accumulation must be accounted for. The following techniques have been applied:

- Coamplification of a standard template can be used for quantitation of a PCR product (52, 53).
- Competitive PCR (54).

Different thermostable polymerases

- Taq polymerase has 5'- to 3'-exonuclease activity, but no 3'- to 5'-exonuclease activity.
- The "Stoffel fragment" of Taq polymerase has no 5'- to 3'-exonuclease activity.
- VENT DNA polymerase (New England Biolabs) exhibits both, 5'- to 3'-exonuclease activity and 3'- to 5'-exonuclease activity.
- A DNA Polymerase of *thermus thermophilus* in the presence of manganese can synthesize DNA from both RNA and DNA templates

References

1. Mullis KB, Faloona (1987) Specific synthesis of DNA in vitro via a polymerase-catalysed chain reaction. Methods Enzymol 155:335-350
2. Erlich HA (ed) (1989) PCR technology. Stockton, New York
3. Innis MA, Gelfand DH, Sninsky JJ, White TJ (eds) (1990) PCR protocols: a guide to methods and applications. Academic press, San Diego
4. Erlich H (1991) Recent advances in the polymerase chain reaction. Science 252:1643-1651
5. Saiki RK, Gelfand DH (1989) Introducing AmpliTaq DNA polymerase. Amplifications 1:4-6
6. Innis MA, Myambo KB, Gelfand DH, Brow MAD (1988) DNA sequencing with Thermus aquaticus DNA polymerase and direct sequencing of polymerase chain reaction-amplified DNA. Proc Natl Acad Sci U.S.A. 85:9436-9440 7. Saiki RK et al. (1988) Primer-directed enzymatic amplification of DNA with a thermostable DNA polymerase. Science 239: 487-491

7. Saiki RK et al. (1988) Primer-directed enzymatic amplification of DNA with a thermostable DNA polymerase. Science 239: 487-491
8. Longo MC, Berninger MS, Harley JL (1990) Use of uracil DNA glycosylase to control carry-over contamination in polymerase chain reactions. Gene 93:125-128
9. Handyside AH, Kontogianni EH, Hardy K, Winston RML (1990) Pregnancies from biopsied human preimplantation embryos sexed by Y-specific DNA amplification. Nature 344:768-770
10. Pääbo S, Huguchi RG, Wilson AC (1989) Ancient DNA and the polymerase chain reaction. J Biol Chem 264:9709-9712
11. Persing DH et al. (1990) Detection of Borrelia burgdorferi DNA in museum specimens of ixodes dammini ticks. Science 249:1420-1423
12. Impraim CC, Saiki RK, Erlich HA, Teplitz RL (1987) Analysis of DNA extracted from formalin-fixed, paraffin-embedded tissues by enzymatic amplification and hybridization with sequence-specific oligonucleotides. Biochem Biophys Res Comm 142:710-716
13. Ochman H, Gerber AS, Hartl DL (1988) Genetic applications of an inverse polymerase chain reaction. Genetics 120:621
14. Triglia T, Peterson MG, Kemp DJ (1988) A procedure for in vitro amplification of DNA segments that lie outside the boundaries of known sequences. Nucleic Acids Res 16:8186
15. Silver J, Keerikatte V (1989) Novel use of polymerase chain reaction to amplify cellular DNA adjacent to an integrated provirus. J Virol 63:1924-28
16. Jeffreys AJ, MacLeod A, Tamaki K, Neil DL, Monckton DG (1991) Minisatellite repeat coding as a digital approach to DNA typing. Nature 354:204-209
17. Weber JL, May PE (1989) Abundant class of human DNA polymorphisms which can be typed using the polymerase chain reaction. Am J Hum Genet 44:338-396
18. Ledbetter SA, Nelson DL, Warren ST, Ledbetter DH (1990) Rapid isolation of DNA probes within specific chromosome regions by interspersed repetitive sequence polymerase chain reaction. Genomics 6:475-481
19. Phadnis SH, Huang HV, Berg DE (1989) Tn5supF, a 264-base-pair transposon derived from Tn5 for insertion mutagenesis and sequencing DNAs cloned in phage l. Proc Natl Acad Sci U.S.A. 86:5908-5012
20. Strausbaugh LD, Bourke MT, Sommer MT, Coon ME, Berg CM (1990) Probe mapping to facilitate transposon-based DNA sequencing. Proc Natl Acad Sci U.S.A. 87:6213-17in press
21. Welsh J, McClelland M (1990) Fingerprinting genomes using PCR with arbitrary primers. Nucleic Acids Res 18:7213-7218
22. Williams JGK, Kubelik AR, Livak KJ, et al. (1990) DNA polymorphisms amplified by arbitrary primers are useful as genetic markers. Nucleic Acids Res 18:6531-6535
23. Olson M, Hood L, Cantor C, Botstein D (1989) A common language for physical mapping of the human genome. Science 245:1434-1435
24. Mueller PR, Wold B (1989) In vivo footprinting of a muscle specific enhancer by ligation mediated PCR. Science 246:780-786
25. Meltzer PS, Guan X-Y, Burgess A, Trent JM (1992) Rapid generation of region specific probes by chromosome microdissection and their application. Nature Genet 1:24-28
26. Rappolee DA, Wang A, Mark D, Werb Z (1989) Novel method for studying mRNA phenotypes in single or small number of cells. J Cell Biochem 39:1-11

27. Loh EY, Elliott JF, Cwirla S, Lanier LL, Davis MM (1989) Polymerase chain reaction with single-sided specificity:analysis of T cell receptor d chain. Science 243:217-220
28. Frohman MA, Dush MK, Martin GR (1988) Rapid production of full-length cDNAs from rare transcripts: amplifacation using a single gene-specific loigonucleotide primer. Proc Natl Acad Sci U.S.A. 85:8998-9002
29. Lee C.C. et al. (1988) Generation of cDNA probes directed by amino acid sequence: cloning of urate oxidase. Science 239:1288-1291
30. Varadi A, Gilmore-Hebert M, Benz EJ (1989) Amplification of the phosphorylation site - ATP-binding site cDNA fragment of the Na^+, K^+-ATPase and the Ca^{2+}-ATPase of Drosophila melanogaster by polymerase chain reaction. FEBS let 258:203-207
31. Rappolee DA, Mark D, Banda M, Werb Z (1988) Wound macrophages express TGF-a and other growth factors in vivo: analysis by mRNA phenotyping. Science 241:708-712
32. Tecott LH, Barchas JD, Eberwine JH (1988) *In situ* transcription: specific synthesis of complementary DNA in fixed tissue sections. Science 240:1661-1664
33. Vigilant L, Pennington R, Harpending H, Kocher T, Wilson AC (1989) Mitochondrial DNA sequences in single hairs from a southern African population. Proc Natl Acad Sci U.S.A. 86:9350-9354
34. Williams SD, Kwok S (1991) In: Lenette EH (ed) Laboratory diagnosis of viral infections. Dekker, New York, pp 147-173
35. Goodenow M et al. (1989) HIV-1 isolates are rapidly evolving quasispecies: evidence for viral mixtures and preferred nucleotide substitutions. J Acquired Immune Def Syndr 2:344-352
36. Lathe R (1985) Synthetic oligonucleotide probes deduced from amino acid sequence data. J Mol Biol 183:1-12
37. Scharf SJ, Horn GT, Erlich HA (1986) Direct cloning and sequence analysis of enzymatically amplified genomic sequences. Science 233:10761078
38. Higuchi R (1989) In: Erlich H (ed) PCR technology: principles and applications for DNA Amplification. Stockton, New York, pp 61-70
39. Sakar G, Sommer SS (1989) Access to a messenger RNA sequence or its protein product is not limited by tissue or species specificity. Science 244:331-334
40. Martin et al. (1990) The GAP-related domain of the neurofibromatiosis type 1 gene product interacts with ras p21. Cell 63:843-849
41. Erlich H (1991) Recent advances in the polymerase chain reaction. Science 252:1643-1651
42. Gyllenstein UB, Erlich HA (1988) Generation of single-stranded DNA by the polymerase chain reaction and its application to direct sequencing of the HLA-DQA locus. Proc Natl Acad Sci U.S.A. 85:7652-7656
43. Saiki RK, Bugawan TL, Horn GT, Mullis KB, Erlich HA (1986) Analysis of enzymatically amplified β-globin and HLA-DQα DNA with allele-specific oligonucleotide probes. Nature 324:163-166
44. Wong C, et al. (1986) Characterization of β-thalassaemia mutations using direct genomic sequencing of amplified single copy DNA. Nature 330:384-386
45. Saiki RK et al. (1988) Diagnosis of sickle call anemia and β-thalassemia with enzymatically amplified DNA and nonradioactive allele.specific oligonucleotide probes. N Engl J Med 319:537-541
46. Conner BJ et al. (1983) Detection of sickle cell b^S-globin allele by hybridization with synthetic oligonucleotides. Proc Natl Acad Sci U.S.A. 80:278-282

47. Erlich H et al. (1991) HLA-DR, DQ and DP typing using PCR amplification and immobilized probes. Eur J Immunogen 18:33-55
48. Fisher SG, Lerman LS (1983) DNA fragments differing by single base-pair substitutions are separated in denaturing gradient gels: correspondence with melting theory. Proc Natl Acad Sci U.S.A. 80:1579-83
49. Myers RM, Sheffield VC, Cox DR (1988) Detection of single base changes in DNA: ribonuclease cleavage and denaturing gradient gel electrophoresis. In: Davies K (ed) Genomic analysis: a practial approach. IRL Press, Oxford
50. Orita M, Iwahana H, Kanazawa H, Hayashi K, Sekiya T (1989) Detection of polymorphisms of human DNA by gel electrophoresis as single-strand conformation polymorphisms. Proc Natl Acad Sci U.S.A. 86:2766-2770
51. Nickerson DA et al. (1990) Automated DNA diagnostics using an ELISA-based oligonucleotide ligation assay. Proc Natl Acad Sci U.S.A. 87:8923-8927
52. Brow MD (1990) In: Innis MA, Gelfand DH, Sninsky JJ, White TJ (eds) PCR protocols: a guide to methods and applications. Academic Press, San Diego, CA, p 189
53. Wang AM, Doyle MV, Mark DF (1989) Quantitation of mRNA by the polymerase chain reaction. Proc Natl Acad Sci U.S.A. 86:9717-9721
54. Gilliland G, Perrin S, Blanchard K, Bunn HF (1990) Analysis of cytokine mRNA and DNA: detection and quantitation by competitive polymerase chain reaction. Proc Natl Acad Sci U.S.A. 87:2725-29
55. Siebert PD, Larrick JW (1992) Competitive PCR. Nature 359:557-558

Chapter 15

Cloning Vectors

ANNETTE DEICHMANN AND KLAUS DEICHMANN

Introduction

For investigation of the sequence, structure and function of DNA fragments a large number of identical copies is required. The production of large amounts of DNA from a small number of copies is called cloning (Levin 1985; Winnacker 1985). For cloning, one can make use of the cell's replication mechanisms. A question arises, which is how to transport a DNA fragment into the cell and how to initiate its replication. For this purpose, cloning vectors have been designed. These vectors are descendants of naturally-occuring, autonomous DNA-species in bacterial cells - the bacteriophage and plasmids. Insertion of foreign DNA fragments into the genomes of phages and plasmids, produces a piece of recombinant DNA called a hybrid or a chimera. The ability of bacteriophage and plasmids to be amplified in bacterial cells is not altered as long as the inserted DNA does not exceed a certain size. The inserted DNA sequences does not disrupt a vector sequence that is essential for its function. Furthermore, there should be a growth advantage for bacterial cells containing these hybrids and easy selection of chimeric vectors over native vectors. In short, vectors should provide us with:

- Simple methods for multiplication of specific DNA fragments
- High reliability regarding the identity of the copies
- Simple sequencing procedures to be carried out directly on the recombinant vector

Annette Deichmann, *Correspondence to* Klaus Deichmann, Universitäts-Kinderklinik, Mathildenstrasse 1, Freiburg, 79106, Germany (*phone* +49-7612704300; *fax* +49-7612706372; *e-mail* deichman@kkl200.ukl.uni-freiburg.de)

- The transcription of isolated genes into corresponding mRNA as a first step of in vivo or in vitro synthesis of their encoded proteins
- The possibility of further subcloning
- A high rate of biological securityModern vectors are designed by DNA engineering to fit a certain range of the demands listed above. We will focus on typical sequencing vectors (M13) as well as typical expression vectors.

Principles of cloning

The steps of cloning are shown in Figure 1 using a plasmid. First, the circular plasmid is linearized using a restriction endonuclease that cuts at a unique recognition sequence producing a single cleavage. The sticky ends or blunt ends formed are characteristic for each of the restriction enzymes (see Chap. 2). The DNA insert to be cloned into the vector is produced by digestion with the same restriction enzyme or an isoschizomer to produce compatible ends. Ligation is performed by joining the ends of foreign DNA and linearized plasmid using a ligase (see Chap. 2), resulting in a circular chimeric plasmid. Ligation is followed by transformation of bacterial host cells (when using bacteriophage, this step is called transfection). The bacterial host cell is made competent' to receive the plasmid through alteration of the bacterial cell wall using $CaCl_2$ or an electrical impulse. As a safety measure, natural transformation of laboratory plasmids is inhibited by removing the genes coding for this ability. Once inside the cell, replication of the plasmid is initiated from an origin of replication (ori). Using multicopy plasmids, all cells originating from the transformed cell will bear several copies of the recombinant plasmid. Following in vitro cell lysis, these plasmids can be isolated and the insert, now present in several thousands of identical copies, can be excised, usually using the same restriction enzymes as for initial ligation. Some phages will cause lysis of the transfected cells (temperate phage). These can be isolated from the supernatant after the lytic cycle has been completed.

Plasmids

Plasmids are small, circular, double-stranded DNA elements (Figure 2). They replicate autonomously in bacterial cells independently of the bacterial chromosome. They are known to the physician as transmitters of bacterial antibiotic resistence. Plasmid vectors are composed of:

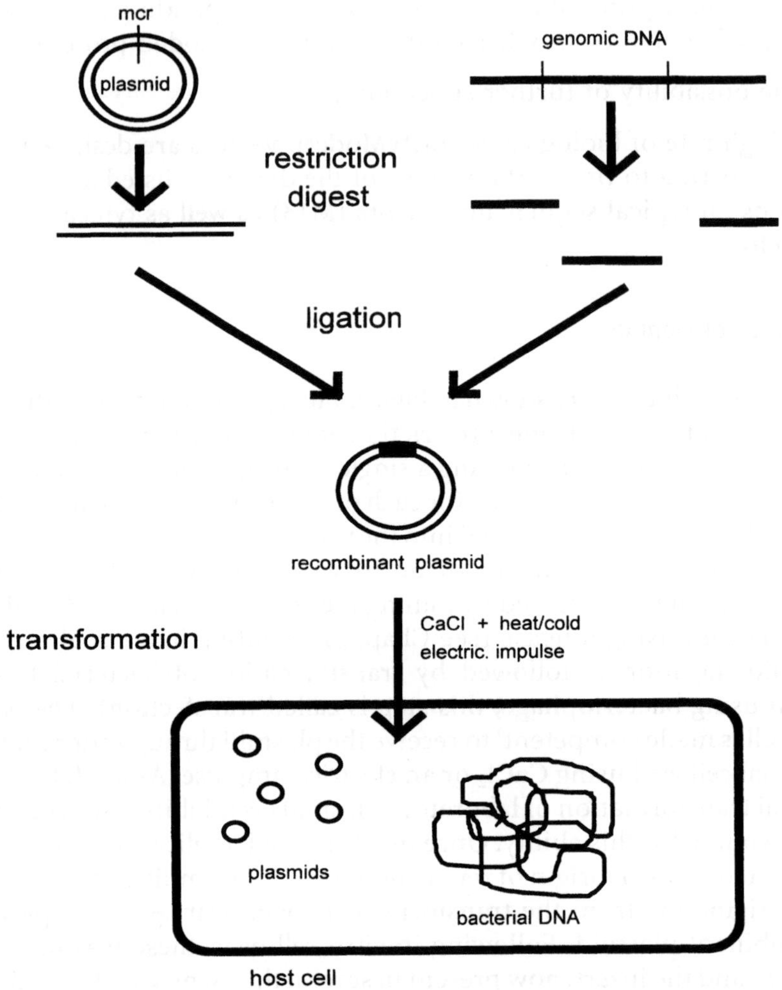

Fig. 1. Fragments of genomic DNA are cloned into a plasmid to form a recombinant plasmid. These are transformed into *E. coli* host cells (mcr = multiple cloning region).

- An origin of replication which initiates their replication and keeps the number of copies within a cell at a characteristic level.
- An optional multiple cloning region (MCR), which is a short sequence bearing unique recognition sites for many restriction enzymes. The MCR allows site-directed cloning of an insert.
- Antibiotic resistance genes providing a selective advantage to transformed bacterial host cells.

- Alternative sequences which might be used for selection, see below.
- Optional (phage) promotors flanking the insertion site of the insert and allowing in vitro transcription of cloned genes.

To keep plasmids stable (i.e., to prevent loss of part or all of the insert) and to assure efficient transformation into bacterial cells, an insert length of 5000 bp should not be exceeded.

- pBR322 (Bolivar 1977, Sutcliffe 1978)
  A double-stranded DNA plasmid cloning vehicle of about 4.4 kb in length. pBR322 carries an origin of replication and resistance genes for ampicillin and tetracycline. Several unique cloning sites within the resistance genes allow for simple differentiation between plasmids bearing inserts and wild-type plasmids.
- pUC (Vieira 1982)
  A double-stranded DNA plasmid cloning vehicle of about 2.7 kb in length. pUC plasmids share in common with pBR322 the ampicillin resistance gene and the origin of replication. In addition they share in common with the M13 derived vectors the lacZ α-peptide gene of *E. coli* and its multiple cloning region (MCR) (see below). This permits color-selection between plasmids bearing inserts and wild-type plasmids. As with M13, pUC plasmids are available in pairs with reversed order of restriction sites in the MCR.

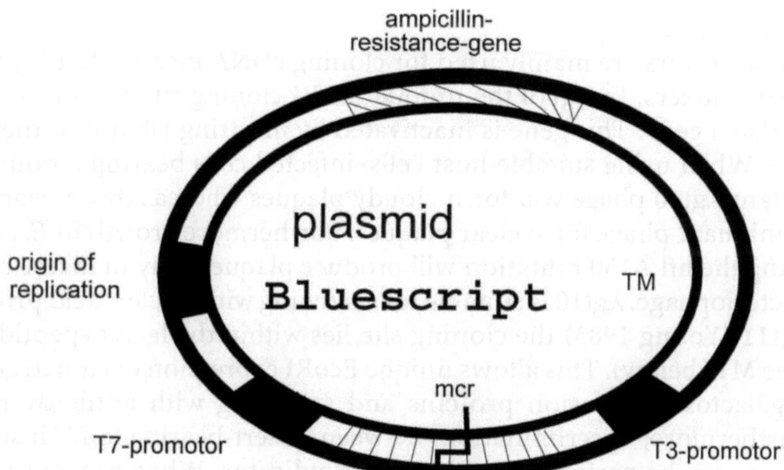

Fig. 2. The plasmid pBluescript. Indicated are characteristic sequences responsible for the function of the vector (mcr = multiple cloning region).

- pSP6/T7 (Melton 1984)
  This vector shares many homologies with pUC. In addition, the plasmid has been supplemented by 2 bacteriophage promotors flanking the MCR. This allows the plasmid to be used as a cloning and in vitro transcription vector.There are many other plasmids constructed to suit special tasks (see catalogues of the main suppliers of products for molecular biology).

Bacteriophages

The use of plasmids is restricted to an insert size of up to about 5 kb. When cloning larger fragments, bacteriophage should be used.

λphage Lambda phage DNA is a 43 kb double-stranded DNA, that is introduced into bacterial cells by transfection (Figure 3). The DNA may be covalently incorporated into the bacterial genome and replicated with it as a prophage. However, during the lytic cycle, hundreds of phage DNA copies are formed and packaged by special enzymes into newly-formed phage particles. Packaging enzymes and head and tail proteins are encoded by the phage DNA. The bacterial cells are broken open (lysis), and phages are set free. This mechanism also applies to other temperate phages such as phi 80, 82, 21, 434, P2 and 299.

- λgt10 / λgt11
  These vectors are mainly used for cloning cDNA inserts after ligation to EcoRI linkers. In **λgt10** the unique EcoRI cloning site lies in the cI (repressor) gene. This gene is inactivated by inserting DNA into the EcoRI site. When using suitable host cells, infected cells bearing a nonrecombinant λgt10 phage will form cloudy plaques whereas those bearing recombinant phage form clear plaques. Furthermore growth in *E. coli* carrying the hfl A150 mutation will produce plaques only in insert-bearing bacteriophage. λgt10 is suitable for screening with nucleic acid probes. In **λgt11** (Young 1983) the cloning site lies within the lacZ α-peptide gene (see M13 below). This allows unique EcoRI expression of cloned genes as ß-galactosidase fusion proteins and screening with antibody probes. Furthermore, discrimination between insert-bearing and insert-free phage can be performed with a color indicator. When grown on plates containing X-Gal, recombinant phage plaques are clear, whereas non-recombinant phage plaques appear blue (mechanism see M13 below).

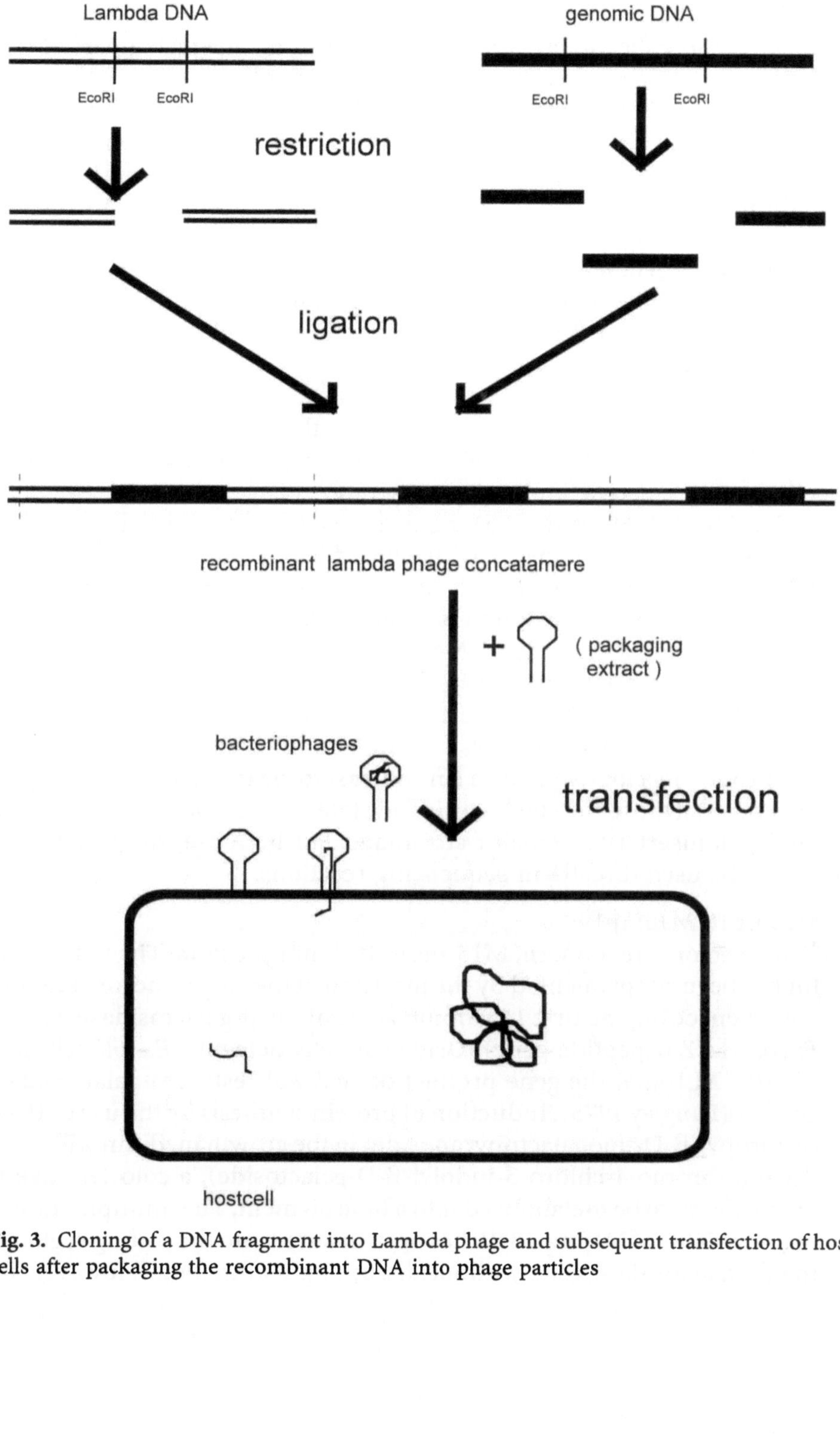

Fig. 3. Cloning of a DNA fragment into Lambda phage and subsequent transfection of host cells after packaging the recombinant DNA into phage particles

- EMBL 3 and 4
  These are so-called lambda replacement vectors, in which a part of that does not contain any genes essential for DNA replication and lytic development has been removed (Frischauf 1983). Thereby, the recombinant phage lose the ability to form prophage. These vectors accomodate inserts of about 15-23 kb in length. Polylinker sequences flanking both sites of the insert region simplify cloning and recovery. Inserts that are outside of the range of 15-23 kb in length will prohibit packaging of the recombinant phage. These vectors are suitable for genomic libraries, especially those produced by partial digestion of genomic DNA using the restriction enzyme Mbo I. EMBL 3 and 4 contain restriction enzyme sites in reverse order.

M13 derivatives M13 is a filamentous, male-specific coliphage, that is about 6.4 kb in length. A so-called male *E. coli* strain is capable of forming F-pili (F'or F^+) which are encoded by an episome and are necessary for phage adherence (Messing 1977, Sambrook 1989, Wu 1983). M13 shares extensive homology and similiar replication mechanisms with other filamentous phage such as fd and f1. During infection of *E. coli* host cells, single-stranded phage DNA (+strand) is converted into double-stranded replicative forms and then amplified. These double-stranded DNA intermediates are used in cloning reactions. Normally they are used in the cell to produce progeny as single-stranded DNA. The DNA molecules are packaged into protein coats and extruded without causing cell lysis. However, infected cells tend to grow slower producing cloudy plaques (in contrast to clear plaques following lytic infections). Due to the single-stranded state, M13 carriers exhibit greater flexibility of insert as one major advantage. Furthermore, single-stranded DNA can be used directly in sequencing reactions.

- M13mp18/M13mp19
  These vectors are modern, M13-derived cloning vectors. The native vector has been supplemented by the insertion of the lac promotor/operator region encoding the first 145 amino acids of the β-galactosidase gene of *E. coli* (lacZ α-peptide gene). Using suitable, defective *E. coli* cell lines (JM103, XL1blue) the gene product of lacZ will restore β-galactosidase activity (Langley 1975). Induction of protein synthesis by the use of IPTG (isopropyl-ß-D-thiogalactopyranoside) in the growth medium will cause X-gal (5-bromo-4-chloro-3-indolyl-ß-D-galactoside), a colorless dye in the medium, to be metabolized into a blue pigment. The multiple cloning region (MCR) of M13mp18/19 has been inserted into the coding region of the β-galactosidase gene without disrupting its function. However, in-

sertion of foreign DNA into the MCR will cause the infected cells to lose their β-galactosidase activity thus forming clear plaques. This allows for a color distinction between cells infected by wild-type phage versus recombinant phage, forming clear or blue plaques, respectively. M13mp18 and M13mp19 differ with respect to the order of restriction sites within the MCR.

Cosmids

Cosmids originate from the plasmid pBR322, but are inserted into cells by transfection. To understand their function, we have to focus once again on the replication of lambda phage (Figure 4).

It is a typical feature of λ-replication, that the phage DNAs are formed as one long, so-called concatemeric molecule. Multiple genomes are joined end-to-end, separated by cohesive sites (cos sites). Cos sites are single-stranded, complementary regions that are recognized by the packaging enzymes. The DNA between cos sites is packaged into a phage particle as long as the distance between the two sites is in the range of 35-45 kb. This mechanism provides the functional basis for cosmids (cos-site; plas-mid) (Collins 1979). pBR322 was changed so that a cos site was inserted into the tetracycline-resistance gene. Linearized plasmids will form concatameres similiar to those described for λ-replication. These cosmids are not infectious, since they lack the phage's own coding sequences. However, in the presence of enzymes ("packaging extract") and lambda particles, in vitro packaging of the recombinant DNA is efficient as long as the insert does not exceed[K.D.6] 35-45 kb in length. Cosmids replicate in the cell via the origin of replication of pBR322 and may be isolated using the techniques described for plasmids. Cosmid vectors are useful for genomic library construction (Meyerowitz 1980).

Host cells

All of the host cells (Bachmann 1972, Bachmann 1980, Sambrook 1989) described below are derived from *E. coli*. Most of the vectors require specific bacterial strains. For example, the use of vectors expressing β-galactosidase gene activity requires the use of lac^-, pro^- bacterial strains, such as JM103 (see M13 derivaties). A modern bacterial strain fulfilling this task is XL1-blue™ (Stratagene). Furthermore this strain can be grown on complete medium, forms sex-pili and carries tetracycline-resistance-gene.

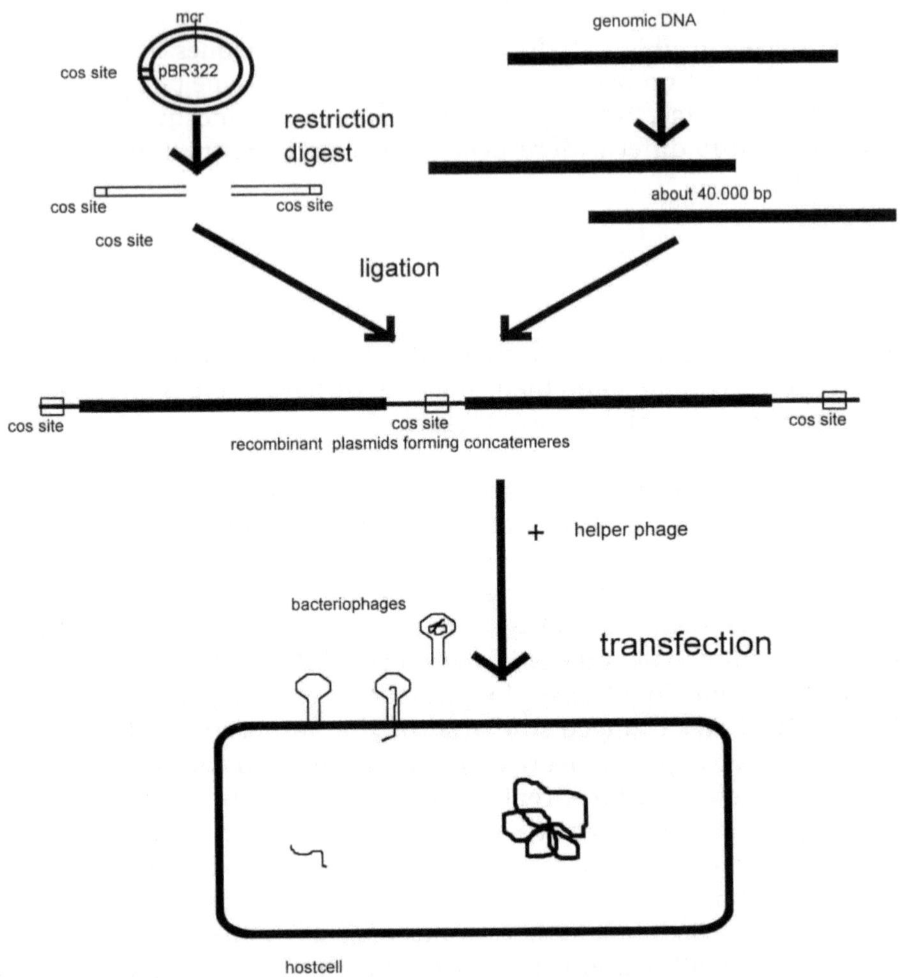

Fig. 4. Cosmids use the transfection mechanisms of bacteriophage to enter cells. Once they are intracellular they are amplified like plasmids (mcr = multiple cloning region).

Subprotocol 1
Transformation of Bacteria with Recombinant Plasmid Using the Ca^{++}-Precipitation Method (Mandel 1970)

The bacterial strain to be transformed depends on the selected vector. For pBR322 and pBR322-derived vectors, *E. coli HB101* is a commonly used bacterial strain. The following transformation procedure refers to this strain. If you wish to use a different strain, the principles of the method

Common Vectors and Host Cells for Molecular Cloning

Vector	Host Cell	Growth Medium	Supplements
pBR322	HB101 (Boyer 1969)	LB	antibiotics
	DH1 (Hanahan 1983)		
pUC	JM103 (Messing 1981)	LB	ampicillin
	XL1-blue™		
pSP6/T7	JM103	LB	ampicillin
	XL1-blue™		
EMBL3	LE392	LB	$MgSO_4$
Cosmids	BHB2688 (Hohn 1979)	LB	
	BHB2690		
M13	JM103 ; JM107	minimal medium	
λgt10	C600 (Appleyard 1954)	LB	$MgSO_4$
λgt11	Y1090	LB	ampicillin; $MgSO_4$

remain the same, but appropriate medium, antibiotics or different reagents for selective growth must be used.

Streak *E. coli HB101* on a stock plate to obtain single colonies. This stock plate can be stored at 4°C for about 1 month. After that you have to replace the stock plate.

Materials

Plates for bacterial growth

Reagents

- Bacto-tryptone
- Bacto-yeast extract
- NaCl
- Agar

Solutions

- 1 M NaOH
- Ampicillin, 50 mg/ml, sterile-filtered

- Tetracycline, 15 mg/ml in 50% ethanol, sterile-filtered

Supplies

- Autoclave
- Petri dishes

Casting plates

1. Prepare **LB Medium:**

H_2O	950 ml
Bacto-tryptone (1%)	10 g
Bacto-yeast extract (0.5%)	5 g
NaCl (1%)	10 g
Adjust pH to 7.3 with 1 M NaOH	
Add H_2O to make 1 liter.	

2. Add 15 g agar to 1 liter of LB-medium.
3. Sterilize in the autoclave, cool to 50°C.
4. Add antibiotics if desired:
   - 1 ml of a 50 mg/ml solution of ampicillin
   - 1 ml of a 15 mg/ml solution of tetracycline
5. Disperse carefully avoiding air bubbles.
6. Pour into Petri dishes. Stack one upon the other and let cool to solidify. Use approximately 25 ml of LB agar per 100 mm Petri dish.
7. Store the plates at 4°C until used.

Transformation

Reagents

- Recombinant plasmid
- LB agar plates containing tetracycline and ampicillin (see above)
- A colony of HB101

Solutions

- LB medium (see above)
- 50 mM $CaCl_2$ (sterile and ice-cold)
- Glycerol sterilized by autoclaving (ice-cold)

Supplies

- Shaker or shaking water bath

- Spectrophotometer
- Centrifuge
- Water bath
- Incubator
- Sterile toothpicks

Procedure

In advance

1. Prepare an overnight culture of HB101: Add 10 ml of sterile LB medium sterile to a culture flask. Transfer one colony of HB101 from the stock plate to the culture flask. Place the culture flask in a shaker at 37°C and let the cells grow overnight.

Standardreaction

Making cells competent

2. Add 45 ml of LB medium and 0.5 ml of the HB101 overnight culture to a sterile culture flask and place it in the shaker at 37°C.
3. Grow the cells to an OD_{600} of 0.3-0.5 (log-phase). This requires approximately 1.5-2 hours.
4. Incubate the cells for 15-30 minutes on ice.
5. Centrifuge the cells at 2500 × g for 5 min at 5°C.
6. Gently resuspend the pellet in 25 ml of ice-cold 50 mM $CaCl_2$. The cells are now very fragile.
7. Place on ice for 15 min.
8. Centrifuge at 2500 × g for 5 min at 5°C.
9. Gently resuspend the pellet in 3.3 ml of ice-cold 50 mM $CaCl_2$. The cells are now competent.

Storage of competent cells

10. The competent cells can be stored overnight at 4°C. For long-term storage mix 600 µl of competent cells with 150 µl of sterile ice-cold glycerol in an Eppendorf tube. Store at -70°C. Thaw on ice before use.

Transformation

11. In a sterile Eppendorf tube mix 0.1-0.5 µg of plasmid or recombinant plasmid DNA with 200 µl of competent cells.

12. Incubate on ice for 30 min.
13. Place the tube in a water bath at 42°C for exactly 2 min (heat shock).
14. Add 1 ml of LB medium and incubate at 37°C in a shaker or shaking water bath for 1 hr.
15. Spread different amounts of this culture (e.g. 50, 100 and 200 µl) over the surface of LB agar plates containing the appropriate antibiotic (depending on the plasmid) to obtain single colonies. Store the remainder of the culture at 4°C for further plating if necessary.
16. As a negative control, spread 100 µl of the untransformed competent cells on a LB agar plate containing the appropriate antibiotic.
17. Place the plates inverted in an incubator at 37°C and let the cells grow overnight.

Identification of the recombinant clones

18. Pick several colonies, using sterile toothpicks, to both an ampicillin and a tetracycline plate. Use a grid so that you can identify the colonies belonging together.
19. Incubate the plates inverted overnight at 37°C in an incubator. Recombinant clones will grow only on one plate (ampicillin or tetracycline, depending on the plasmid).

Subprotocol 2
Transformation of Bacteria with Recombinant Plasmid by electroporation (Dower 1988)

The electroporation apparatus used for this experiment can be purchased from different companies. We describe here the use of the Gene Pulser system from Bio-Rad-Laboratories. The recombinant DNA to be transformed must be in a low ionic-strength buffer such as TE. The ligation mixture, which can be used directly for transformation following the Ca^{++}-precipitation method, is not suitable for electroporation. To circumvent this difficulty different methods such as dilution (Willson 1988), precipitation (Zabarovsky 1990), dialysis (Jacobs 1990) or heat inactivation (Ymer 1991) have been described. Recently a new method which combines several of these steps has been reported (Kobori 1993).

Materials

Reagents

- One colony of an *E. coli* strain (e.g. LE392, MC1061, HB101)
- Recombinant plasmid in a low-salt buffer
- LB agar plates containing the appropriate antibiotics

Solutions

- LB medium (see Subprotocol 1, Materials)
- 10% glycerol solution, sterile-filtered (ice-cold)
- SOC medium (2% Bacto-tryptone, 0.5% Bacto-yeast extract, 10 mM NaCl, 2.5 mM KCl, 10 mM $MgCl_2$, 10 mM $MgSO_4$, 20 mM glucose)

Supplies

- Shaker
- Centrifuge
- Spectrophotometer
- Gene Pulser system (Bio-Rad Laboratories)
- 0.2 cm electroporation cuvette (Bio-Rad Laboratories)
- sterile toothpicks

Procedure

In advance

1. Prepare an overnight culture of the chosen *E. coli* strain: To a sterile culture flask add 10 ml of LB medium, and inoculate the medium with a single colony of the *E. coli* strain. Incubate the culture flask in a shaker at 37°C overnight.

Cell preparation

2. Add 990 ml of LB medium to a sterile culture flask and inoculate the medium with 10 ml of the overnight culture from step 1. Incubate the culture at 37°C shaking vigorously.
3. Grow the cells to an OD_{600} of 0.5-0.7 (log-phase). This requires approximately 3 hrs.
4. Place the cells briefly on ice.
5. Centrifuge the cells at 4000 $\times$ g for 15 min at 4°C.
6. Discard the supernatant and resuspend the cells in 300 ml of ice-cold 10% glycerol solution.

7. Centrifuge the cells at 4000 × g for 15 min at 4°C.
8. Resuspend the cells in 100 ml of ice-cold 10% glycerol solution.
9. Centrifuge the cells at 4000 × g for 15 min at 4°C.
10. Resuspend the cells in 50 ml of ice-cold glycerol solution.
11. Centrifuge the cells at 4000 × g for 15 min at 4°C.
12. Discard the supernatant and resuspend the cells in 1 ml of ice-cold 10% glycerol solution.
13. The cells are now ready for use in electroporation. As well they can be stored in small aliquots at -70°C (before use gently thaw the cells on ice).

Transformation

14. To an ice-cold Eppendorf tube add 40 μl of the *E. coli* cell suspension and 1-2 μl of DNA in a low ionic-strength buffer.
15. Mix well and incubate on ice for 1 min.
16. Set the Gene Pulser apparatus (Bio-Rad Laboratories) at 2.5 kV, 25 μF and 200 Ohms parallel resistance. The time constant achieved by the above settings equals approximately 4.7 msec.
17. Transfer the mixture of cells and DNA to the bottom of an ice-cold 0.2 cm electroporation cuvette.
18. Apply one pulse at the above settings. This should result in a time constant of approximately 4.5-5 msec.
19. Immediately after the pulse, add 1 ml of SOC medium to the cuvette and mix gently.
20. Transfer the transformed cells to a culture tube and incubate at 37°C for 1 hr with vigorous shaking.
21. Spread different amounts of the culture (e.g. 50, 100, and 200 μl) over the surface of LB agar plates containing the appropriate antibiotic. Incubate the plates inverted in an incubator at 37°C and let the cells grow overnight.
22. Identify the recombinant clones of e.g. pBR322 as described in Subprotocol 1.

Subprotocol 3
Transformation of Bacteria with Recombinant Bacteriophage DNA

For transformation with bacteriophage DNA a male *E. coli* strain such as JM101, JM103 or JM107 must be used. The procedure is the same for all of these strains. We describe here the transformation with recombinant M13 DNA. You can also use other filamentous phages such as fd or f_1.

Materials

For the plates

Reagents

- Bacto-tryptone
- Bacto-yeast extract
- NaCl
- $Na_2HPO_4 \cdot 2\ H_2O$
- KH_2PO_4
- NH_4Cl
- Agar
- sterile H_2O

Solutions

- 1 M NaOH
- 20% glucose solution, sterile-filtered
- 10 mM $CaCl_2$, autoclaved
- 1 M $MgSO_4$, autoclaved
- Thiamine (100 (g/ml), sterile-filtered

Supplies

- Autoclave
- Petri dishes
- Sterile filters

Media

2 × YT medium

H_2O	950 ml
Bacto-tryptone (1.6%)	16 g
Bacto-yeast extract (1%)	10 g
NaCl (1%)	10 g
Adjust pH to 7.3 with 1 M NaOH.	
Add H_2O to make 1 liter.	
Sterilize in the autoclave.	

For casting 2 × YT plates follow steps 2-7 in Subprotocol 1, Materials, without adding antibiotics.

Agar

2 × YT top-agar
To 1 liter 2 × YT medium add 7 g agar. Sterilize in the autoclave and cool to 48°C if you need the top-agar the same day. Otherwise you can store the autoclaved top-agar at room temperature. Before use it must be melted in the autoclave or microwave oven and cooled to 48°C.

Casting plates

Casting minimal agar plates

1. Prepare the following salt solution:

$Na_2HPO_4 \cdot 2\ H_2O$ (0.7%)	7 g
KH_2PO_4 (0.3%)	3 g
NaCl (0.05%)	0.5 g
NH_4Cl (0.1%)	1 g
Add H_2O to 950 ml.	
Adjust pH to 7.4 with 10 M NaOH.	

2. Add 15 g agar, autoclave and cool to 60°C.
3. Add the following solutions:

20% glucose (sterile-filtered)	10 ml
10 mM $CaCl_2$ (autoclaved)	10 ml
1 M $MgSO_4$ (autoclaved)	1 ml
Thiamine (100 (g/ml; sterile-filtered)	50 µl
Add sterile H_2O to make 1 liter.	

4. Disperse carefully to avoid air bubbles.
5. Pour into Petri dishes. Stack them one upon the other and let cool to solidify.
6. Store the plates at 4°C until used.

For transformation of JM103 with recombinant M13 DNA (BRL-focus 1980)

Reagents

- Recombinant M13 DNA
- A minimal agar plate (see above)
- 2 × YT agar plates (see above)
- 2 × YT top-agar (see above)
- *E. coli* JM103 grown on minimal medium

Solutions

- 2 × YT medium (see above)
- 50 mM $CaCl_2$ (sterile and ice-cold)
- Glycerol sterilized by autoclaving (ice-cold)
- 100 mM IPTG (isopropyl-ß-D-thiogalactopyranoside), dissolved in sterile H_2O
- 2% X-gal solution (5-bromo-4-chloro-3-indolyl-ß-D-galactoside), dissolved in dimethylformamide (DMF)

Supplies

- Shaker or shaking water bath
- Spectrophotometer
- Centrifuge
- Water bath
- Incubator

Procedure

In advance

1. Streak JM103 cells on a minimal agar plate so that single colonies can be obtained. Incubate the plate inverted at 37°C overnight.
2. Inoculate 10 ml of 2 × YT medium sterile with a single colony of JM103 grown on a minimal agar plate. Incubate the culture in a shaker at 37°C and let the cells grow overnight.

Standard reaction

3. To make the JM103 cells competent, follow the procedure described in Subprotocol 1, Procedure, steps 2-10. In place of LB medium use 2 × YT medium. In step 4 remove 2 ml of the log-phase cells and store them in a sterile tube at 4°C.
4. Mix 10-100 ng of recombinant M13 DNA with 200 μl of competent JM103 cells.
5. Place on ice for 40 min.
6. Heat shock the cells at 42°C for exactly 2 min. Place them on ice immediately.
7. For the following steps, sterile glass tubes, 2 × YT agar plates and 2 × YT top-agar must be prewarmed at 48°C.
8. Divide the transformation mixture by two: To one prewarmed glass tube add 20 μl to another 180 μl of the transformation mixture. Keep the tubes at 48°C.
9. To each of the glass tubes add:

Log-phase JM103 cells from step 3	200 μl
100 mM IPTG	10 μl
2% X-gal solution	50 μl
2 × YT top-agar	3 ml

10. Mix well and pour each tube immediately on to a 2 × YT agar plate prewarmed at 48°C. Swirl the plate so that the surface of the 2 × YT agar is covered.
11. Allow the top-agar to harden which takes approximately 15 min.
12. Incubate the plates inverted overnight at 37°C. Clear plaques correspond to transformed cells bearing an insert, blue plaques to those without insert.
13. It may be necessary to prepare 5-10 plates with IPTG, X-gal solution, JM103 cells and top-agar to streak out the recombinant plaques the next day.

Subprotocol 4
"Mini-Prep" Method for Plasmid Isolation

This method is recommended for rapid isolation of plasmids from a high number of clones (Birnboim 1979, Ish-Horowicz 1981). The isolated DNA is partially purified and is useful primarily for analytical restriction analysis. The yield is about 1-5 μg. If the clones bear a low copy-number plasmid, use 5 ml of bacterial culture and double the volumes of the corresponding solutions.

Materials

Reagents

- Lysozyme
- RNase A (1 mg/ml in 5 mM Tris-HCl (pH 8.0), heated 10 min at 70°C)

Solutions

- LB medium (see Subprotocol 1, Materials)
- Solution A (25 mM Tris-HCl (pH 8.0), 50 mM glucose, 10 mM EDTA, 2 mg/ml lysozyme). Add the lysozyme just before use.
- Solution B (0.2 M NaOH, 1% SDS). Make fresh each time.
- Solution C (3 M Na-acetate (pH 4.8)): Dissolve 40.8 g sodium-acetate · 3 H_2O in a small volume of H_2O, adjust the pH with glacial acetic acid and add water to 100 ml).
- Phenol: Dissolve crystallin phenol carefully at 37°C, add 10 mM Tris, 1 mM EDTA solution (pH 7.5) and store at room temperature. The phenol will be in the bottom layer.
- Chloroform/isoamyl alcohol (24/1 v/v)
- 95% ethanol
- 70% ethanol
- Sterile distilled water

Supplies

- 10 ml tubes with caps
- Sterile toothpicks
- Eppendorf centrifuge
- 1.5 ml tubes for the Eppendorf centrifuge

- Incubator
- Vortex

Procedure

In advance

1. Pick a single colony from the agar plate using a sterile toothpick. Let the toothpick drop in 2 ml of LB medium containing the appropriate antibiotic. Repeat this for as many colonies as you wish to work up. Grow the cells with shaking overnight at 37°C.
2. The following day prepare solution B and add the lysozyme to solution A.

Standard reaction

3. Fill a 1.5 ml microcentrifuge tube with the overnight culture from step 1. Store the remainder of the culture at 4°C and use later to make a frozen bacterial stock of the positive clones.
4. Centrifuge the cells in the Eppendorf centrifuge for 20 sec.
5. Resuspend the pellet in 100 μl of solution A. Incubate for 5 min at room temperature.
6. Add 200 μl of solution B. Place on ice for 5 min.
7. Add 150 μl of solution C. Mix carefully and place on ice for 5 min.
8. Centrifuge for 2 min.
9. Save the supernatant in a new microcentrifuge tube and discard the pellet.
10. Add 150 μl of phenol and 150 μl of chloroform/isoamyl alcohol (24/1). Vortex and centrifuge for 10 sec.
11. Pipet the supernatant containing the plasmids into a new microcentrifuge tube. Add 1/10 volume of solution C and 2 volumes of 95% ethanol. Mix well and incubate at room temperature for at least 2 min.
12. Centrifuge in the Eppendorf centrifuge for 2 min.
13. Discard the supernatant carefully and wash the pellet in 1 ml of 70% ethanol.
14. Centrifuge for 2 min.
15. Discard the ethanol and air dry the pellet for 5 min.

16. Resuspend the pellet in 36 µl of distilled water.
17. Add 4 µl of RNase A (1 mg/ml) and incubate 30 min at 37°C. If you wish to digest the DNA with restriction enzymes, the restriction buffer and restriction enzyme can be added together with the RNase A and the reaction carried out at the same time.

Subprotocol 5
Chromatographic Method for Plasmid Preparation

Another plasmid isolation method makes use of the low molecular weight and different binding properties of plasmid DNA compared to chromosomal DNA (Birnboim 1979, Birnboim 1983). The DNA solution is passed through columns packed with different absorbents (e.g. Sepharose, Dextrane) that reversibly bind DNA. In a second step the DNA is eluted in fractions thereby separating plasmid DNA from other forms of DNA, RNA and protein. In the past several years kits have become commercially available making use of the above-described technique. The absorbent material has been further developed enabling large-scale plasmid preparations in high purities. One of those kits used by the authors will be described in the following.

Materials

Reagents

- A single colony of an *E. coli* strain (e.g. XL1-Blue, DH5α)
- QIAGEN-tip 500 (by QIAGEN Inc.)
- RNase A (10 mg/ml in 5 mM Tris-HCl (pH 8.0), heated to 70°C for 10 min or lyophilized RNase A from the Qiagen kit)

Solutions

- LB medium (see Subprotocol 1, Materials)
- Buffer P1 (50 mM Tris/HCl, 10 mM EDTA (pH 8.0))
- Buffer P2 (0.2 M NaOH, 1% SDS)
- Buffer P3 (3.0 M potassium-acetate (pH 5.5))
- Buffer QBT (750 mM NaCl, 50 mM MOPS, 15% ethanol (pH 7.0), 0.15% Triton X-100)
- Buffer QC (1.0 M NaCl, 50 mM MOPS, 15% ethanol (pH 7.0))

- Buffer QF (1.25 M NaCl, 50 mM Tris/HCl, 15% ethanol (pH 8.5))
- Isopropanol
- 70% ethanol (ice-cold)
- TE buffer (10 mM Tris; 1 mM EDTA (pH 8.0))

Supplies

- Centrifuge
- Shaker
- Sterile toothpicks
- Spectrophotometer

Procedure

In advance

1. With a sterile toothpick inoculate 5 ml of LB medium containing the appropriate antibiotic with a single colony of the transformed *E. coli.* Let the cells grow overnight with shaking at 37°C.
2. Inoculate 148.5 ml of LB medium containing the appropriate antibiotic with 1.5 ml of the overnight culture from step 1. Grow the cells with shaking overnight at 37°C.

Standard reaction

3. Harvest the bacterial cells by centrifugation at 4°C and 6000 × g for 15 min. Discard the supernatant.
4. Dissolve 1 mg of RNase A in 10 ml of buffer P1 and resuspend the bacterial cells in this solution.
5. Add 10 ml of buffer P2 and mix very gently. Incubate at room temperature for 5 min.
6. Add 10 ml of ice-cold buffer P3, mix very gently and incubate on ice for 20 min.
7. Centrifuge at 30,000 × g for 30 min at 4°C. Save the supernatant.
8. If the supernatant is not clear, a second centrifugation (30,000 × g, 4°C, 15 min) is recommended. The turbid supernatant can also be filtered over a folded filter.
9. Apply 10 ml of buffer QBT to a QIAGEN-tip 500 and allow the column to empty by gravity flow (equilibration).

10. Apply the supernatant from step 7 or 8 on the QIAGEN-tip and allow it to enter by gravity flow.
11. Wash the column twice with 2 × 30 ml of buffer QC.
12. Elute the DNA with 15 ml of buffer QF.
13. Precipitate the DNA with 0.7 volumes of isopropanol. Centrifuge the DNA immediately at 15,000 × g for 30 min at 4°C.
14. Discard the supernatant carefully.
15. Wash the DNA pellet with 15 ml of cold 70% ethanol. Centrifuge at 15,000 × g for 30 min at 4°C.
16. Air dry the pellet for 5 min and dissolve the DNA in a suitable volume of TE buffer. The maximum binding capacity for plasmid DNA of the QIAGEN-tip 500 is 500 μg.
17. Determine the DNA concentration with the spectrophotometer at an OD_{260}.

Subprotocol 6
Large-Scale Preparation of Bacteriophage DNA (Replicative Form (RF))

This method describes the preparation of double-stranded DNA from the replicative form (RF) of bacteriophage M13. The procedure is also suitable for other filamentous bacteriophage such as fd or f_1. Furthermore, the preparation method for closed circular DNA described in steps 7-27 can be used for a large scale plasmid preparation to yield plasmids of high purity. However, you can also use the column chromatographic method (Subprotocol 5) to isolate M13-RF-DNA.

Materials

Reagents

- A single colony of *E. coli* JM103 grown on a minimal agar plate
- An M13 phage plaque (see Subprotocol 3, Materials)
- CsCl
- Lysozyme

Solutions

- LB medium (see Subprotocol 1, Materials)

- 2 × YT medium (see Subprotocol 3, Materials)
- Chloramphenicol, 30 mg/ml in ethanol, sterile-filtered
- Solution A: 100 mM Tris-Cl (pH 8.0), 50 mM NaCl; 15 mM EDTA (pH 8.0)
- Solution B: Add 1.6% Triton X-100 (v/v) to solution A
- Ethidium bromide, 5 mg/ml
- CsCl solution (1 g/ml) in TE buffer
- TE buffer: 10 mM Tris; 1 mM EDTA (pH 8.0)
- Isopropanol, saturated with CsCl (dissolve 15 g CsCl in 15 ml of TE buffer and add approximately 100 ml of isopropanol; two phases will appear)
- Solution for dialysis tubing: 2% Na_2HCO_3, 1 mM EDTA
- 95% ethanol
- 70% ethanol
- 1 mM EDTA (pH 8.0)
- 3 M Na-acetate (pH 7.0)

Suppliess

- Shaker or shaking water bath
- Spectrophotometer
- Centrifuge
- Ultracentrifuge (e.g. Beckman, with VTi50 and VTi65 rotors)
- Quick-seal ultracentrifuge tubes for the selected rotors and sealing apparatus (Beckman)
- UV-lamp
- Dialysis tubing
- Dialysis clips
- Vortex
- Magnetic stirrer
- Eppendorf centrifuge
- Vacuum evaporator

Procedure

1. Preparation of dialysis tubing: Heat the dialysis tubing (cut in pieces of approximately 25 cm) in 2% Na_2HCO_3, 1 mM EDTA (pH 8.0) and boil for 1 min. After cooling rinse the tubing first in distilled water then in ethanol. Rinse the dialysis tubing in 1 mM EDTA (pH 8.0) and store it in this solution at 4°C. **In advance**
2. Inoculate 100 ml of LB medium with a single colony of JM103 (grown on a minimal agar plate). Grow the cells with shaking at 37°C to an OD_{600} of 1.0, then store these cells at 4°C. **Day 1**
3. Inoculate 100 ml of LB medium with a M13 phage plaque. Grow with shaking at 37°C overnight.
4. To 900 ml of 2 × YT medium add 100 ml of JM103 culture from step 2. Grow the cells with shaking to an OD_{600} of 1.5. **Day 2**
5. Add 100 ml of the phage stock from step 3 and shake 15 min at 37°C.
6. Add 0.5 ml of chloramphenicol solution. Shake 1-2 hours at 37°C (accumulation of RF DNA).

Preparation of closed circular M13 RF DNA (Triton-lysozyme method)

7. Centrifuge the cells at 4000 × g for 10 min at 4°C.
8. Discard the supernatant and resuspend the cells in 30 ml of solution A.
9. Add 1 ml of a freshly made lysozyme solution (30 mg/ml).
10. Incubate at room temperature for 30 min.
11. Add 3 ml of solution B.
12. Incubate at room temperature for 30 min.
13. Centrifuge at 27,000 × g for 20 min at 4°C.
14. Pour the supernatant in a graduated cylinder. Add 1 g CsCl for each ml of the supernatant and dissolve the CsCl.
15. In Quick-seal ultracentrifuge tubes, add ethidium bromide solution (1/20 volume of the DNA/CsCl solution) and the DNA/CsCl solution from step 14. If there is a remaining volume, fill with CsCl solution to the constriction. Balance the tubes and dry the inlets. Seal the tubes with the prewarmed sealer and allow the sealing caps to cool completely

before removing. Check the tubes for leakage by pressing gently in the middle of the tube. Place them in the rotor (e.g. VTi50 from Beckman).

Note: Be careful when handling ethidium bromide because it is a very carcinogenic agent. Furthermore, UV light in the presence of ethidium bromide causes nicks in the closed circular DNA. Therefore take care of light conditions and expose the gradients as briefly as possible to the UV light.

16. Centrifuge at 225,000 × g for 16 hrs at 20°C (e.g. VTi50 at 48,000 rpm).

Day 3

17. Stop the centrifuge without brake.
18. Remove a tube carefully and place it in a clip. Using a UV-lamp you will see 2 bands. The upper band contains nicked circular and linear M13 DNA and linear fragments of the bacterial chromosome. The lower band contains the desired closed circular M13 RF DNA.
19. Pierce the top of the tube with a 20-gauge needle to allow air to enter. With a second 20-gauge needle pierce the tube just below the lower band and immediately collect the solution dripping out of the needle which contains the closed circular DNA. After the lower band is collected, discard the rest of the gradient.
20. Place the collected material in smaller Quick-seal tubes and add 1/20 volume of ethidium bromide solution. Fill with CsCl solution. Seal the tubes and place them in the rotor (e.g. VTi65 from Beckman).
21. Centrifuge at 278,000 × g for at least 8 hrs at 20°C (e.g. VTi65 at 54,000 rpm).

Day 4

22. Stop the centrifuge without brake.
23. Remove a tube carefully and visualize the DNA under UV light. You should see one thick band which contains the desired phage DNA and only a small upper band or no upper band. Collect the DNA as described in step 19.
24. To the collected material add an equal volume of CsCl- saturated isopropanol (use only the isopropanol layer). Vortex and discard the isopropanol layer which is now colored pink and contains the ethidium bromide. Repeat this step four more times until the isopropanol remains clear.
25. Dialyze the DNA against TE buffer for 4 hours: Remove a piece of dialysis tubing from the stock (see step 1) and clip one end. Insert the DNA solution and clip the other end of the dialysis bag. Place the bag in 1-2 l

of TE buffer and stirr with a magnetic stirrer. Change the TE buffer 3 times.

26. Remove the DNA from the dialysis bag and determine the DNA concentration with the spectrophotometer at an OD_{260} (dilute an aliquot of the DNA solution 1/20).

27. If you wish a higher DNA concentration, precipitate the DNA:
    - In Eppendorf tubes mix the DNA solution with 2 1/2 volumes of 95% ethanol and 1/10 volume of 3 M Na-acetate (pH 7.0).
    - Incubate at -20°C for 1-2 hrs.
    - Centrifuge at 4°C in the Eppendorf centrifuge for 10 min.
    - Discard the supernatant carefully and wash the pellet with 70% ethanol (add 1 ml of 70% ethanol, vortex).
    - Centrifuge at 4°C in the Eppendorf centrifuge for 5 min.
    - Discard the supernatant and let the pellet dry in the vacuum evaporator.
    - Dissolve the pellet in the desired volume of TE buffer.

References

Appleyard R K (1954) Segregation of new lysogenic types during growth of a doubly lysogenic strain derived from *Escherichia coli K12*. Genetics 39:440-452

Bachmann B J (1972) Pedigrees of some mutant strains of *Escherichia coli K12*. Bacteriological Reviews 36:525-557

Bachmann B J, Low K B (1980) Linkage map of *Escherichia coli K12*, Edition 6. Microbiological Reviews 44:1-56

Birnboim H C (1983) A rapid alkaline extraction method for the isolation of plasmid DNA. Methods Enzymol 100:243-255

Birnboim H C, Doly J (1979) A rapid alkaline extraction procedure for screening recombinant plasmid DNA. Nucleic Acids Res 7:1513-1523

Bolivar F, Rodriguez R L, Greene P J, Betlach M C, Heyneker H L, Boyer H W (1977) Construction and characterization of new cloning vehicles. II. A multipurpose cloning system. Gene 2:95-113

Boyer H W, Roulland-Dussoix D (1969) A complementation analysis of the restriction and modification of DNA in *Escherichia coli*. J Mol Biol 41:459-472

BRL-focus (1980) M13 Cloning/Didesoxy-sequencing manual

Collins J, Hohn B (1979) Cosmids: A type of plasmid gene-cloning vector that is packable in vitro in bacteriophage lambda heads. Proc Natl Acad Sci U.S.A. 75:4242-4246

Dower W J, Miller J F, Ragsdale C W (1988) High efficiency transformation of *E.coli* by high voltage electroporation. Nucleic Acids Res 16:6127-6145

Frischauf A-M, Lehrach H, Poustka A, Murray N J (1983) Lambda replacement vectors carrying polylinker sequences. J Mol Biol 170:827-842

Hanahan D (1983) Studies on transformation of *Escherichia coli* with plasmids. J Mol Biol 166:557-580

Hohn B (1979) In vitro packaging of Lambda and cosmid DNA. Methods Enzymol 68:299-309

Ish-Horowicz D, Burke J F (1981) Rapid and efficient cosmid cloning. Nucleic Acids Res 9:2989-2998

Jacobs M, Whendt S, Stahl U (1990) High-efficiency electro-transformation of *Escherichia coli* with DNA from ligation mixtures. Nucleic Acids Res 18:1653

Kobori M, Nojima H (1993) A simple treatment of DNA in a ligation mixture prior to electroporation improves transformation frequency. Nucleic Acids Res 21:2782

Langley K E, Villarejo M R, Fowler A V, Zamenhof P J, Zabin I (1975) Molecular basis of β-Galactosidase α-Complementation. Proc Natl Acad Sci U.S.A. 74:560-564

Levin B (1985) Genes II. John Wiley & Sons, New York (USA)

Mandel M, Higa A (1970) Calcium dependent bacteriophage DNA infection. J Mol Biol 53:154-162

Melton D A, Krieg P A, Rabagliate M R, Maniatis T, Zinn K, Green M R (1984) Efficient in vitro synthesis of biologically active RNA and RNA hybridization probes from plasmids containing a bacteriophage SP6 promotor. Nucleic Acids Res 12:7035

Messing J, Gronenborn B, Muller-Hill B, Hofschneider PH (1977) Filamentous coliphage M13 as a cloning vehicle. Insertion of a HincII fragment of the lac regulatory region in M13 replicative form in vitro. Proc Natl Acad Sci U.S.A. 74:3642-3646

Messing J, Crea R, Seeburg P H (1981) A system for shotgun DNA sequencing. Nucleic Acids Res 9:309-321

Meyerowitz E M, Guild G M, Prestidge L S, Hogness D S (1980) A new higher capacity cosmid vector and its use. Gene 11:271-282

Sambrook J, Fritsch EF, Maniatis T (eds) (1989) Molecular Cloning, a laboratory manual. 2nd edition. Cold Spring Harbor Laboratory Press

Sutcliffe J G (1978) Complete nucleotide sequence of the *Escherichia coli* plasmid pBR322. Proc Natl Acad Sci U.S.A. 75:3737-3741

Vieira J, Messing J (1982) The pUC plasmid, an M13mp7-derived system for insertion mutagenesis and sequencing with synthetic universal primers. Gene 19:259-268

Willson T A, Gough N M (1988) High voltage *E.coli* electro-transformation with DNA following ligation. Nucleic Acids Res 16:118-20

Winnacker E-L (1985) Gene und Klone: Eine Einführung in die Gentechnologie. VCH Verlagsgesellschaft, Weinheim (Germany)

Wu R, Grossmann L, Maldave K (eds), Messing J (1983) Methods Enzymol 101 (part C): Recombinant DNA. Academic Press, New York

Ymer S (1991) Heat inactivation of DNA ligase prior to electroporation increases transformation efficiency. Nucleic Acids Res 19:6960

Young R A, Davis R W (1983) Efficient isolation of genes by using antibody probes. Proc Natl Acad Sci U.S.A. 80:1194-1198

Zabarovsky E R, Winberg G (1990) High efficiency electroporation of ligated DNA into bacteria. Nucleic Acids Res 18:5912

Part II

Strategies for Gene Isolation and Characterization

Chapter 16

Cloning Strategies: An Overview

FRIEDHELM HILDEBRANDT

Introduction

When trying to clone a gene for a genetic disease, different approaches can be followed, depending on the availability of prior information about the function or nature of the gene or its product. Figure 1 gives an overview of the different cloning strategies that are available. The relevant methods are discussed in the subsequent chapters.

Post-purification cloning

If the relevant gene product is known, antibodies against the protein can be generated and used for cDNA library screening. Alternatively, protein purification can be attempted in order to obtain partial amino acid sequence. cDNA sequence can be inferred from the amino acid sequence, and oligonucleotides can be synthesized for library screening or for use in a PCR-based cloning approach. Due to the degeneracy of the genetic code, the cDNA sequence will not be unequivocal. This problem is discussed below.

Cloning by sequence similarity

In a situation where there is some prior sequence available of a related gene in a different species, or of an isoform of a protein, this sequence information can be used to design oligonucleotides, that represent a "best guess" with respect to the target cDNA sequence. When designing oligonucleotides

Friedhelm Hildebrandt, Universitäts-Kinderklinik, Mathildenstrasse 1, Freiburg, 79106, Germany (*phone* +49-761-270-4301; *fax* +49-761-270-4533; *e-mail* hildebra@kkl200.uk-l.uni-freiburg.de)

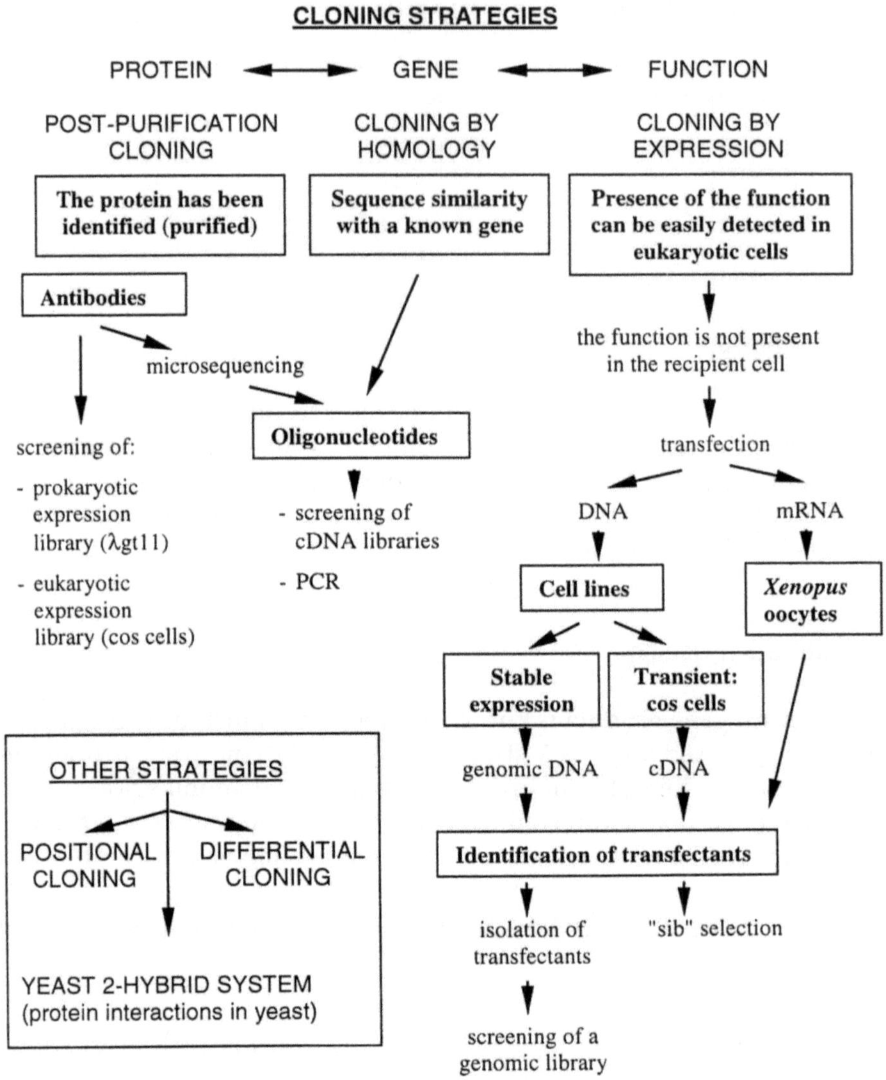

Fig. 1. Summary of cloning strategies (modified after C. Sardet, In: Molecular Medicine, Freiburg, 1994)

based on sequence similarity, the following should be considered, especially when the sequence is derived from amino acid sequence.

Choosing the region to derive a sequence from

- A portion of the amino acid sequence should be chosen in which there are amino acids that have only a few codons (e. g. Trp (one codon) rather than Ser (six codons)).
- A region should be selected, which is likely to have been conserved during evolution and which is likely not to be interrupted by an intron (if cDNA is to be used in a PCR).

Choosing codons for amino acids

- All possible codons for each amino acid should be written down below the amino acid sequence, and their relative codon usage in the target species (as published in codon preference tables (Pesole, Attimonelli, Liuni 1988) should be indicated behind each codon. An oligonucleotide mixture can be designed that includes only the codons most frequently used in the target species. The number of different molecules in the mixture is the product of the number of choices in each position in the sequence. Ideally, this number should be less than 256, although probes with higher degeneracy have been used successfully for cloning by hybridization as well as by PCR.
- Inosine, which base pairs with adenine, cytosine and uracil can be subsituted for the nucleotide in the third codon position if this nucleotide is T, G or A.

Expression cloning

If the gene product in question can be identified by a functional assay, and if there exists a eukaryotic cell line that is deficient in this function, then stable or transient transfection with, e.g. human genomic DNA can be performed. Transfected cells can be assayed for reconstitution of the desired function. Human DNA fragments retained in the cell line expressing the reconstituted function can then be identified and the respective gene can be isolated, e.g. by screening of a genomic library prepared from the transfected cell line.

Alternatively, *Xenopus laevis* oocytes may be microinjected with pools of mRNA and assayed for expression of the desired protein. Pools of mRNA are then divided into smaller and successively smaller fractions until a single clone is identified, that gives rise to the assayed function.

Positional cloning, differential cloning, etc.

If there is no prior knowledge available concerning protein sequence or function, positional cloning (see Chap. 22) can be employed. In positional cloning, cytogenetic abnormalities or linkage analysis point to a small genomic region in which the desired gene resides. This region is then defined more narrowly by physical mapping techniques, and mutations in the responsible gene are identified by sequencing.

If cell lines are available that express functional differences of interest, differential cloning by subtractive hybridization or differential display can be used.

References

Pesole G, Attimonelli M, Liuni S (1988) A backtranslation method based on codon usage strategy. Nucl Acids Res 16:1715-28

Chapter 17

Genomic Libraries

LUIZ F. ONUCHIC AND GREGORY G. GERMINO

Introduction

Library construction and screening are essential techniques that permit the isolation and manipulation of a DNA sequence of interest. A library of clones, whether derived from genomic DNA or mRNA, is a set of molecules that, ideally, faithfully represents the entire population of molecules that are in the source material from which the library was made. DNA fragments and DNA copies of mRNA (complementary DNA, also called cDNA), however, cannot propagate without assistance. Therefore, these molecules are cloned into biological vectors which provide the necessary cellular machinery required for DNA replication and clonal amplification. The principal types of vectors that are used are plasmid, cosmid (hybrid plasmid cloning vector containing the lambda *cos* site), lambda bacteriophage, bacteriophage (P1) and yeast artificial chromosomes (YACs). The limiting factors involved in library construction are basically related to the biological characteristics of these vectors.

This chapter will review the main features of genomic libraries, as well as the basic principles essential to the efficient use of libraries: library complexity, redundancy, chimerism and unclonable elements. It will also review some of the uses of genomic libraries, emphasizing how the anticipated use of the library affects the design of its construction and the selection of an appropriate host and vector. Basic features of the methods used to construct genomic libraries, such as partial digestion, first and second strand synthesis, size selection, and packaging, will also be discussed in this chapter. The features and methods of cDNA libraries will be described in Chap. 18.

Luiz F. Onuchic, Johns Hopkins University, School of Medicine, Division of Nephrology, 720 Rutland Avenue, Ross 970, Balitmore, Maryland, USA, *Correspondence to* Gregory G. Germino, Johns Hopkins University, School of Medicine, Division of Nephrology, 720 Rutland Avenue, Ross 970, BalitmoreMarylandUSA (*phone* +01-410-614-1650; *fax* +01-410-955-0485; *e-mail* ggermino@welchlink.welch.jhu.edu)

Vector and host selection

Recombinant DNA strategies require the ability to distinguish and select specific fragments of interest. Finding a specific fragment buried in a large population of similarly sized molecules and then obtaining enough of it to allow its manipulation would be a daunting challenge if not for the availability of cloning techniques and libraries. Rodent-human somatic cell hybrids are one way of propagating very large human DNA fragments, but the fraction of total DNA that is of human origin is relatively small, and the human fragments are often too large to be easily characterized or manipulated. Another drawback is that the human and rodent chromosomes cannot be separated except by flow-sorting or by preparing metaphase spreads. Thus, a variety of vectors and hosts have been constructed and derived to allow more efficient and manageable propagation of DNA fragments: YACs that are capable of propagating very large fragments spanning several hundred kilobases (kb) to megabases; and plasmids, lambda bacteriophage and cosmids for smaller fragments of 1 – 50 kb. Larger fragments contain more information but are more difficult to characterize and separate from host DNA. There are a number of factors that determine selection of a vector (the cloning vehicle) and host (the cell line infected/transfected or transformed with the recombinant molecule) for a given library, which are presented and discussed below.

Desired size of cloned inserts

In chromosome "walking", larger inserts are desirable in order to reduce the number of "walks" required. Each walk yields, on average, "new DNA" equal to one half the insert size of the vector being used: e.g., lambda bacteriophage with 20 kb inserts yield 10 kb, cosmids with 40 kb yield 20 kb, P1s with 100 kb yield 50 kb and YACs with 1000 kb yield 500 kb (Figure 1). It is obvious why YACs and P1 clones have replaced cosmids and lambda bacteriophage in most physical mapping and chromosome walking projects. Smaller insert sizes are preferable, however, for many other projects requiring genomic fragments. Examples include mapping intron-exon boundaries and transcriptional regulatory sequences or developing constructs for gene knock-outs via homologous recombination. Smaller inserts are more easily characterized, isolated and subcloned. Lambda bacteriophage and cosmids are the vectors most commonly used to generate such multipurpose libraries. They are highly efficient cloning vehicles and carry inserts in the appropriate range of sizes.

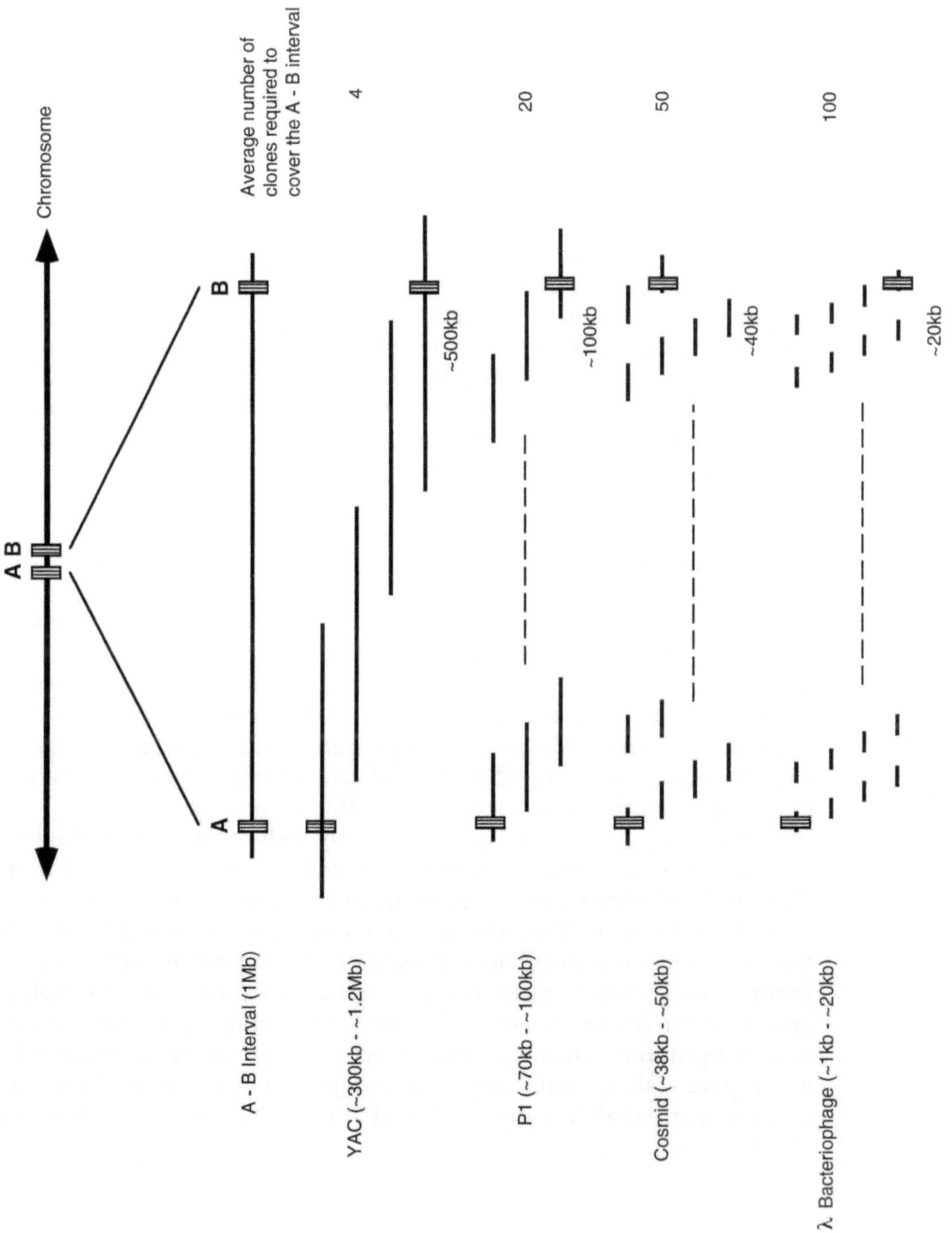

Fig. 1. DNA inserts clonable by different vectors: sizes and walking efficiencies

Tolerance of 'difficult to clone' sequences

Some hosts have restriction systems that cleave methylated DNA or DNA containing unusual restriction sites, thereby preventing their propagation. A library grown in such a host would be incomplete because genomic fragments having susceptible restriction sites would not be represented. Fortunately, a number of mutant strains have been isolated that allow cloning and stable propagation of these previously unclonable segments. Genomic fragments with unusual sequence properties such as inverted repeats, long stretches of repetitive DNA, or sequences with homology to host or vector sequences, may also be unclonable or undergo partial deletion or rearrangement in some hosts. Lastly, there appear to be segments that are repeatedly unclonable in cosmid libraries and yet can be reproducibly isolated in lambda, YAC or P1 libraries (and vice versa). Thus, failure to isolate a clone from one library does not mean that the segment is unclonable, and the use of additional libraries increases the likelihood of success.

Quantity and complexity of starting material

Single cos site vectors (i.e., pcos2EMBL, PWE series) are less efficient in packaging than newer double cos site vectors such as sCos1 which can be used with extremely small quantities of starting DNA. Likewise, YAC transformation efficiencies are orders of magnitude lower than those achieved with plasmids or bacteriophage/cosmid transfections. Vectors with the highest cloning efficiencies (bacteriophage and/or plasmid) are preferred when the quantity of source DNA is limiting.

Occasionally, only small and reasonably well-defined DNA fragments are required. The source of fragments could be an entire or partial human chromosome (propagated in a rodent cell line) or a YAC, P1, cosmid or bacteriophage insert. Virtually any fragment can be subdivided and propagated as smaller subunits. These libraries, constructed from only part of the genome, are known as subgenomic libraries. They are useful in the isolation and characterization of segments present in multiple copies within the genome. Subgenomic libraries, however, are most commonly constructed for use in large scale genomic sequencing projects. YACs or cosmids are more easily sequenced after being subcloned into plasmids which contain much smaller inserts.

Future uses of the genomic clone

Many vectors have features incorporated into their design that facilitate their use in subsequent studies. For example, one may wish to rapidly map an insert, identify its ends and then cleave it free of vector sequences for use in chromosome walking. The PWE and sCOS1 series of cosmid vectors are examples of two vectors with features that are useful for such studies. They contain a *Not*I restriction site, which occurs infrequently in genomic DNA, flanking the *Bam*HI cloning site. Between the *Not*I and *Bam*HI sites are short segments of known sequences for which primers are commonly available (T3, T7). Hybridization of Southern blots of digested clones with radiolabeled T3 or T7 primers can rapidly identify the respective end fragments and, when combined with partial digests, can assist in the construction of a restriction map of the clone.

On occasion, large-insert genomic clones are useful for *in vivo* expression studies. This approach often involves transfection of an intact genomic clone into an eukaryotic cell line. The process is not very efficient, and transfectants will usually make up a small fraction of the final total cell population. The fraction will be greatly increased if a selectable marker such as antibiotic resistance is encoded by the vector. While many cosmid vectors contain these markers, most lambda vectors do not.

Complete vs. partial digest libraries

Intact human chromosomes are too large to be easily manipulated and propagated in prokaryotic cell hosts. All the currently available hosts require fragmentation of the source DNA. This can be accomplished by either random shearing or complete or partial digestion using a restriction endonuclease. Each of these methods has certain advantages and limitations.

Complete digest libraries

As the name implies, in this approach the source DNA is cleaved to completion with an enzyme that produces fragments with ends that are compatible with the cloning site of the vector. This is the most efficient method for cloning a specific restriction fragment if its size is known to be within the acceptable range for the vector being used, since each chromosomal copy in the source material is potentially clonable. If digestion is truly complete, however, no clones with overlapping fragments could ever be isolated from such a library, making it impossible to do sequential chromosome walks (Figure 2). This problem can be overcome by using multiple libraries that are prepared from DNA cleaved with different endonucleases.

Another major problem associated with one-enzyme complete digest libraries is that they lack restriction fragments of the original source material that are not within the acceptable cloning limits of the vector (Figure 2). Complete restriction enzyme digestion of complex genomes typically yields fragments that range greatly in size. The proportion of the total that will be

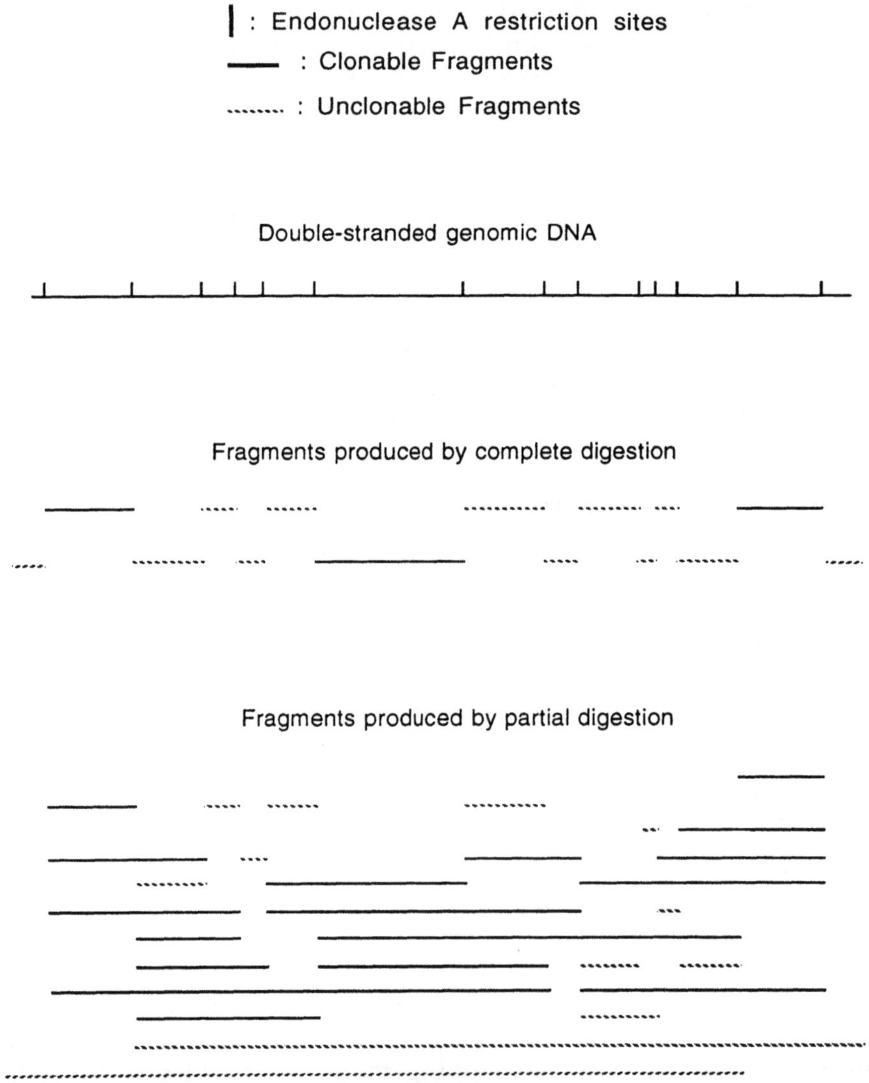

Fig. 2. Partially and completely digested libraries

clonable will depend on the enzyme and vector selected. Plasmid libraries generally lack fragments greater than 5 – 10 kb in length and preferentially contain small inserts. Libraries constructed using either lambda bacteriophage or cosmids have both upper and lower limits of insert size since recombinant vectors >50 kb or <38 kb cannot be packaged.

Partial digest libraries

One can avoid many of the aforementioned limitations by using partial digest libraries. Consider the genomic segment in Figure 2. There are 14 restriction sites for enzyme A in this example. Complete restriction with enzyme A yields 15 fragments, most of which are below the minimum insert size of the vector. Reducing the amount of enzyme or decreasing the duration of the reaction results in incomplete digestion and yields a ladder of fragments. Segments that were previously below the cloning limit of the vector are now clonable, and the total fraction of the original source material that is represented in the library is greatly improved. The ratio of clonable/unclonable segments is optimized by careful selection of the reaction conditions. The cloning efficiency is highest when the size of the most abundant fragments equals the vector capacity. This is achieved by using reduced quantities of dilute enzyme, brief restriction times, and an enzyme that cuts DNA frequently and without bias in selection of its restriction sites (Seed, Parker and Davidson 1982). The enzyme *Sau*3A or *Mbo*I is usually selected because it recognizes a 4-bp sequence and yields fragments that can be ligated to vectors cleaved with *Bam*HI. One empirically determines the optimal conditions in a test digest prior to scaling up for a preparative digest (see Procedure). Aliquots of the test digest are removed at various times after addition of the enzyme and evaluated by gel electrophoresis. The time point and enzyme concentration that yield the most clonable fragments are selected for the preparative reaction.

There is a second major advantage of partial digest libraries. A well-made library typically is comprised of clones containing overlapping inserts (Figure 2). A "deep" or representative library (see Comments, Library Complexity) often includes between 5 and 10 copies of a locus, each cloned as a unique, overlapping fragment. It is this feature that makes such libraries especially suitable for chromosome walking.

One disadvantage of partial digest libraries is that they are technically more challenging to construct. The depth of a library correlates directly with the quality of the DNA used for its construction. For optimal results, the average size of the source DNA before partial digestion must be at least three times greater than that required by the vector. The highest cloning yields are obtained when the partial-digestion products are size fractionated prior to ligation to the vector. This step, however, can result in significant losses of DNA if the method of size fractionation is not very efficient. Excessive manipulation of the DNA can further reduce cloning efficiency by shearing molecules or producing ragged ends that are incompatible with the vector. Lastly, inhibitors of ligation can be introduced if the reagents are not carefully prepared and purified. Sucrose gradient purification and preparative gel electrophoresis are the most commonly used fractionation methods. Preparative gel electrophoresis offers better resolution and is capable of resolving larger quantities of DNA. The sucrose gradient method, however, is very simple, reliable and fast and is the method most commonly used in genomic library construction. The basis for fractionation is that DNA fragments migrate through a linear sucrose gradient at a rate dependent on their size. This technique can resolve fragments in the range of 5 – 60 kb, which are suitable for generating cosmid and bacteriophage libraries. Such a procedure is also used for purification of lambda bacteriophage arms. The steps are decribed in the Procedure.

Randomly sheared libraries

Although partial-digest libraries generally lack fewer sequences than complete-digest libraries, there often are some cloning gaps. Figure 2 illustrates one reason why this occurs. Restriction sites are variably sensitive to cleavage. Some sites that always cut may be clustered together, always yielding very small, unclonable fragments. Other sites may be very resistant and yield fragments that are too large to clone under conditions that yield an optimal size for the majority of fragments. Partial-digest libraries lack fragments that are bounded by restriction sites that are either "hypersensitive" or "hyper-resistant" to the enzyme. There are several solutions to this problem. One approach is to clone the DNA into vectors with different insert size limits, eg. bacteriophage and cosmid. Another solution is to make several libraries, using different enzymes for each. A final approach is to avoid the use of enzymes altogether and resort to random shearing of the source DNA. This process is thought to be less biased in its inclusion of fragments. Although this premise is unlikely to be completely true, it undoubtedly is more random than restriction digestion. The major drawback to this approach is that sheared DNA has either ragged or blunt ends, which require further manipulations that result in less efficient cloning (Meyerowitz et al. 1981).

Procedure

Partial enzyme digestion

1. Transfer 100 μg of genomic DNA (0.1 – 0.5 mg/ml) to a microcentrifuge tube. Dilute to 900 μl and add 100 μl of the appropriate 10 x restriction enzyme buffer. Remove a 40 μl aliquot, add to 10 μl of stop solution, and place on ice. Incubate the remainder at 37°C for 10 minutes.
2. Add to the main reaction mix sufficient enzyme to digest completely only 20% of the DNA in 60 minutes. Remove 40 μl aliquots every 10 minutes and add each to 10 μl of stop solution as in Step 1. Follow this procedure for 90 minutes.
3. Electrophorese the aliquots of digested DNA on a 0.5% agarose gel, being certain to include markers that span a range of sizes from 10 kb – 100 kb. Determine the time interval that yielded the highest concentration of fragments that equals the optimal insert size range of the selected vector. Between 10 and 20% of the total amount of DNA should be within this size range. One may need to repeat the experiment if the previous results were unsatisfactory, adjusting the enzyme concentration accordingly (usually 10 x more or 10 x less).
4. Once suitable conditions are determined, scale up the reaction to 1 mg of DNA in a 10 ml final reaction volume. Remove 3.3 ml of the digest 5 minutes before the optimal time point, at the optimal time point and 5 minutes after. Combine the aliquots, add to 10 mls phenol-chloroform, and mix well. Separate the phases by centrifugation at 5000 – 10000 x g for 5 minutes. Remove the supernatant and safely dispose of the organic phase. Add 1 ml of 5 M NaCl and 30 ml 100% ethanol to the supernatant, chill in a dry ice/ethanol slurry for at least 5 minutes, and centrifuge at 20,000 rpm. Dilute the precipitated DNA in 1 ml STE buffer (10 mM Tris-Cl (pH 7.5), 10 mM NaCl, 1 mM EDTA (pH 8.0)) and check 10 μl on a 0.5% agarose gel.

Size fractionation

Sucrose density gradient centrifugation for fractionating DNA according to size. This technique is also used to purify the arms of lambda bacteriophage.

1. Dissolve the digested DNA in STE buffer at a concentration of 1 mg/ml.

2. Construct a linear 10% to 40% sucrose density gradient according to the manufacturer's guidelines (several gradient makers are commercially available) using either a SW-28 or SW-41 centrifuge tube.
3. Heat the digested DNA at 65°C for 5 min to dissociate DNA aggregates. Gently pour no more than 0.5 mg (SW-28) or 0.2 mg (SW-41) of heated genomic DNA solution onto the top of the sucrose density gradient. Lambda vector DNA should not be used in amounts exceeding 50 μg (SW-28) or 20 μg (SW-41).
4. Centrifuge the gradient tubes for 16–18 hr for cosmid-sized inserts and 24 h for bacteriophage-sized inserts. Spin at 113,000 x g (SW-28 rotor) or 154,000 x g (SW-41 rotor) at 20°C.
5. After centrifugation is completed, 750 μl aliquots are collected from the bottom of the tube using a capillary tube. The size range of each portion is evaluated by gel electrophoresis.
6. The desired fractions are subdivided into 4 microcentrifuge tubes containing ~180 μl/ each. 320 μl of water and 1 ml ethanol are added to the samples, and the tubes are placed at –20°C overnight to precipitate the samples.
7. At this point, the samples can be stored indefinitely at –80°C for future use or redissolved in 100 μl of TE buffer (10 mM Tris-Cl (pH 7.5), 1 mM EDTA, (pH 8.0)) or autoclaved deionized water.

Additional information related to these and other protocols can be found in Ausubel et al. 1991.

Comments

Library complexity

A library is described as being "deep" or "representative" when it contains all fragments from a given DNA source. No library has yet been produced which includes a complete set of overlapping clones. As discussed previously, some of the reasons for this are:

- repetitive sequences that are unstable in prokaryotic or yeast hosts or that are prone to recombination (rearrangements)

- chromosome telomeres, since they do not end in convenient restriction sites, and
- fragments flanked by restriction sites that either cleave too infrequently to allow cloning in currently available vectors or conversely, cleave too often, below the minimum size requirement for most vectors.

The cloning gaps can be minimized by careful DNA preparation coupled with careful vector and host selection.

The size of the library, defined as the number of recombinants produced prior to its amplification, is another parameter that is useful in defining the depth of a library. Consider a 40 kb partial *Mbo*I human genomic library in the cosmid PWE 15. Since the total length of the human genome is ~3.0×10^6 kb, end-to-end it would be equivalent to ~7.5×10^4 clones. Partial libraries, however, are comprised of overlapping rather than end-to-end segments. Thus, one aims for 5 – 10 fold redundancy (i.e., 5 – 10 genome equivalents in the number of primary recombinants) to have a 95% chance of finding a particular genomic segment in a partial digest library (Seed, Parker and Davidson 1982). In the example above, ~$4.0 - 8.0 \times 10^5$ primary recombinants would be necessary. The size of the library can be improved by maximizing the efficiency of packaging and transfection/transformation, and by increasing the quantity of starting material. If a set of restriction sites cleaves only 5% of the time, increasing the number of molecules digested and cloned will increase the likelihood of finding a clone that contains a very short fragment, as presented in Figure 2.

Primary libraries generally are more representative than amplified ones since clones that have a growth advantage become over-represented when libraries are amplified. Clones with inserts near either the minimum or maximum size limits may propagate less efficiently than others. Likewise, clones that include repetitive sequences or encode genes that are transcribed by the host also can have a detrimental effect on growth. As a consequence, a mimimum of 5x the number of primary recombinants should be evaluated before concluding that a genomic library lacks a particular genomic clone.

Chimerism

A problem common to all libraries is the coincidental cloning of more than one fragment in the same clone (some YAC libraries have a rate of chimerism as high as 30%). Chimeric clones isolated from complete digest libraries are easily recognized since their digestion with the enzyme used to prepare the library releases more than one band. Chimeric clones isolated from partial digest libraries are much more difficult to detect. Careful comparison of the restriction maps of the clone with the source DNA may yield clues, but hybridization studies are often required to detect small, non-contiguous fragments that are incorporated during packaging. If such end-fragments remain undetected, the next chromosome walk will isolate a non-contiguous set of clones elsewhere in the genome, leading to a fatal mapping mistake.

Effort required to construct a representative library compared with the availability of previously prepared libraries - considerations and recommendations

Unless one has specific requirements or has ample experience making libraries, the effort required to construct a genomic library in lambda bacteriophage or cosmids is considerable and may ultimately cost much more in terms of time and money than purchasing a commercial library. YAC and P1 libraries are even more difficult to construct. A host of lambda bacteriophage and cosmid libraries are commercially available, and their quality has greatly improved in recent years. A number of commercial and not-for-profit organizations offer YAC and P1 library screening, and gridded chromosome-specific libraries are also available from a number of sources (eg. Los Alamos National Laboratory, USA). There are, nevertheless, circumstances in which one may require a genomic library prepared from a specific source of DNA, such as when one wishes to clone a translocation breakpoint associated with tumor formation in a particular cell type. Such "do-it-yourself" libraries usually are constructed using lambda bacteriophage as the vector since this system is the most efficient and least tedious to prepare.

References

Ausubel FM, Brent R, Kingston RE, Moore, DD Seidman JG, Smith JA and Struhl K (1991) Current Protocols in Molecular Biology, John Wiley and Sons. New York, New York

Meyerowitz EM, Guild GM, Prestidge LS and Hogness DS (1981) A new high-capacity cosmid vector and its use. Gene 11:271282

Seed B, Parker RC and Davidson, N (1982) Representation of DNA sequences in recombinant DNA libraries prepared by restriction enzyme partial digestion. Gene 19:201209

Chapter 18

cDNA Libraries

LUIZ F. ONUCHIC AND GREGORY G. GERMINO

Introduction

Proteins are the work horses of the cell: they comprise the structural and regulatory molecules, the motors and pumps, enzymes and receptors. Determining the complete sequence of a specific protein, however, or producing it in sufficient quantity to allow its manipulation and characterization are extremely tedious and expensive procedures if one is limited to the use of traditional biochemical approaches. Gene cloning and genetic engineering have greatly contributed to the field of biochemistry. Determining protein sequences would still be a daunting task, however, if one was restricted to the use of genomic DNA as the source material since only ~2% is thought likely to encode protein. The remainder includes structural and regulatory elements, introns, untranslated exons, and repetitive, non-coding sequences without known function ("junk DNA"). The effort required to determine a gene product's sequence is greatly reduced if the analysis is restricted to coding sequences. Since one of the principle tenets of molecular biology is that mRNA serve as the template for protein synthesis, mRNA molecules are the ideal substrates for determining the sequence of proteins. Unfortunately, none of the commonly available cloning vectors accept RNA molecules as inserts. Therefore, an essential step in the generation of an expressed sequence library is the conversion of mRNA molecules into double-stranded DNA. DNA copies of mRNA molecules are called cDNA, and a library comprized of DNA copies of the mRNA species from a cell or tissue is called a cDNA library.

Luiz F. Onuchic, Johns Hopkins University, School of Medicine, Division of Nephrology, 720 Rutland Avenue, Ross 970, BalitmoreMarylandUSA, *Correspondence to* Gregory G. Germino, Johns Hopkins University, School of Medicine, Division of Nephrology, 720 Rutland Avenue, Ross 970, BalitmoreMarylandUSA (*phone* +01-410-614-1650; *fax* +01 410-955-0485; *e-mail* ggermino@welchlink.welch.jhu.edu)

An ideal cDNA library is a faithful and complete representation of all mRNA molecules present in the tissue from which it was prepared. Many of the principles reviewed in the preceding sections are also applicable to cDNA libraries:

- the choice of vector and host are determined by the ultimate use of the library;
- the strategy used to generate inserts depends on the optimal size of insert desired;
- some expressed sequences confer significant growth advantages/disadvantages and thereby distort their representation within the library;
- chimeric clones are common; and
- one must consider the relative cost and effort of constructing one's own library versus the availability of previously prepared libraries.There are a number of features unique to expressed sequence libraries, however, which are reviewed in the following sections.

mRNA purification

The first step in constructing a cDNA library consists of isolating the RNA molecules that are expressed within a given tissue or cell type. mRNAs comprise only 1 – 5% of the total RNA present within a cell, so additional purification techniques are required to enrich for mRNA. Selection of mRNA from total cellular RNA is based on the observation that virtually all eukaryotic mRNAs have a long stretch (>20 bp) of adenosines at their 3' ends. This sequence, called the poly(A) tail, is unique to mRNA molecules and is absent in the other major RNA species, ribosomal RNAs (rRNAs) and transfer RNAs (tRNAs). The poly(A) tail usually is long enough to hybridize to its complement, a synthetic oligonucleotide made of deoxythymidines [oligo(dT)], and it is this property that allows selection of mRNAs from a mixture of RNA molecules. Although there are numerous methods available for selecting mRNA molecules, they primarily differ in how the oligo(dT)-mRNA complex is separated from the total RNA mixture. The protocols fall basically into three groups:

- Oligo(dT) affinity chromatography: The original protocol by Aviv and Leder (1972) used affinity chromatography with an oligo(dT) (12 – 18 mer) linked to cellulose. In this procedure, the total RNA mixture is loaded onto the top of a column of the matrix, and mRNAs anneal to the the oligo(dT) as the solution slowly drips through the column under gravity. Non-poly(A)+ RNA does not bind and is easily washed off the column. The bound mRNA is rapidly eluted by washing the column with a low-salt buffer. In a modification of this technique [oligo(dT) spin columns], centrifugation replaces gravity as the force separating the components. Consequently, the time allowed for annealing and washing is shortened. The total time required to isolate mRNA is reduced from several hours to less than 1 h, but yield is also reduced.

- Solution hybridization: This protocol also uses oligo(dT)-cellulose but the matrix is added directly to the total RNA solution rather than a column. The annealing and washing steps are performed in solution with the matrix-bound RNA pelleted by centrifugation between steps. This method benefits from the rapidity and efficiency of solution kinetics.
- Biotinylated oligo(dT) and streptavidin-coated magnetic beads: In this protocol, a biotinylated oligo(dT) primer is annealed in solution to total RNA. Magnetic beads which have been coated with streptavidin are then added, and the bead-mRNA complex is separated from the bulk solution using a magnet. The magnet is removed, wash solutions are added, and the magnetic isolation is repeated. The oligo(dT) primer dissociates from mRNA in the absence of salt. This is an extremely fast, inexpensive and efficient method for the isolation of mRNA. In an alternative method, the magnetic beads are directly bound to oligo(dT).

cDNA synthesis and cloning

Since mRNAs can not be propagated as RNA molecules, they are copied into single-stranded cDNAs by an enzyme called reverse transcriptase prior to cloning (see Procedure). Single-stranded DNAs, however, are also unsuitable for cloning since most ligases will only join together double-stranded molecules. Thus, complementary strands are synthesized by a DNA polymerase, and the double-stranded products are cloned in cDNA libraries (Figure 1). The efficiency in generating complete, double-stranded molecules is variable, and very short fragments often contaminate the mixture. Moreover, shorter molecules are preferentially cloned and, if not removed, will result in a library that is poorly representative of its source mRNA. This problem can be avoided by size-fractionation of the cDNA prior to cloning. This is accomplished by either agarose gel electrophoresis or centrifugation through sizing columns that retain small molecules.

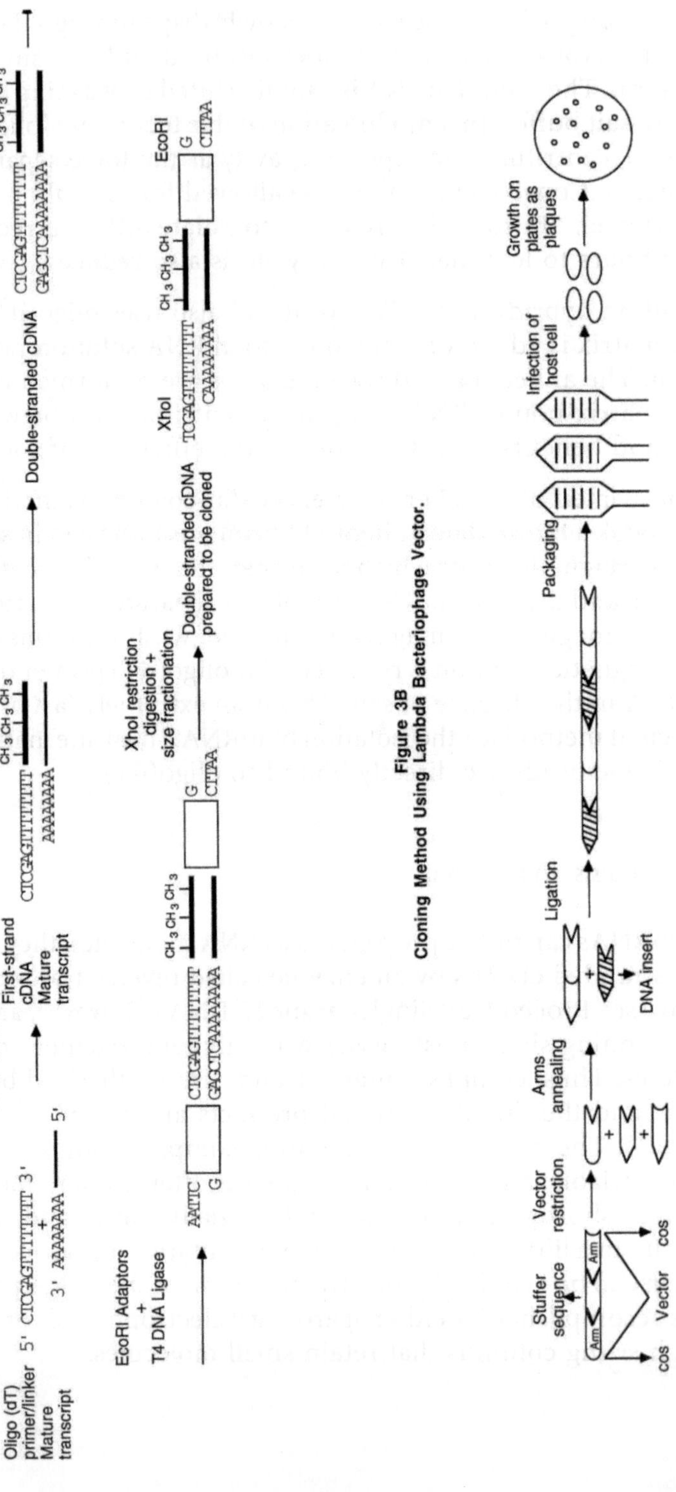

Figure 3B
Cloning Method Using Lambda Bacteriophage Vector.

Previous chapters have discussed the relative cloning efficiencies of cohesive vs. blunt ends. Double-stranded cDNA molecules synthesized using DNA polymerase I have blunt ends which ligate very inefficiently. A large fraction of the cDNA molecules would be lost from a library if one tried to ligate them to a blunt-ended vector. Therefore, virtually all protocols for synthesis of cDNA libraries include a step that adds cohesive ends which are compatible with the cloning site of a vector. This is accomplished by ligating short, double-stranded DNA molecules, called linkers. A unique feature of these synthetic oligonucleotides is that they contain restriction sites which are compatible with the vector's cloning sites. A very high molar ratio of linkers to cDNA molecules is easily achieved, and this compensates for the inefficiency of blunt end ligations. Excess linkers are removed by size fractionation prior to cloning. The preparation of cDNA described under Procedure as shown in Figure 1A presents one of the many approaches in current use (a modification of Gubler and Hoffman, 1983).

As previously discussed, there are methodologic problems common to both genomic and cDNA libraries. cDNA libraries also have several unique features that require consideration:

- Most cDNA libraries are constructed with the goal of obtaining full-length cDNA copies of mRNAs. Careful selection of the enzymes (eg. Superscript (GibcoBRL, Gaithersburg, MD) for the reverse transcriptase, etc.) and reaction conditions are important but by far the most important determinant of success is the quality of the original mRNA. Libraries generated from low quality mRNA (i.e., partially degraded) will lack full-length copies of larger mRNAs as well as rare transcripts.
- A high frequency of chimeric clones results when the concentration of linkers used to construct the library is too low.
- After packaging or transformation, most libraries are amplified for storage and re-use. Ideally, each clone present in an original library is replicated to the same degree so that its fractional representation remains

Fig. 1. **A** Full preparation of a cDNA insert. **B** Cloning method using lambda bacteriophage vector. The vector is digested with restriction enzymes that produce ends compatible with the cDNA linker adaptors. Size fractionation is used to remove the stuffer fragments. The phage arms are allowed to reanneal at their *cos* sites prior to ligation to cDNA inserts. This results in long concatamers of alternating phage arms and inserts. Included in packaging extract are proteins that cleave at *cos* sites and then package the 38–50 kb DNA fragments into phage heads. Phage tails, also included in packaging extract, self-attach to the filled heads. The resultant particles are infectious and are incubated with bacteria for propagation.

unchanged after amplification. Unfortunately, library amplification often results in the loss of clones which grow poorly. Amplifying plasmid libraries in a semi-solid suspension culture, such as in a low-melting temperature agarose mixed with culture broth, minimizes this problem.

- Linkers that ligate to the ends of cDNA molecules must be cleaved with a restriction enzyme that generates cohesive ends prior to ligation to a similarly-prepared vector. Protocols that rely on linkers to enhance cloning efficiency must include a step that protects the restriction sites within the cDNA from cleavage. This is accomplished in one of four ways. In older protocols, the double-stranded cDNA was treated with a methylase specific for the restriction enzyme sites prior to the ligation of linkers. The linkers were then ligated to the methylated cDNA, digested with the appropriate restriction enzyme, and finally ligated to a compatible vector. Incomplete methylation, however, resulted in restriction of internal cDNA sites and produced clones that were of incomplete length. An alternative method uses linkers that contain restriction sites such as *Not*I that are rarely found in cDNAs. A significant advantage of this approach is that methylation, which is often an inefficient process, is no longer required. An additional benefit is a reduction in the number of steps required to synthesize a library. A drawback is that some cDNAs do include rare restriction sites within their sequences (for example, there are at least three genes within the polycystic kidney disease 1 (PKD1) region that have *Not*I sequences within their cDNAs). A third approach is to add 5-methyl-dCTP to the reaction mix for the synthesis of either the first or both strands (Figure 1A). The hemi-methylated DNA will resist restriction with many enzymes. A newer approach that has even fewer steps involves the use of linker-adapters. These synthetic oligonucleotides have been designed with one end blunt, suitable for ligation to the cDNA molecules, and one end that contains an overhang suitable for ligation to a compatible vector.
- Oligod(T)-primed libraries often lack 5' end sequences. When primed by oligo(dT), reverse transcriptase begins cDNA synthesis at the poly-A tail and may "fall off" the molecule before copying its full length. Second strand synthesis can also be an incomplete process. These problems can be circumvented by constructing a library in which reverse transcription is primed using random primers, mixtures of short oligonucleotides that have, on average, multiple annealing sites distributed along an mRNA molecule. Inserts constructed using random primers, however, are unlikely to be full-length, and average-insert size tends to be shorter than in oligo-d(T) primed libraries.

- One can screen cDNA libraries for gene sequences using many more approaches than can be used to screen genomic libraries. Standard hybridization techniques using labeled oligonucleotide, cDNA, RNA or genomic fragments as probes are similar for the two types of libraries. cDNA libraries, however, by virtue of being comprised of expressed sequences, can also be screened by testing for their gene products. Antibodies can be used to identify clones expressing particular antigens. Ligands can be used to identify clones expressing particular receptors (or vice versa), etc. Genes can also be identified on the basis of the functional properties of the encoded proteins (eg. transporters, channels).
- Inefficient ligation, packaging, transfection or transformation results in incomplete representation of the transcribed sequences from the tissue that was the source for the mRNA.

Theoretical considerations for cloning cDNA sequences

Determining the number of clones that must be screened to ensure >95% coverage of the genome is relatively straightforward, as was previously explained. Each genome equivalent is expected to have one copy of a chromosome locus. These calculations, however, are much more difficult for cDNA libraries since not all genes are expressed in equal abundance in a cell. Some cells have a large number of copies of mRNA (~1500/cell) derived from a handful of genes whereas other mRNA molecules are present at less than 10 copies/cell. Consider a chicken red blood cell (RBC) library: 3 mRNA species (~1500 copies/cell) comprise over 90% of the total mRNA molecules present in chicken RBCs. One plate of less than 100 clones has a probability >95% of having at least one copy of each of the most common genes. Alternatively, over 1000 fold more plaques would be required to find a message that is rarely expressed in this cell type. The vast majority of chicken genes will never be found no matter how many millions of plaques one screens. Only approx. 1000 genes, or $\approx$ 1 – 5% of the total, are expressed in this cell type. Thus, even a perfect cDNA library derived from chicken RBCs will lack copies of most genes.

In sum, there are four major principles that determine the number of clones that must be screened.

- The relative abundance of the transcript in the tissue or cell type that is the source of the RNA is one of the most important factors. Northern blot analysis or a quantitative RNA protection assay can help one estimate expression levels.

- Next in importance is the total number of primary recombinants in a library. In an unamplified library, each recombinant represents a unique mRNA species. Genes with very low levels of expression (1/50,000 or 1/100,000) are unlikely to be present in a library containing less than 250,000 recombinants. The number of primary recombinants must be much higher if one uses expression screening approaches (see below, Principles of Vector Selection: Expression vs. Non-Expression Libraries).
- Amplification of a library often results in under-representation of some clones. Unfortunately, there is no accurate way of predicting how many additional clones must be screened to compensate for this effect. We favor initially evaluating the same number of clones as was present in the unamplified library. If this screen is negative, we recommend making/borrowing a primary library or using a library derived from a different tissue rather than plating an additional number of clones.
- Random-primed libraries often are more representative than oligo(dT)-primed libraries. Short oligonucleotides of random sequence can prime synthesis from multiple sites along an mRNA molecule whereas oligo(dT) usually primes synthesis from only one site. In the latter case, each mRNA molecule has only one chance of becoming a cloned cDNA, and rare transcripts are easily lost if any of the steps are inefficient.

It is not uncommon for investigators seeking rarely expressed genes in tissues that express a complex population of genes (i.e., brain, estimated at >20,000 different expressed genes) to screen more than 1×10^6 primary recombinants. This can rapidly become an expensive and arduous task. A novel solution to this problem has recently been proposed. This technique seeks to normalize the number of copies of each gene in the final library (Patanjali, Parimoo and Weissman 1991). Rather than 3000 copies of genes A, B and C and only two copies of D, E and F genes as in a conventional library, a normalized library would ideally have two copies of A-F. In a normalized library, screening of a total of 100 colonies would be required to ensure a high likelihood of finding a copy of genes A-F whereas over 100-fold more clones would be required using a conventional library. Normalized libraries are constructed by annealing single-strand cDNA to a limiting amount of genomic DNA, removing the excess unbound copies, and cloning only those cDNAs that bind to genomic DNA. In practice, the normalization protocol is only moderately efficient and has the drawback that the inserts are usually very short.

Principles of vector selection: expression vs. non-expression libraries

The number of vector-host systems available for cloning genes is constantly increasing. Since this process is likely to continue, it is more appropriate for a review such as this to discuss some of the basic principles of vector selection for cDNA libraries. It is anticipated that the reader will then be able to use his/her knowledge to evaluate the advantages of new systems.

Some of the factors that govern vector/host selection for cDNA library construction are similar to those for genomic libraries. High transfection/transformation efficiencies, stable propagation and selectable markers are desirable. Unlike genomic libraries, the size of the insert is less often a problem since most full length cDNAs can be easily cloned into either plasmid or lambda bacteriophage, and virtually all cDNA libraries are constructed using one of these vectors. Historically, bacteriophage were preferred because packaging and transfection efficiencies far exceeded plasmid transformation rates and yielded a higher number of primary recombinants per microgram of starting cDNA. Technical improvements (electroporation and transformation of plasmids into high-efficiency competent cells) have since made this difference negligible, and other features are now used to select an appropriate system. Bacteriophage libraries are easier to screen and are more likely to include longer gene sequences. Furthermore, one can use a modified bacteriophage which requires a minimum insert size that is greater than the average transcript size (eg. >5kb) to select for longer cDNAs. On the other hand, plasmid libraries have some other advantages. They are easier to amplify in a way that maintains the relative proportion of clones within them. Plasmid DNA also is more easily prepared and transfected into eukaryotic cells for expression studies.

The most important factor that determines selection of a vector is how the library is to be screened. As discussed previously, clones encoding genes of interest can be identified in cDNA libraries using standard hybridization protocols. A major advantage of cDNA libraries is that one requires neither the gene sequence nor a sub-fragment to isolate its full length clone. Because cDNA clones are derived from mRNA, they potentially encode the protein products of the genes from which they were derived. This feature allows an investigator to identify genes on the basis of functional properties rather than sequence. Antibodies, ligands, and functional attributes such as transport activity or ability to rescue function become additional tools that are useful in identifiying clones derived from genes of interest. A yeast-based technique, known as the two-hybrid system, has been developed to screen for genes which encode proteins capable of interacting with a protein of interest (bait). The assay is based on the ability to separate eukaryotic transcriptional activators such as GAL4 into two modular domains: a DNA-binding domain (BD) and a transcriptional activation domain (AD). While the first localizes the transcription factor to specific DNA sequences positioned in regulatory sequences immediately 5' of a reporter gene, the second domain contacts other transcription-related elements necessary for the initiation of transcription. Both domains must be in close proximity for normal transactivation to occur. Normally, this occurs because the two domains are segments of the same gene product. In the yeast two hybrid technique, the domains are separately encoded as portions of hybrid proteins containing other protein domains. Transactivation only occurs when the AD and BD domain are brought together because of an interaction between the other domains of their respective hybrid proteins. Proteins that interact with the bait protein are identified by generating hybrids of the BD and the bait protein (H1), and the AD and a library of proteins (H2). A specific interaction between H1 and H2 localizes both the BD and the AD to a reporter gene, leading to its transcriptional activation and identifying a candidate binding partner for the target protein. This system is sensitive enough to detect weak and transient protein-protein interactions, generally observed in large native complexes.

Expression Libraries. Libraries that are constructed to optimize expression of their inserted cDNA sequences are called expression libraries. The vectors used in construction of cDNA expression libraries have special features. One essential element is an RNA promoter sequence adjacent to the cloning site. This is used to transcribe RNA from the cDNA insert in a bacterial host. Expression must be regulated, however, since the products of many cloned genes are toxic to the host cell. If left unregulated, clones expressing deleterious products would be at a growth disadvantage and would be rapidly lost from a library. Most vectors used in constructing expression libraries contain inducible promoters that allow propagation of a clone before inducing expression of its insert. Expressed proteins can be fixed to nitrocellulose membranes in a manner similar to that used to make DNA filters and then screened with labeled antibodies, ligands, etc. An added feature of expression libraries is that they also can be plated so as to generate DNA filters suitable for hybridization with nucleic acid probes. Thus, expression libraries are the most versatile since they can be used for multiple purposes.

There are, however, several drawbacks to their use. First, expression libraries without full-length copies of mRNA molecules are the most useful. The extreme 5' end of mRNA molecules is non-coding and its inclusion could disrupt expression of the remainder of the message. Furthermore, smaller foreign polypeptides are more stably expressed in prokaryotic hosts. Second, expression cloning is an inefficient process. Half of the clones will have inserts cloned in the wrong orientation (3' – 5'), which is unsuitable for expression. Moreover, the cDNA will be inserted "in frame" with the gene encoded by the vector and thereby expressed only one out of three times. Taken together this means that six times as many clones must be screened if one is searching for a gene based on its expressed product rather than on its DNA sequence.

The inefficiency of this process can be halved by directionally cloning the insert into the vector. This is achieved by synthesizing the cDNA with a modified oligo(dT) primer that includes the sequence of a unique restriction site. A second unique restriction site is added to the 5' end by the ligation of a linker-adapter (Figure 1A). The asymmetric sites allow insertion of the sequence in the vector in the correct orientation, thus preserving the 5' – 3' orientation of the gene. Nonetheless, a minimum of three times more primary recombinants are still required for expression screening than are necessary for nucleic hybridizations.

A number of expression vectors have also been designed that allow expression in eukaryotic hosts. The requirements for regulating expression in prokaryotic and eukaryotic hosts are different, and the expression cassettes included in eukaryotic expression vectors reflect these differences. Another feature of eukaryotic expression vectors is that they often encode markers that allow selection of successfully transfected cells.

Non-Expression Libraries. Nucleic acid hybridization is the method most commonly used to identify cDNA clones of interest in gene cloning projects. There are numerous suitable bacteriophage and plasmid vectors available. Distinguishing features include:

- limits of acceptable insert size;
- variety of cloning sites;
- ease of recovering inserts;
- positive selection for recombinants; and
- flanking RNA promoter sequences that allow generation of sense and anti-sense transcripts. Lambda gt10 traditionally has been one of the most commonly used lambda vectors since it accepts inserts from 0 – 5.0 kb, yields large and uniform plaques with high titers, and has a positive selection system.

The cloning site has been placed in the *cI* gene (lambda repressor) and successful ligation of an insert disrupts its expression. *cI+* bacteriophage will not form plaques whereas insert-containing bacteriophages form plaques when cultured in an *hfl*$^+$ host. Important considerations regarding lambda gt10 cDNA libraries are reviewed by Huynh et al. (Huynh, Young and Davis 1984).

Many of the additional features listed above have been developed to overcome some of the limitations of lambda gt10:

- The original vector has an *Eco*RI site as its only cloning site. As discussed previously, there are numerous circumstances where one would prefer to have alternative or additional cloning sites.
- Subcloning inserts from lambda gt10 can often be a frustrating task. Newer vectors such as Lambda ZAPII (Stratagene) or λSHlox (Novagen) have novel excision systems that allow direct rescue of the insert as a plasmid.
- Many investigators seek to identify genes that are differentially expressed between two tissues or cell types. A variety of subtraction protocols have been developed that allow enrichment for clones that differ in their level of expression. Typically, pools of sense and anti-sense single stranded molecules derived from each of the libraries are required. Vectors such as Lambda ZAPII and λSHlox have multiple cloning sites that allow unidirectional cloning (to preserve 5' – 3' orientation) and flanking RNA promoter sequences. The pools are generated using RNA polymerases that prime from the respective RNA promoter sequences that flank the cloning sites.

Procedure

Preparation of cDNA inserts for unidirectional cloning - guidelines

1. First strand cDNA synthesis is performed using *Xho*I primer-adaptors, 5-methyl-dCTP, dATP/dTTP/dGTP, and reverse transcriptase. This reaction yields single-stranded cDNA molecules that are methylated at all deoxycytosines. Many investigators choose to simultaneously perform a tracer reaction using an aliquot of the final mix which allows them to determine the efficiency of the reaction. A small amount of $\alpha[^{32}P]$dCTP is added to a fraction of the total reaction, and the two samples are incubated for 60 minutes at 42°C. A portion of the labeled reaction is removed for gel electrophoresis, another sample is allowed to dry to completion (to determine total number of counts in sample), and a third fraction is used to determine incorporated counts (number of

counts precipitated). The first strand yield is calculated using the following equations:

$$\% \text{ incorporation} = \frac{\text{incorporated cpm x 100\%}}{\text{total cpm}}$$

$$\text{nmoles dNTP incorporated} = \text{dNTP [nmoles/}\mu\text{l] x reaction vol. [}\mu\text{l] x \% incorp./100}$$

$$\text{cDNA synthesized [ng]} = \text{nmoles DNTP incorporated x 330 ng/nmole}$$

$$\% \text{ mRNA converted to cDNA} = \frac{\text{ng cDNA synthesized}}{\text{ng mRNA in reaction}}$$

2. Second strand synthesis is performed in two steps:
a) RNAse H is used to partially degrade the RNA components of the cDNA/RNA hybrids. The remaining RNA fragments serve as primers for second strand synthesis by DNA polymerase I.
b) *E. coli* DNA ligase is added to covalently link the newly synthesized cDNA, and T4 DNA polymerase is used to create blunt ends. Chimeric clones are not produced by this reaction because *E. coli* DNA ligase will **not** ligate blunt end molecules.
The second strand is synthesized without methylated dNTPs so that the resultant products are hemi-methylated. This is accomplished by one of two methods. One approach is to use a minimal concentration of methyl-dCTP/other dNTPs in the first strand reaction and a much larger concentration of unmethylated dNTPs for the second reaction. The reaction kinetics are such that the probability of methylating the *Xho*I primer-adapter during synthesis of the second strand is very low. Another approach is to remove the methyl-dCTP from the reaction mixture using spin columns or by purifying the cDNA by ethanol precipitation after completing the first strand synthesis. Hemi-methylation is important because it protects internal *Xho*I restriction sites from subsequent restriction. By removing the methyl-dCTP from the second reaction, the *Xho*I primer-adapter remains unmethylated and susceptible to cleavage.
Calculations of second strand yield are similar to those described above. The concentration of the dNTPs, however, will depend on the method used to synthesize an unmethylated second strand. dNTPs from the initial reaction must be included in the calculation if they are not removed by precipitation or centrifugation.Aliquots of the first and second strand reaction products are electrophoresed to determine average fragment length.

$$\% \text{ second strand incorporatiion} = \frac{\text{incorporated cpm}}{\text{total cpm}}$$

nmoles dNTP incorporated = {(dNTP [nmoles/μl] x reaction vol. [μl]) – nmoles incorporated in the first strand reaction} x % second incorp./100

second strand cDNA synthesized [ng] = nmoles dNTP incorporated x 330 ng/nmole

$$\% \text{ conversion to double-strand cDNA} = \frac{\text{second strand cDNA synthesized[ng]}}{\text{first strand cDNA synthesized [ng]}}$$

3. *Eco*RI adaptors are ligated to the blunt-ends by T4 DNA ligase. Excess linkers are removed by centrifugation through a spin column.
4. The products are prepared for ligation to the vector by cleavage with *Xho*I. The *Xho*I site contributed by the 3' end primer-adapter is the only unmethylated site in each molecule, and its digestion yields molecules with vector-compatible *Eco*RI-*Xho*I cohesive ends. Small end-fragments are removed from the mix by precipitation or spin column centrifugation, and the pool is now ready for ligation. 10 μg of mRNA yields on average 14 μg of double-stranded product suitable for cloning. 100 ng of vector-ready cDNA yields 1 x 10^6 primary recombinants.

Cloning cDNA into lambda bacteriophage

1. Digest a suitable bacteriophage vector (ZAP Express [Stratagene], Lambda Zap II [Stratagene]) with *Eco*RI and *Xho*I. Treat with alkaline phosphatase and remove the stuffer fragments by size fractionation. Resuspend the arms at a concentration of 1 μg/μl. Pre-prepared vector arms are also commercially available. Add 100 ng of cDNA containing *Eco*RI and *Xho*I ends (5 μg/ml final), 10 μl of vector, 2 μl of 10 x ligase buffer and adjust with H_2O to a final volume of 19 μl. Pipet gently and add 1 μl (400 U) of T4 DNA ligase (20 U/μl). Pipet again gently and incubate overnight at 4°C. Inoculate appropriate medium supplemented with $MgSO_4$ and maltose with a suitable *E. coli* host (eg. XL1-Blue MRF' [Stratagene]).
2. Inoculate 50 ml of $MgSO_4$- and maltose-supplemented broth with 0.5 ml of the overnight host culture and allow to grow to an OD_{600} of 0.5 – 1.0.

Pellet the cells by centrifugation at 2000 x g for 10 min. Resuspend the cells in half the original volume of cold 10 mM $MgSO_4$. The cells should be stored at 4°C until use (within 24 h).

3. Incubate the ligation mix for 30 min at -20°C, thaw at room temperature and then add to the highest efficiency packaging extract **immediately** after thawing the latter. Incubate at room temperature for 2 h.
4. Dilute the packaging mix with phage dilution buffer (100 mM NaCl, 20 mM Tris-HCl (pH.7.4), 10 mM $MgSO_4$) to a final volume of 2 ml, add a few drops of chloroform, gently mix (~3 s) and place on ice. The library is now ready for titering.
5. The library should be kept at 4°C until plating. Packaged bacteriophage are unstable in vitro and will suffer a significant drop in titer if not plated within the next 24-48 hours.
6. Order of magnitude dilutions are prepared for titering. Aliquots of diluted extract are adsorbed to 200 µl of host cells for 15-20 minutes at 37°C, added to 7 ml of melted top agar (cooled to below 50°C), and poured onto dry, pre-warmed agar plates.
7. The plates are incubated at 37°C overnight. Cloning efficiency is expressed as the number recombinants/µg cDNA and is calculated using the following formula: (number of recombinant plaques) x (dilution factor) / (number of µg cDNA).
8. Plate the remainder of the library at a dilution adequate to yield ~50×10^4 plaques/130 mm plate.

References

Aviv H and Leder P (1972) Purification of biologically active globin messenger RNA by chromatography on oligothymidylic acid-cellulose. Proc Natl Acad Sci USA 69:1408 1412

Gubler U and Hoffman BJ (1983) A simple and very efficient method for generating cDNA libraries. Gene 25:263 269

Huynh TV, Young RA and Davis RW (1984) Construction and screening cDNA libraries in lambda gt10 and lambda gt11. In DNA Cloning, Vol 1:49 78. IRL Press, Oxford

Patanjali SR, Parimoo S and Weissman SM (1991). Construction of a uniform-abundance (normalized) cDNA library. Proc Natl Acad Sci USA 88:1943 1947

cDNA Library Screening

JOHANN PETER HOSSLE

Introduction

cDNA libraries are collections of recombinant DNA sequences, derived from and representing entire mRNA populations present at a given stage of development or differentiation in certain tissues or cell types. cDNA libraries can be self-made and tailored according to individual experimental needs. Alternatively, they can be purchased from a commercial distributor, if the library representing the desired mRNA source is available. In any case, the collection of cDNA sequences is cloned into a vector, a plasmid or phage, allowing the manipulation of the cloned sequences and propagation of the whole library in suitable host bacteria.

The major purpose of cDNA libraries is to serve as a source for isolation and cloning of individual recombinant DNA sequences of interest. Individual recombinants within libraries are identified by a number of possible screening methods. The most obvious method is to screen for homology with a known nucleic acid sequence, thereby using labeled complementary nucleic acid sequences as probes. Expression libraries, which reflect the protein biosynthesis potential of the original mRNA source, allow for functional and antibody screening.

Vector and host selection

cDNA libraries can be constructed using either plasmid or bacteriophage vectors. Plasmid libraries offer the advantage of ease of manipulation but are somewhat limiting if large numbers of recombinant clones need to be screened. They are suitable for screening up to about 10,000 to 50,000 re-

Johann Peter Hossle, University Children's Hospital, Division of Immunology/Hematology, Steinwiesstraße 75, Zürich, 8032, Switzerland (*phone* +41-1-266-73-41; *fax* +41-1-266-71-71; *e-mail* hosslejp@kispi.unizh.ch)

combinants. For screening of larger numbers of recombinants (>100,000), phage libraries are more suitable, because of library size and efficiency of screening. Introduction of recombinant DNA into host bacteria for propagation is more efficient by phage infection than by conventional plasmid transfection procedures. This permits construction of libraries containing larger numbers of recombinants. Furthermore, screening of phage libraries can be carried out at very high plaque densities thereby allowing for screening of much larger numbers of recombinants at the same time. Advantages of both systems have recently been combined in so-called excision vector systems in which initial screening is phage-based but later cDNA inserts are excised in vivo and can be further propagated and screened as phagemids (see below).

Expression libraries can be plasmid or bacteriophage based. They offer versatility over conventional plaque hybridization screening, since antibody recognition and phenotypic expression can also serve as screening tools. However, isolation of a particular sequence from an expression library requires screening of at least 6 times the number of recombinants needed in standard hybridization screening. This is due to possible orientation (factor two) of the cDNA insert relative to its integration site as well as its frame relative to the initiation codon from which translation starts (factor three). One potential problem of cDNA expression screening is the synthesis of a protein product that might be toxic to the host bacteria. This is one of the reasons why phage expression libraries are often favored over plasmid expression libraries. Phage vectors such as λgt11 can easily be grown to high numbers under lysogenic conditions, before protein expression is induced for detection.

The choice of host bacteria is primarily determined by the vector used. The most common vectors used are bacteriophage λgt10 for plaque hybridization screening and λgt11 (Jendrisak et al. 1987) and pUC plasmid (Helfman and Hughes 1987) for cDNA expression screening. Among the most widely used excision vectors is the λZAP expression vector (Short and Sorge 1992). If the biological activity of a peptide that is expressed from an isolated clone is to be verified later in an eukaryotic cell, prokaryotic to eukaryotic shuttle vector systems can be used (Kato et al. 1994).

Library size and status

The size of a cDNA library and the percentage of independent recombinants it contains are of critical importance, for determing the probability that a specific RNA sequence is represented in the library. To find a specific sequence during screening, it will be necessary to screen a certain number of individual recombinants. As a rule of thumb, the number of individual recombinants in the library should exceed 5 times the number expected, given the relative abundance of the desired mRNA in the original RNA population (Kimmel and Berger 1987). It is preferable to use unamplified libraries, because the relative frequency of recombinant clones contained in the library may be distorted after amplification due to uneven growth rates of the individual recombinants.

Cloning of cDNA into most vectors is performed by insertion into the β-galactosidase (lacZ) gene that is present in the vector molecule. Inserted DNA interrupts the coding sequence in lacZ, thereby causing the phenotype of the vector to change from blue (parental) to colorless (recombinant) in a simple blue plaque test. In this fashion, the number of recombinants present in the library can be determined easily. Average size λ-libraries are usually in the range of 10^6-10^8 plaque-forming units with around 60% recombinants.

Number of recombinants needed to be screened

The number of unique recombinants (N) that must be screened to ensure isolation of the desired sequence can be calculated as: $N = \frac{ln(1-P)}{ln(1-n)}$ where P is the probability of isolating the desired sequence (usually set at 99%); and n is the fractional proportion of the total mRNA population represented by the desired single mRNA species (Jendrisak et al. 1987). The abundance of the mRNA encoding the sequence of interest needs to be estimated by, e.g., quantitative RT-PCR. The numbers of recombinants that are required can be us high as 10^6 for rare mRNA species.

Screening based on sequence homology

Homologous nucleic acid probes

The most direct method for screening a cDNA library is to use nucleic acid probes for hybridization. This involves the following steps (Wahl and Berger 1987):

1. Aliquots of the library to be screened (phage or plasmid transfected bacteria) are grown on master plates (or filters) from which replica filters are made.
2. The masters are preserved while the replica filters are processed.

3. Phage particles are disrupted or bacteria are lysed in situ on the filters.
4. DNA is bound to the filter while RNA is hydrolyzed.
5. Filters are washed and non-specific binding sites are blocked during a prehybridization step. The bound partially denatured DNA is hybridized to a labeled complementary nucleic acid sequence probe that is capable of binding to the immobilized target sequences at the experimentally chosen hybridization stringency.
6. The pattern of the DNA on the filter replicas is aligned with the plated recombinants on the masters. The image of positive signals obtained, therefore, serves as a map for picking positive recombinants from the master.
7. Positive colonies or plaques are purified and reamplified for further analysis.

To avoid false positive clones, which are quite common using this method, it is highly recommended that duplicate filters of each master plate are screened in addition to negative and positive controls. Hybridization screening requires information about the sequence being sought. Often part of the sequence to be isolated is already known and the procedure is used to search for additional flanking sequence regions. In other cases, closely related sequences may cross-hybridize to the desired cDNA and can be used as probes. For optimal screening results, careful consideration of hybridization parameters is required. For DNA-DNA interactions these are related as follows (Wahl et al. 1987): $T_m = 81.5^\circ + 16.6 \log M + 41$ (%[G+C] - 500/L-0.62 (%formamide) where T_m is the melting temperature, defined as the temperature at which half the duplex molecules have dissociated into their constituent single strands; M is the monovalent cation concentration in moles per liter; (G+C) is the base composition expressed as mole fraction of G and C residues; L is the length of the shortest chain in the duplex and %formamide is the relative concentration of the helix-destabilizing agent. The maximum rate of hybridization occurs at 25° below Tm with a broad range of nearly ideal conditions between T_m -32° and T_m -18°. Standard conditions used are T_m -25°, 0.18M cation concentration, and a single stranded fragment length of 500 nucleotides. For most experiments this results in optimal hybridization temperatures between 65°C to 68°C, or approx. 42°C in the presence of 50% formamide.

Base mismatches decrease the stability of hybrid duplexes according to mismatch number and location. For hybrids longer than 150 bp, the T_m of a DNA duplex decreases by approx. 1° for every 1% of base pair mismatch. In order to assure specificity, hybridization reactions must be performed under the most stringent conditions possible. Stringency can be adjusted either during hybridization or afterwards in the post-hybridization washes. Labeling of probes can be radioactive or nonradioactive. There are many different protocols for labeling nucleic acids with ^{32}P, which is the most widely used isotope for this purpose. Since the introduction of ^{33}P, a low energy β-emitter labeling has become even more convenient. Colony and plaque hybridization screening has also been successfully performed using non-radioactive systems, e.g. digoxigenin (Höltke et al. 1995).

Screening using synthetic oligonucleoid probes

If there is not enough information available about the nucleotide sequence of interest, it may be possible to design a synthetic oligonucleotide probe based on partial amino acid sequence. Alternatively, one may use synthetic oligonucleotide probes to check the presence of a particular known mutation. In most cases where oligonucleotide probes are derived from protein sequences it will be necessary to design an oligonucleotide mixture based on codon usage due to redundandency of the genetic code. When synthesizing such probes, the number of different oligonucleotides (complexity) present in the mixture should be taken in account, and should as a rule of thumb not exceed 128 (Wallace and Miyada 1987). If more complex combinations must be tested, oligonucleotide mixtures should be hybridized as several separate pools.

For calculating the appropriate hybridization temperature (T_h) the following equations are commonly used: $T_h = T_d - 5^\circ$ with T_d (the temperature at which half of the duplexes is dissociated) determined as: $T_d = 2^\circ$ (number of A+T nucleotides) + 4° (number of G+C nucleotides) (applies to duplexes of 11 to 23 bases in 1 M Na^+).

A very rapid colony hybridization screening method for oligonucleotide probes has been developed (Paddock 1993), in which a special filter paper (Whatman 541) is used rather than conventional nitrocellulose or nylon membranes.

It should be noted that acid depurination of the DNA, which facilitates blot transfer, should be avoided in oligonucleotide screening since oligonucleotides do not hybridize efficiently to DNA treated in this manner (Wallace and Miyada 1987).

PCR amplified probes using degenerate primers

An approach for identification and subcloning of related cDNA sequences (e.g. gene families) that is similar to the synthesis of complex oligonucleotide mixes for screening makes use of the polymerase chain reaction (PCR). DNA derived from pools of cDNA clones serves as a template in PCR amplifications using degenerate primers. Primer pairs are derived from conserved amino acid sequences bracketing more diverse sequences. Amplification makes such sequences selectively and directly accessible for subcloning and comparative sequence analysis (Watson et al. 1992).

Differential hybridization

DNA libraries themselves can serve as probes in screening of cDNA libraries. Such strategies are used for isolation of differentially expressed genes or extremely rare sequences. Libraries of DNA clones derived from sequences that are differentially expressed can be prepared by subtractive hybridization (Lee et al. 1991) or differential display of mRNA (Liang and Pardee 1992).

Screening based on antigenicity

In cases where no nucleic acid or protein sequences are available, expression vectors offer the alternative of looking for expression of specific eukaryotic cDNAs in host bacteria. Although both plasmid and phage vectors are used, the most frequently used expression vector is probably λgt11 (Young and Davis 1983a, Young and Davis 1983b). This and other widely used expression vectors provide sequence elements that direct the synthesis of the proper protein product from a cloned cDNA expressed under the control of a bacterial promoter. Usually, such proteins are expressed as fusion proteins containing portions of prokaryotic proteins. Many expression vectors permit insertion into the bacterial β-galactosidase gene (lac Z). As a consequence, cDNA inserts are expressed as β-galactosidase fusion proteins. It has been shown that fusion of foreign sequences to the carboxy terminus of β-galactosidase gives rise to relatively stable fusion proteins (Stanley 1983, Mierendorf et al. 1987). Proper expression of integrated cDNA coding information depends on the reading frame and the orientation of the insert. Therefore, only one sixth of the recombinant clones will have the potential to express the desired protein product. The host strains used for expression (e.g. *E. coli* Y1090) are deficient in certain proteases which prevents premature degradation of expressed proteins. These strains also contain the lac repressor which prevents fusion protein synthesis until isopropyl-β-D-thiogalactopyranoside (IPTG) is added to the medium. This allows the induction of potentially toxic foreign proteins to be delayed.

Manipulation of expression libraries is similar to conventional libraries. However, in addition to recombinant DNA bound to filter replicas, expressed fusion proteins will also bind to the filter membranes and can be detected immunologically using an antibody directed against the peptide product of the sequence to be identified. Some further points are worth considering:

- Antibody choice:
  Both monoclonal and polyclonal antibodies have been used successfully. Ideally, the use of several monoclonal antibodies, which recognize distinct epitopes, would be most advantageous. Because expressed fusion proteins may contain only portions (preferentially C-terminal) of the desired protein, utilization of a single monoclonal antibody carries the risk of missing positive clones. In practice polyclonal antisera rather than a mix of monoclonal antibodies are often used. In any case it is of paramount importance to test beforehand, whether partially denatured antigen bound onto filter can be recognized by the antibody that will be used.

- Antigen detection:
  Sensitivity is critically dependant on the quality (titer, binding characteristics and preparation) of the antibody used. Antibody dilutions must be optimized individually. A dilution 1/100 of the primary antibody is useful to start with. In most cases, non-radioactively labeled secondary antibodies (e.g. peroxidase or phosphatase-conjugated) work as well as radiolabeled antibodies. The amount of antigen bound to the filter replica after plaque lift-off is in the range of 50 to several hundred picograms, depending on expression and stability of the fusion protein in the host cells (Mierendorf et al. 1987).

- Preadsorbtion of antibodies:
  Since IgG often binds non-specifically to *E. coli* proteins it may be necessary to preadsorb the primary as well as secondary antibody using immobilized host cell extracts (e.g. bound to a cyanogen bromide activated matrix).

- Plaque density:
  affects the ability to detect specific recombinants with antibody. Plaque density should not exceed 3 to 4 x 10^4 plaques per 90 mm petri dish.

Screening based on function

In addition to immunological screening other approaches can be used to screen cDNA expression libraries. Possibly the most appealing are screening strategies that are based on the biological function of the protein of interest. For example, expression of receptors can be detected by screening with a radiolabeled ligand (Sikela and Hahn 1987). Screening by antigen binding permits identification of monoclonal Fab fragments in combinatorial lambda libraries (Huse et al. 1989). Binding activities expressed on the surface of released phagemids have been identified by capturing on to immobilized ligand (biopanning) (Hogrefe et al. 1993). Protein-protein interactions can be studied in the so-called two-hybrid system (Fields and Sternglanz 1994). Screening for binding activity is not restricted to protein-protein interactions, but can also be used for identification of DNA binding proteins (e.g. transcription factors) as originally described by Singh et al. (1988) and as modified since (Messier et al. 1993). cDNA libraries may be propagated in *E. coli* then transferred into eukaryotic cells where they are screened for expression. Shuttle vectors permit growth in prokaryotic as well as eukaryotic cells. In this fashion, exciting screening strategies such as identification of expressed receptor functions have become possible (Spengler et al. 1993).

Subcloning of desired cDNA insert - excision vectors

An additional vector system that deserves special notice is the λZAP expression vector (Short and Sorge 1992, Snead et al. 1997). The central distinguishing feature of this class of vectors, called excision vectors, is the ability to automatically excise the cDNA insert in the host cell in vivo and to convert it into so-called phagemid sequences. This in vivo excision eliminates the need to subclone cDNA inserts and significantly reduces the time and effort required for their analysis. Briefly, the principle is as follows: cDNA sequences cloned into the λZAP bacteriophage are flanked by the initiator and terminator sequences derived from the f1 filamentous phage origin. Upon coinfection with M13 helper phage the M13 gene II protein is expressed which recognizes the initiator sequence of the f1 origin of replication. Subsequently single-stranded DNA is synthesized from the initiator to the terminator including the cDNA insert, and is then circularized. The DNA circle is packaged as filamentous phage particles using structural proteins provided by the M13 helper phage. The packaged phagemids are extruded into the culture supernatant where they and can infect *E.coli* and are propagated as plasmids. Therefore the λZAP expression vector system combines the advantages of lambda phage screening with the ease of manipulating plasmid vectors.

Subprotocol 1
Plating of a λgt11 cDNA Library

Materials

Media

- LB (Luria-Bertani) medium (per 1 liter):
  10 g Bacto-tryptone
  5 g yeast extract
  10 g NaCl
  adjust pH with NaOH to 7.5; autoclave immediately
- LB plates: Add 7.5 g Bacto-agar (DIFCO) to 0.5 liter LB medium and autoclave. Cool hot agar solution by placing in a water bath until the temperature is below 60°C. Alternatively store agar solutions on the shelf at room temperature and remelt (melting temperature of agar: 47°C) in the microwave oven before use. Pour melted agar into petri dishes (0.5 liter agar solution for 20 to 25 90 mm dishes). Cool the plates in a laminar flow work bench for at least 30 minutes. Let plates dry further by leaving them at room temperature overnight. Dry plates are stored in the cold room at 4°C until used.
- LB top-agar: 0.72 to 0.75% (w/v) Bacto-agar (DIFCO) in LB medium
- λgt11 cDNA library: titrated in λ-dil
- λ - dil: 10 mM Tris-Cl, pH 7.5; 10 mM $MgSO_4$; autoclaved

Procedure

Preparation of host bacteria

1. 2 x 5 ml LB-medium are each inoculated with a single colony of *E. coli* host strain (Y1088 for plaque-hybridization screening, Y1090 for immuno-screening) and incubated at 37°C overnight with constant agitation.
2. Centrifuge overnight cultures for 5 minutes at 3,000 x g
3. Resuspend cell pellets in 2 ml λ - dil each.
4. Cell suspensions can be stored at 4°C for up to one week.

Preparation of ready to plate phage suspensions

Starting with cDNA libraries or other phage suspensions of known titer (i.e., known number of phage particles per ml), dilutions in λ-dil are prepared to achieve the desired number of phage per plate. For initial screening 5 x 10^3 phages per 100 µl are ideal for each 90 mm plate.

Plating

1. The required number of LB plates are placed in a 42°C incubator for prewarming.
2. LB top-agar (minimum of 2.5 ml for each plate) is melted in a microwave oven and then placed in a water bath at 49°C.
3. For each of the plates in the incubator one 12 ml screw cap tube is placed in a suitable rack at room temperature.
4. 100 µl of the host bacteria suspension in λ-dil is pipetted into each of the screw cap tubes.
5. 100 µl of diluted phage suspension is added to each tube and mixed with the bacteria by brief vortexing.
6. Tubes containing bacteria and phages (infection mixes) are incubated at 37°C for 25 minutes, then placed at room temperature.
7. Using a sterile 10 ml glass pipette which has been briefly prewarmed in the flame of a Bunsen burner, 2.5 ml melted top-agar is pipetted from the stock at 49°C to one of the infection mixes.
8. The contents of the tube are immediately mixed by rolling several times between the palms and are then poured evenly onto a prewarmed LB plate.
9. The poured plate is left to stand upright on a flat surface at room temperature until the top-agar has solidified (at least 5 minutes).
10. Steps 7 to 9 are repeated for each of the infection mixes until all of the mixtures of bacteria and phages are plated.
11. Plates are placed in an incubator at 42°C (for λgt11) until plaques develop.

Subprotocol 2
Hybridization Screening - cDNA Sequences as Probes

Materials

Solutions

- Nitrocellulose membranes: Schleicher & Schüll (82 mm; 45 μm)
- Denaturation solution: 0.5 M NaOH; 1.5 M NaCl
- Neutralization solution: 0.5 M Tris-Cl (pH 7.5); 1.5 M NaCl
- SSC: 3 M NaCl; 0.3 M Na-citrate; autoclaved
- Sodium dodecyl sulfate (SDS): 20%
- Prewash solution: 50 mM Tris-Cl (pH 8.0); 1 M NaCl; 1 mM EDTA; 0.1% SDS
- 20 x SSPE: 3 M NaCl; 0.2 M NaH_2PO_4; 25mM EDTA (pH 7.4); autoclaved
- 100 x Denhardt's solution: 2% Ficoll; 2% Polyvinylpyrrolidone; 2% BSA; store at -20°C
- Calf thymus DNA (cT-DNA): 10 mg/ml; sonicated and stored at -20°C
- Hybridization solution: 5 x SSPE; 1 x Denhardt's; 100 μg/ml cT-DNA; 0.1% SDS
- Radiolabeled ds cDNA probe: e.g. ^{32}P-labeled nick-translated cDNA probe with a specific activity of 5×10^7 to 2×10^8 cpm/μg DNA
- X-ray film: Kodak XAR5 35 x 43cm
- Whatman 3MM: filter paper

Procedure

1. Plate library as described above. For plaque hybridization experiments use *E. coli* host strain Y1088.
2. Plaques develop at 42°C overnight.
3. Remove plates from the incubator and cool at 4°C for at least 1 hr.
4. Membrane lifts: First, the dry nitrocellulose membrane is marked with the number of the plate that is to be replicated using a soft pencil.

Second, the dry filter membrane is carefully placed onto the bacterial lawn starting from the center of the plate towards the rim. From the time when the membrane has become completely wet (color turns gray) wait another 30 seconds (1.5 minutes in case of a second lift). Meanwhile, mark the position of the membrane on the plate by three asymmetrically placed needle sticks. Carefully remove the filter from the plate and place contact side up on a clean sheet of Whatman 3MM. Step 4 is repeated for all of the plates before proceeding to the next step.

5. Float membranes contact side up on top of denaturation solution for 5 minutes.
6. Neutralize for 5 minutes.
7. Wash in 2 x SSC for 5 minutes.
8. Place membranes on a fresh sheet of Whatman 3MM and wrap in aluminum foil together with the supporting paper.
9. Bake for 2 hrs at 80°C in a vacuum oven.
10. Briefly prewet membranes in 6 x SSC then prewash in prewashing solution at 68°C for at least 1 hr. Use a tray placed in a shaking water bath.
11. For prehybridization, 25 filter membranes are transferred into a preserving jar (90 mm at least) filled with 50 ml of hybridization solution. Prehybridization is performed at 68°C for 2 hrs.
12. For each ml of the hybridization solution 2 times10^5 to $1x10^6$ cpm of double-stranded probe are denatured by boiling in a water bath for 10 minutes followed by rapid cooling on ice.
13. The denatured probe is then immediately added to the prehybridization mix.
14. Hybridization is performed at 68°C overnight in the sealed preserving jar under constant agitation to prevent adherence of filter membranes to each other.
15. After hybridization, filter membranes are washed 3 times for 1 hr in 2 x SSC, 0.01% SDS at room temperature.
16. Nitrocellulose membranes are then mounted while still damp (keep from drying!) onto sheets of stiff paper or X-ray film and the needle marks are labeled with radioactive ink.

17. Filters are wrapped in plastic wrap (e.g. Saran) then exposed with intensifying screen to X-ray film overnight at -70°C.

Subprotocol 3
Immuno-Screening

Materials

Buffer

- Coupling buffer: 0.1 M $NaHCO_3$; 0.5 M NaCl
- HCl: 0.001 M
- Saturation buffer: 0.1 M glycine; 0.1 M $NaHCO_3$; 0.5 M NaCl
- Washing buffer (pH 4.0): 0.1 M Na-acetate; 1 M NaCl
- Washing buffer (pH 8.0): 0.1 M boric acid; 1 M NaCl
- PBS: 0.1 M NaCl; 2.6 mM KCl; 6.4 mM Na_2HPO_4; 1.4 mM KH_2PO_4; (pH 7.0)
- 10 x TBS: 0.5 M Tris-Cl (pH 8.0); 1.5 M NaCl
- TBT: TBS + 1% (w/v) BSA; 0.1% (v/v) Triton X-100; store at 4°C
- Enzyme-reaction buffer: 24.3 ml 0.1 M citric acid; 25.7 ml 0.2 M Na_2HPO_4; 50 ml H_2O; pH adjusted to 5.0; store at 4°C
- CNBr-activated sepharose 4B: (Pharmacia) 5 g (1 g 3.5 ml gel) are preswollen on a glass sinter filter using cold (4°C) HCl and are then washed with total 1 liter cold HCl.
- Isopropyl-β-D-thiogalacto-pyranoside (IPTG): 16 mM in H_2O, sterile filtered
- 5-bromo-4-chloro-3-indolyl-3-D-galactopyranoside (X-Gal): 2% (w/v) in dimethylformamide
- Thimerosal: 1%
- 4-chloronaphthol: 0.3% (w/v) in methanol; store at -20°C
- H_2O_2: 30%

Host strains

- Lysogenic *E. coli* strain: BTA282(λgt11amp3) or BNN97
- *E. coli* host strain: Y1088

Procedure

Preparation of Sepharose-coupled bacterial lysate

1. 2 x 7.5 ml LB media are each inoculated with a single colony of the lysogenic temperature-sensitive *E. coli* strain and grown overnight at 32°C.
2. The overnight cultures are diluted 100-fold in LB and grown at 32°C to an OD_{600} of 0.5.
3. Phage expression is induced by incubation at 45°C for 15 minute.
4. Further incubate cultures at 39°C for approx. 2 hrs.
5. To test phage content and readiness for harvesting, $CHCl_3$-tests are performed: 3 drops of $CHCl_3$ are added to a 1 ml test sample of culture and the mixture is then incubated at 37°C. A control aliquot of culture to which no $CHCl_3$ had been added is incubated in parallel. After 5 minutes of incubation the test sample is compared to the control. If the tested cell suspension has cleared upon addition of $CHCl_3$ the bacteria are fully packed with phage and just about to lyse. At this stage the culture is ready for harvesting.
6. Harvest bacteria by centrifugation at 4,000 x g for 10 minutes at 4°C.
7. Cell pellets from both original 7.5 ml cultures are resuspended separately in cold coupling buffer and respun.

Note: coupling buffer must be cold (4°C), otherwise premature lysis of cells will occur.

8. Cells are resuspended in each 5 ml cold coupling buffer and pooled.
9. Sonicate cell suspension until the increased viscosity caused by liberation of nucleic acids disappears.
10. Couple the bacterial lysate to the previously prepared CNBr-activated Sepharose 4B overnight at 4°C.
11. Recover the coupled material by centrifgation at 3,000 x g for 5 minutes.
12. Wash pellet once by resuspension in coupling buffer and recentrifugation.
13. For saturation of unoccupied reactive groups of the CNBr-activated Sepharose the pellet is resuspended in cold (4°C) saturation buffer and incubated for 1 hr at 4°C on a rotating wheel.

14. Non-adsorbed proteins are removed by three successive washing cycles, starting with washing buffer (pH 4.0), then switching to pH 8.0 and a final wash at pH 4.0. Before and after this treatment and also in between the individual washing steps the material is recovered by centrifugation at 2'500 x g for 5 minutes at 4°C.

15. The Sepharose-coupled bacterial lysate can be stored as a suspension in PBS (+0.01% thimerosal) at 4°C for several weeks.

Preadsorbtion of the antibodies used for screening

1. The undiluted primary antibody as well as the secondary antibody (e.g. peroxidase-coupled goat anti-rabbit, Calbiochem) at a protein concentration of 0.5 to 1 mg/ml are each mixed with 1.7 volumes of Sepharose-coupled bacterial lysate and incubated overnight at 4°C on a rotating wheel. For each screening cycle approx. 300 µl of treated primary antibody (final working dilution: 1:100) and 15 µl of treated secondary antibody (final working dilution: 1:2000) are required.

2. The suspensions are filled into separate pipette tips that are plugged with a small amount of glass wool. Non-retarded antibodies are eluted with cold blocking buffer (at least 5 x times the column volume).

3. The concentration of the eluted antibody solutions is adjusted with cold blocking buffer to 1/20 of the original concentration. Thimerosal is added (0.01% final).

Note: Thimerosal is used instead of Na-azide, if mold contamination of peroxidase-conjugated antibody solutions is to be prevented. Antibody solutions are now ready to use and can be stored at 4°C.

Plating

1. Plating is performed as described above, with the following changes:
   a. One additional 12 ml screw-cap tube containing 2.5 ml top-agar is placed in the water bath at 49°C for plating of an induction control.
   b. In one infection mix, 100 µl of a phage suspension containing a known percentage of wild-type phage with an intact β-galactosidase gene is used to serve as an induction control.
   c. For plating of the induction control 20 µl X-Gal is added to the 2.5 ml of top-agar immediately before adding to the corresponding infection mix.

2. Incubation at 42°C is performed for 3-4 hrs (using *E. coli* Y1090) until plaques become visible.

3. Plates are removed to a laminar flow hood where they are overlaid with a nitrocellulose membrane that has been soaked in IPTG to induce expression from the phage lac Z gene. For each of the plates, a dry nitrocellulose membrane is labeled with a soft pencil then soaked in the IPTG solution in a petri dish. Excess fluid is at the rim of the dish drained off and the membrane is then carefully lowered onto the bacteria lawn starting from the center of the plate towards the rim. Filter membranes are always handled with flat forceps.

4. Plates are placed inverted in an incubator overnight at 37°C.

5. Some plaques on the induction control plate will turn blue overnight, due to expression of an intact β-galactosidase gene, indicating successful IPTG induction.

6. Before lifting the nitrocellulose membranes from the plates the position of the membrane should be marked with three asymmetrically placed needle sticks.

7. Nitrocellulose membranes are then carefully lifted off the plates and placed contact side up on Whatman 3MM until all membranes have been removed.

Immunological staining

1. In order to wash off top-agar residue and cell debris a maximum of 5 nitrocellulose membranes at a time are transferred into a tray (10 x 10 cm) placed on a shaker and washed with 25 ml TBS for 2 x 10 minutes at room temperature.

2. Non-specific binding sites on the filter membranes are then blocked by incubating the filters for 30 minutes in 25 ml blocking buffer (at least 5 ml per one 90 mm membrane) at room temperature. This step is performed with constant agitation to prevent filters from adhering to each other.

Primary antibody incubation

3. Each membrane is placed in a petri dish and individually incubated with 3 ml blocking buffer containing the preadsorbed primary antibody. A typical working dilution of a polyclonal primary antibody is 1:100. Such working dilutions of antibody can be reused up to 5 times and are stored at 4°C. Incubation with primary antibody is performed at

room temperature for at least 2 hrs. To ensure that the filters remain completely soaked with antibody solution, the petri dishes are placed on a rocking platform during incubation.

4. Unbound primary antibodies are removed by washing the membranes (5 at a time) for 3 x 10 minutes in 25 ml TBT at room temperature while shaking.

Secondary antibody incubation

5. This step is performed as in step 3. The working dilution of the preadsorbed secondary antibody typically is approx. 1:2000 in 3 ml blocking buffer (e.g. peroxidase-coupled goat anti-rabbit, Calbiochem). Secondary antibody working solutions should not be reused more than once. After the second usage, the capacity of the secondary antibody working solution seems to be exhausted and often results in a sudden loss of positive signals.

6. Unbound secondary antibodies are removed by washing the membranes 3 x 10 minutes in 25 ml TBS at room temperature while shaking.

Color development for peroxidase conjugated secondary antibodies

7. In a plastic tray 45 ml cold (4°C) enzyme reaction buffer are mixed with 5 ml cold (-20°C) chloronaphthol and 50 µl H_2O_2. A maximum of 5 membrane filters at a time are placed into this color development solution and gently agitated. Within several minutes positive purple plaques should develop.

8. Nitrocellulose filters are placed on a light box, if results are documented by Polaroid photography (film type 665; f-stop 4.5; shutter speed 1/15; orange filter).

9. To assist subsequent picking of positive plaques, it is convenient to copy the pattern of positive plaques with the corresponding needle stick marks onto a transparent overhead film.

Plaque purification

1. Areas corresponding to positive plaque signals are punched out of the top-agar using the wide end of a sterile Pasteur pipette. The thin top-agar disk usually comes easily off the plate and can be suspended into 0.5 ml of λ-dil in an eppendorf tube.

2. Phage particle contained in the punched-out plaques are released from top-agar by vigorous vortexing for at least 1 min. The newly created phage suspension is placed for at least 1 hr at room temperature before plating again.

3. Typically dilutions in the range of 10^{-3} to 10^{-5} of this phage suspension are retransformed into the bacterial host strain and rescreened at lower plaque density for positive signals either with radiolabeled probes or antibodies.

4. Steps 1. to 3. have to be repeated until positive plaques can be punched out of the top agar as single plaques derived from one single original phage infection.

Phage propagation

1. To produce homogeneous phage suspensions of high titer, 100 µl to 250 µl of the individual plaque phage suspensions are plated in *E. coli* Y1088 for amplification.

Note: $CHCl_3$ must be removed from phage suspensions when plating dilutions less than 10^{-2}, since growth of host bacteria is severely inhibited by traces of $CHCl_3$. To remove $CHCl_3$, blow a gentle stream of N_2 into the phage suspension.

2. After overnight incubation at 42°C, one confluently lysed plate per phage clone is overlaid with 7 ml λ-dil, and phage particles are extracted by constant shaking for 5 hrs at room temperature.

3. Supernatants containing phage are removed with a sterile 10 ml pipette while carefully avoiding cellular debris. A few drops of $CHCl_3$ are added, and the supernatants are stored as phage stock solutions at 4°C indefinitely.

4. Phage stock titers should be in the range of 10^9 to 10^{10} phage particles per ml suspension. If a phage stock solution does not contain enough phage, steps 1 to 3 are repeated until sufficient phage titers are achieved.

5. The homogeneity of the phage stocks should be tested by replating a highly diluted aliquot of the stock (several hundred phages at a maximum) and rescreening with the probe. At least 99% of the tested plaques should be positive to consider the tested phage stock "homogeneous".

Subprotocol 4
Purification of Phage Particles, DNA Isolation and Subcloning

Materials

- SM (without gelatin): 0.1 M NaCl; 8 mM $MgSO_4$; 50 mM Tris-Cl (pH 7.5); autoclaved
- RNase A: 1 mg/ml in H_2O
- DNase I: 1 mg/ml in H_2O
- CsCl: Solutions of different densities:ρ=1.45, 1.5, and 1.7
- Solutions for balancing: 5 ml SM + 2.5 g CsCl, 5 ml SM + 3.75 g CsCl
- Refractometer
- Polyethylene glycol (PEG): 6000 (SERVA)

Procedure

1. For each of the phage to be prepared, 10 LB plates with at least 10^5 phage are plated in *E. coli* Y1088 and incubated at 42°C overnight.
2. Supernatants containing phage are prepared as described previously, and the supernatants from all 10 plates (10 x 5 ml) are pooled.
3. Each pool is divided into two 25 ml aliquots in 40 ml polypropylene centrifuge tubes.
4. Bacterial nucleic acid is digested completely by addition of 25 µl RNase A and DNase I, then incubation at room temperature for 30 min.
5. 1.46 g NaCl is added per tube and dissolved completely. After 1 hr at room temperature, cell debris is removed by centrifugation at 10,000 rpm for 10 minutes at 4°C.
6. Supernatants are transferred into a fresh 40 ml polypropylene tube and 2.5 g PEG is added and completely dissolved by constant shaking for 30 min at room temperature. Phage particles are PEG-precipitated overnight on ice.
7. Phage precipitates are recovered by centrifugation at 10,000 rpm for 10 min at 4°C.

8. Supernatants are discarded and the inner walls of the tubes are carefully wiped out using a soft paper.

9. The two pellets derived from the same phage clone are resuspended and pooled in a total of 4 ml of SM. The milky pellet material is homogenized by pipetting up and down with a 5 ml pipette.

10. Extract PEG by adding 4 ml $CHCl_3$ and vortex for 30 sec; centrifuge at 1,600 g for 15 min at 4°C.

11. The aqueous (top) phase is transferred into a fresh 12 ml polypropylene tube, and the volume is adjusted with SM to 4 ml precisely. The jelly-like organic phase and the interphase are discarded.

12. In a 14 ml polyallomer ultracentrifuge tube (Kontron) a CsCl step gradient is prepared containing from bottom to top: 3 ml CsCl with a density of ρ=1.7; 1.5 ml (ρ=1.5); and 1.5 ml (ρ=1.45).

13. 2 g solid CsCl are dissolved in each of the 4 ml phage suspensions in SM.

14. The phage suspensions are layered onto the CsCl step gradient and centrifuged at 30,000 rpm for 2 hrs at 15°C in a Beckman SW40 rotor or equivalent.

15. The gradients are carefully removed from the rotor and placed under a strong lamp. At this point phage particles are clearly visible as a milky band at the interphase between density zones 1.5 and 1.45. The phage particle are harvested using an 18G needle punctured through the side of the tube. Volume is approx. 1.5 ml.

16. The volume of the harvested sample is brought to 6 ml with SM, and solid CsCl (usually approx. 3 g) is added until the refractive index of the solution is precisely 1.380. The volume of the sample is then further expanded with CsCl solution (0.75g/ml) to a final volume of 9.5 ml.

17. Phage particles are re-centrifugated through a continuous density gradient (e.g. 30,000 rpm in a Beckman Ti75 rotor overnight at 20°C).

18. Phage particles are collected as in step 15 and are ready for DNA extraction by standard procedures (Maniatis et al. 1989).

References

1. Fields S, Sternglanz R (1994) The two-hybrid system: an assay for protein-protein interactions (Review). Trends in Genetics 10:286-292
2. Helfman DM, Hughes SH (1987) Use of antibodies to screen cDNA expression libraries prepared in plasmid vectors. In: Berger SL, Kimmel AR (eds) Guide to Molecular Cloning Techniques. Methods in Enzymology, vol 152. Academic Press, London, pp 451-457
3. Hogrefe HH, Amberg J, Hay B, Sorge JA, Shopes B, (1993) Cloning in a bacteriophage lambda vector for the display of binding proteins on filamentous phage. Gene 137:85-91
4. Höltke HJ, Ankenbauer W, Muhlegger K, Rein R, Sagner G, Seibl R, Walter T (1995) The digoxigenin (DIG) system for non-radioactive labelling and detection of nucleic acids – an overview. Cellular & Molecular Biology 41:883-905
5. Huse WD, Sastry L, Iverson SA, Kang AS, Alting-Mees M, Burton DR, Benkovic SJ, Lerner RA (1989) Generation of a large combinatorial library of the immunoglobulin repertoire in phage lambda. Science 240:1275-1281
6. Jendrisak J, Young RA, Engel D (1987) Cloning cDNA into λgt10 and λgt11. In: Berger SL, Kimmel AR (eds) Guide to Molecular Cloning Techniques. Methods in Enzymology, vol 152. Academic Press, London, pp 359-371
7. Kato S, Sekine S, Oh SW, Kim NS, Umezawa Y, Abe N, Yokoyama-Kobayashi M, Aoki T (1994) Construction of a human full-length cDNA bank. Gene 150:243-250
8. Kimmel AR, Berger SL (1987) Preparation of cDNA and the generation of cDNA libraries: overview. In: Berger SL, Kimmel AR (eds) Guide to Molecular Cloning Techniques. Methods in Enzymology, vol 152. Academic Press, London, pp 307-316
9. Lee SW, Tomasetto C, Sager R (1991) Positive selection of candidate tumor-suppressor genes by subtractive hybridization. Proc Natl Acad Sci USA 88:2825-2829
10. Liang P, Pardee AB (1992) Differential display of eukaryotic messenger RNA by means of the polymerase chain reaction. Science 257:967-971
11. Maniatis T, Fritsch E, Sambrook J (1989) Molecular cloning: a laboratory manual. 2nd ed. Cold Spring Harbor Laboratory, New York
12. Messier H, Brickner H, Gaikwad J, Fotedar A (1993) A novel POU domain protein which binds to the T-cell receptor β enhancer. Mol Cell Biol 13:5450-5460
13. Mierendorf RC, Percy C, Young RA (1987) Gene isolation by screening λgt11 libraries with antibodies. In: Berger SL, Kimmel AR (eds) Guide to Molecular Cloning Techniques. Methods in Enzymology, vol 152. Academic Press, London, pp 458-469
14. Paddock GV (1993) Rapid colony hybridization on Whatman 541 paper using oligonucleotide probes. In: Wu R (ed) Recombinant DNA Part H. Methods in Enzymology, vol 217. Academic Press, London, pp 340-346
15. Short JM, Sorge JA (1992) In vivo excision properties of bacteriophage λ ZAP expression vectors. In: Wu R (ed) Recombinant DNA Part G. Methods in Enzymology, vol 216. Academic Press, London, pp 495-509
16. Sikela JM, Hahn W (1987) Screening an expression library with a ligand probe: isolation and sequence of a cDNA corresponding to a brain calmodulin-binding protein. Proc Natl Acad Sci USA 84:3038-3042
17. Singh H, LeBowitz JH, Baldwin Jr. AS, Sharp PA (1988) Molecular cloning of an enhancer binding protein: isolation by screening of an expression library with a recognition site DNA. Cell 52:415-423

18. Snead MA, Alting-Mees MA, Short JM (1997) Clone excision methods for the Lambda ZAP-based vectors. Methods in Molecular Biology 69:53-60
19. Spengler D, Waeber C, Pantaloni C, Holsboer F, Bockaert J, Seeburg PH, Journot L (1993) Differential signal transduction by five splice variants of the PACAP receptor. Nature 365:170-175
20. Stanley KK (1983) Solubilization and immune-detection of β-galactosidase hybrid proteins carrying foreign antigenic determinants. Nucleic Acids Res 12:4077-4092
21. Wahl GM, Berger SL (1987) Screening colonies or plaques with radioactive nucleic acid probes. In: Berger SL, Kimmel AR (eds) Guide to Molecular Cloning Techniques. Methods in Enzymology, vol 152. Academic Press, London, pp 415-423
22. Wahl GM, Berger SL, Kimmel AR (1987) Molecular hybridization of immobilized nucleic acids: theoretical concepts and practical considerations. In: Berger SL, Kimmel AR (eds) Guide to Molecular Cloning Techniques. Methods in Enzymology, vol 152. Academic Press, London, pp 399-407
23. Wallace RB, Miyada CG (1987) Oligonucleotide probes for the screening of recombinant DNA libraries. In: Berger SL, Kimmel AR (eds) Guide to Molecular Cloning Techniques. Methods in Enzymology, vol 152. Academic Press, London, pp 432-442
24. Watson JD, Gilman M, Witkowski J, Zoller M (1992) Recombinant DNA. 2nd ed Freeman & Co, New York
25. Young RA, Davis RW (1983a) Efficient isolation of genes by using antibody probes. Proc Natl Acad Sci USA 80:1194-1198
26. Young RA, Davis RW (1983b) Yeast RNA polymerase II genes: isolation with antibody probes. Science 222:778-782

Chapter 20

Expression Cloning with *Xenopus* Oocytes

DANIEL MARKOVICH, ANDREAS WERNER AND HEINI MURER

Introduction

Expression cloning using *Xenopus laevis* oocytes has been proven to be an excellent tool for the structural/functional identification of proteins of all origins. Due to their great availability, large size and relative ease of handling, *X. laevis* oocytes are optimal tools for the expression and cloning of proteins, when compared to traditional expression systems, such as *Escherichia coli* or eukaryotic cell lines. The *Xenopus* oocyte system, pioneered in 1971 (Gurdon et al. 1971), is able to efficiently transcribe and translate injected genetic information; perform assembly of the foreign protein products; correctly process the nascent polypeptides; and target them to the proper subcellular compartment. Some advantages of the *Xenopus* oocyte expression system over other functional expression systems (e.g. in somatic cells) are: easy and rapid transfer of genetic information by microinjection; simple handling of single cells after transfer of genetic information; high proportion of cells expressing transferred genetic information and good control of the oocyte environment. In this section, we will describe a general strategy which can be successfully used for expression cloning of proteins using *Xenopus* oocytes. The manipulation of oocytes, protein assay (in our case transport measurements), cDNA library screening and cloning by expression, will be covered in detail below.

Correspondence to Daniel Markovich, University of Queensland, Department of Physiology and Pharmacology, Brisbane, 4072, Australia (*phone* +61-7-3365-1400; *fax* +61-7-3365-1766; *e-mail* danielm@plpk.uq.edu.au), Andreas Werner, Max-Planck-Institut für molekulare Physiologie, Rheinlanddamm 201, Dortmund, 44139, Germany, Heini Murer, Universität Zürich, Physiologisches Institut, Winterthurerstrasse 190, Zürich, 8057, Switzerland

Materials

Equipment

- Semiautomatic injector (e.g. Inject+Matic System. J.A. Gabay, Switzerland or Nano-ject microinjector, Drummond, USA)
- Luria Broth plates (LB), 9 cm
- Nitrocellulose or nylon filters (Schleicher and Schuell)
- Sephadex G-50 Quick Spin columns, RNase-free (Boehringer Mannheim or self made)

Reagents

- $^{35}SO_4^{2-}$, 20-40 µCi/ml
- K_2SO_4
- 100 mM NaCl
- 100 mM choline chloride
- 10% SDS
- Scintillation fluid (e.g. Emulsifier Safe, Kontron)
- Ampicillin 100 µg/ml
- ammonium acetate 5 M
- ethanol
- 1% agarose formaldehyde RNA

Solutions

- ethyl m-aminobenzoate solution (1 g/liter in water)
- ORII solution: 82.5 mM NaCl, 2 mM KCl, 1 mM $MgCl_2$, 10 mM Hepes adjusted with Tris to pH 7.5
- modified Barth's solution (MBS): 88 mM NaCl, 1 mM KCl, 0.82 mM $MgSO_4$, 0.41 mM $CaCl_2$, 0.33 mM $Ca(NO_3)2$, 2.4 mM $NaHCO_3$, 10 mM Hepes/Tris (pH 7.4), gentamicin sulfate 20 mg/l
- Solution A: 100 mM Choline chloride, 2 mM KCl, 1 mM $CaCl_2$, 1 mM $MgCl_2$, 10 mM Hepes/Tris (pH 7.5)
- Transcription buffer 1 ×: 40 mM Tris-HCl (pH 7.9), 2 mM spermidine and 6 mM $MgCl_2$; supplied with polymerase (Promega)
- Transcription mixture (final concentrations): transcription buffer 1 ×; 0.5 mM ATP; 0.5 mM CTP; 0.5 mM UTP; 0.5 mM m7G(5')ppp(5')G (RNA cap analog; New England Biolabs); 0.1 mM GTP; 10 mM DTT; 50 units

RNAase inhibitor (RNA guard Pharmacia); 50 units T3 or T7 RNA polymerase (Promega).

- phenol:chloroform:isoamylalcohol (25:24:1)
- collagenase A (or D), low trypsin activity (Boehringer, Mannheim; 2 mg/ml)
- mRNA (of choice)
- Restriction enzyme cutting at the 3' end of the insert (e.g.NotI)
- RNAase inhibitor
- DNAase I, RNAase free (Pharmacia)

Kits

- SuperScript plasmid system, pSPORT-1 vector (Gibco/BRL)

Procedure

Toads, oocytes and injections

1. For removal of oocytes, a *Xenopus* toad is anaesthetized for 15-30 min in half a liter of ethyl m-aminobenzoate solution.
2. The toad is then placed on top of crushed ice covered with wet tissue paper, to avoid the skin from drying out. A small incision is made in the upper skin of about 1 cm length, and a 0.5 cm cut in the inner skin and the muscular layer.
3. Using forceps, small clumps of oocytes (depending on the amount required, usually 500-1500 oocytes) are removed and are treated twice for 90 minutes in collagenase A (or D), having low trypsin activity in a calcium free ORII solution, in order to remove follicular layers. Typically, 10 ml of collagenase solution is placed in a sterile 15 ml tube, and after cutting the ovarian lobes into small clusters of oocytes, up to 3 ml of oocytes are added per tube.
4. The tube is rotated in the dark at 18°C on an orbital rotor spinning at low speed (20 rpm) for 90 min. and then replaced by a new collagenase-ORII solution for another 90 min.
5. The oocytes are then washed thoroughly 5 times with ORII solution, followed by 5 times with modified Barth's solution.

6. The oocytes are sorted for morphologically intact, healthy looking stage V-VI oocytes. The oocytes are subsequently maintained in MBS at 18 °C (kept in the dark).
7. Usually on the next day, healthy looking stage V-VI oocytes are injected with either 50 nl water (control) or mRNA (at concentrations varying from 0 to 1 µg/µl), using a semiautomatic injector. Oocytes are then kept at 18°C in MBS for 1 to 7 days, with changes of MBS solution every 1-2 days.

Protein expression (transport) assay

1. Oocytes (7-10 oocytes per individual data point) are first washed for 1-2 minutes in solution A. This solution is then replaced by 100 µl of solution A supplemented with the desired concentration of substrate (eg. 1 mM K_2SO_4), at the desired specific activity for the isotope. The specific activities for the different substrates used (eg. $^{35}SO_4^{2-}$) are usually between 20-40 µCi/ml. Uptakes are routinely performed in the presence of sodium (100 mM NaCl) or in its absence (100 mM choline chloride) and incubated in a water bath at 25°C for various times (1-60 min.).
2. After the incubation, the uptake solution is removed and the oocytes are washed 3 times with 3 ml of ice cold stop solution (solution A).
3. Each single oocyte is then placed into a separate scintillation vial, dissolved in 200 µl of 1% SDS (allowed to shake vigorously for 45 min at room temp).
4. When the ooyctes have completely dissolved, they are supplemented by 2 ml scintillation fluid (Emulsifier Safe, Kontron). Isotopic uptake (transport assay) is measured by liquid scintillation spectrometry (normally counted for 2 min/vial).

Construction and screening of a cDNA expression library

1. Using a functionally positive unfractionated or size-fractionated mRNA as starting material, a directional cDNA library is constructed using a commercial kit, following the manufacturers protocol (in our case the SuperScript plasmid system, pSPORT-1 vector, Gibco/BRL; SuperScript reverse transcriptase having no RNase H activity). The total library, usually in the range of 10^5-10^6 clones, is subdivided into pools of

1,000 colonies, which are plated onto 9 cm Luria Broth plates (containing ampicillin 100 μg/ml).

2. When the colonies are visible, two replicas are made onto nitrocellulose or nylon filters with both replica filters being placed on top of LB plates (colonies side up) for about 6 hours until colonies are visible (1 mm diameter).

3. One filter is placed on top of an agar/glycerol plate, with the colonies side up and kept as a master plate at -80°C (sealed with parafilm). The second replica is submerged into 10ml Luria Broth containing ampicillin (100 μg/ml) and grown overnight at 37°C with shaking (300 rpm).

4. The cells are pelleted by centrifugation for 1 min and plasmids are isolated using a general plasmid (alkaline lysis) method. About 1-2 μg of plasmid are digested with an appropriate restriction enzyme cutting at the 3' end of the insert (in our case NotI). An aliquot is run through an agarose gel to check if the digestion was complete and the rest is extracted twice with a mixture of 0.1 M Tris pH 8 saturated phenol: chloroform: isoamylalcohol (25:24:1) and then precipitated with 0.1 volume 3 M sodium acetate and 2.5 volume cold 100% ethanol at -20°C for 2 hours. Plasmids are then dissolved in 10 μl of water and used for in vitro transcription, using the appropriate RNA polymerase (in our case T7).

In vitro cRNA-synthesis, including capping

5. For in vitro transcription, the whole 10 μl (approx. 1 μg) of linearized plasmid DNA is used and is added to 40 μl of the transcription mixture. The reaction is incubated at 37°C for 1 hour.

6. Then 50 units of RNAase inhibitor and 10 units of DNAase I, RNAase free, are added to the samples and incubated for 15 minutes at 37°C to digest the remaining DNA.

7. The samples are then extracted twice with phenol:chloroform:isoamylalcohol (25:24:1) and precipitated by adding one volume of ammonium acetate 5 M and 2.5 volumes of ethanol, in order to remove free nucleotides.
Alternatively, after the extractions with phenol: chloroform:isoamyl alcohol, samples can be passed through RNase-free Sephadex G-50 Quick Spin columns to remove free nucleotides, then precipitated with sodium acetate and ethanol as described above.

8. The samples are precipitated at -80°C for at least 3 hours and then are centrifuged for 45 min at 4°C, pellets rinsed with 70% ethanol and vacuum dried for 5 min.

9. The pellet is resuspended in 15μl of water. 1 μl is used to measure the RNA concentration spectrophotometrically at OD_{260} and 1-2 μl is run through a 1% agarose formaldehyde RNA gel, to check for integrity of the transcripts, with the rest being used for injection.
10. The in vitro transcribed RNA (cRNA) is directly injected into oocytes and after 1-6 days post injection, the protein (transport) assay can be measured (as described above).

Troubleshooting

Upon successful construction of a relatively large cDNA library (›50,000 clones from fractionated mRNA), containing sufficiently large cDNA inserts (›1.5kb, checked on an agarose minigel), the library is divided into pools of 1000 colonies (more clones may diffuse the signal). Each plate of 1000 colonies is screened for expression (as described above) and once a positive pool of plasmids is found, the bacteria containing these plasmids are subdivided into smaller pools by one order of magnitude. These new pools are screened, as well as the master plate and then further subdivided by sib selection until a single clone, responsible for the increased expression in oocytes, is found. This part of the work can be very time consuming (several months) and is not always straightforward.

In order to simplify the work, sib selection is used to narrow down the positive plates and eventually find the positive clone. It is performed the following way: Assuming you want to screen 25 plates of 100 colonies each, the way you would do this is to place the plates into a grid of 5 × 5 plates, with 5 columns and 5 rows. The columns you can denote A, B, C, D, E and the rows F, G, H, I, J. Now column A will be made up of plates 1-5, column B of plates 6-10, column C of plates 11-15, etc. The products of rows and columns can now be pooled, either at the plasmid isolation level, or at the level of in vitro transcription, thus reducing the number of reactions from 25 to 10 groups. In addition, once a positive group is found (let's say column A and row F), then you can exactly pin-point the positive plate in one step, by looking at the positive column versus its intersecting positive row (in this case it would be plate 1). This sib selection technique can be applied at the single colony step or at the steps with tens or hundreds of colonies per plate.

Generally, pools of bacteria are only subdivided after the protein assay results are consistently positive (2 or more times). Typical expressed uptakes for different groups of colonies during screening are: master plates

(1000 colonies) show at least a 1.5-3-fold transport activity over background; progressive subdivision of one positive master plate down to plates of 100-200 produce a 5-8-fold increase in transport rate; subdivision down to plates of 10-30 colonies show transport signals of more than 10-fold over background; and finally down to 1 positive colony, stimulation up to 40-fold over background. It is most important, that with each subdivision of colonies, an increase in the stimulated transport rate is necessary before proceeding further. For a general overview used for expression cloning of transport proteins, see Figure 1.

Applications

Examples of expression cloning

Since the discovery of *Xenopus laevis* oocytes as an expression system, numerous laboratories, including our own, have applied this technology towards the cloning of membrane proteins. Below, we wish to list a few transport proteins that were isolated using the *Xenopus* oocyte expression cloning system: SGLT1, coding for an intestinal Na^+/glucose cotransport protein (Hediger et al, 1987); NaPi-1, encoding Na^+/phosphate cotransport (Werner et al. 1991); rBAT-1 and rBAT-2, encoding dibasic and neutral amino acid transport (Bertran et al. 1992; Wells and Hediger, 1992; Markovich et al. 1993b); NaPi-2 and NaPi-3, encoding Na^+/phosphate cotransport (Magagnin et al. 1993); NaSi-1, encoding Na^+/sulfate cotransport (Markovich et. al.,1993a); Na^+/bile acid (taurocholate) transport (Hagenbuch et al. 1992); sat-1, Na-independent sulfate transport (Bissig et al. 1993); GAT-1, encoding a transporter for the neurotransmitter gamma-aminobutyric acid (GABA) protein (Guastella et al. 1990); EAAC1, encoding a glutamate transporter (Kanai and Hediger, 1992).

Despite some difficulties which may be encountered initially with protein expression (depending on the substrate type and assay conditions), this method seems to be reliable in the isolation of functional proteins from all origins. It is currently being introduced and widely used in many laboratories. Expression cloning in *Xenopus* oocytes has proven extensively to be a rewarding technique for the isolation of novel (functional) proteins.

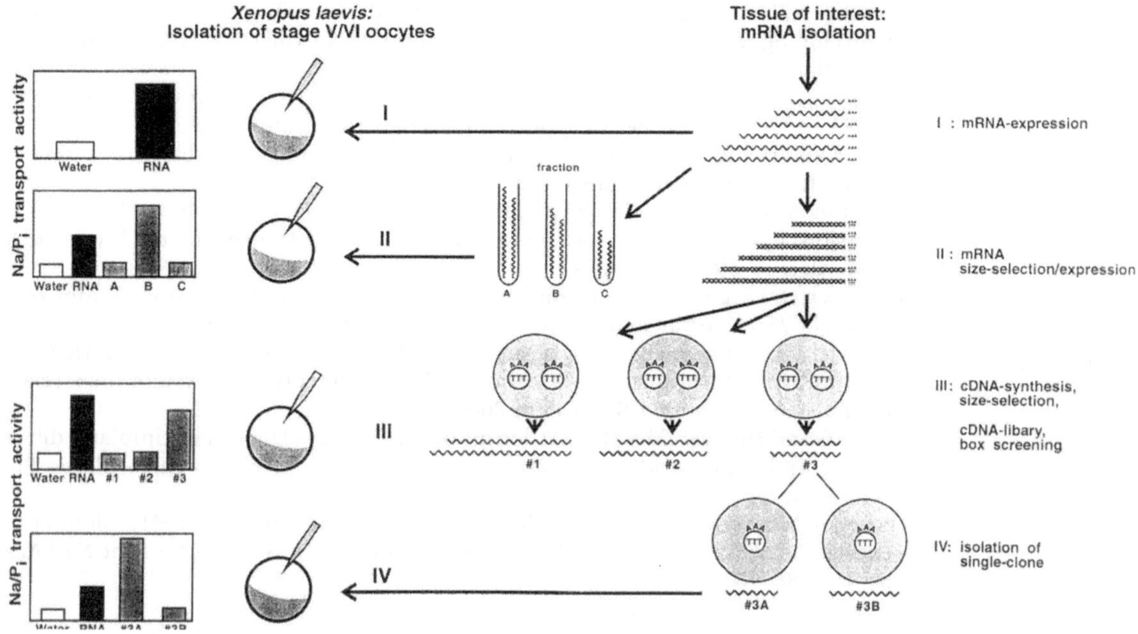

Fig. 1. Schematic outline of the cloning strategy, using functional expression in *Xenopus laevis* oocytes. The right panel shows the different cloning steps graphically. The left panel represents the assay with the expected results at the corresponding stage of the project. The strategy in brief: 1) Isolation of total poly $(A)^+$ RNA from rabbit renal cortex followed by injection and transport assay (I) in oocytes. 2) Fractionation of the mRNA, injection of the fractions and transport assay (II). 3) Construction of a cDNA library from rabbit renal cortex mRNA. Division of the library into different pools of clones. Synthesis of in vitro transcribed mRNA, injection and transport assay (III). 4) Subdivision of the pool(s) as described above down to one single clone (IV).

References

Bertran, J, Werner, A, Moore, ML, Stange, G, Markovich, D, Biber, J, Testar, X, Zorzano, A, Palacin, M and Murer, H (1992) Expression-cloning of a cDNA from rabbit kidney cortex that induces a single transport system for cystine, dibasic and neutral amino acids. Proc Natl Acad Sci USA 89: 5601-5605

Bissig, M, Hagenbuch, B, Stieger, B, Koller, T and Meier, PJ (1993) Functional expression cloning of the canalicular sulfate transport system of rat hepatocytes. Submitted to: J Biol Chem

Guastella, J, Nelson, N, Nelson, H, Czyzyk, L, Keynan, S, Miedel, MC, Davidson, N, Lester, HA and Kanner, BI (1990) Cloning and expression of a rat brain GABA transporter. Science 249: 1303-1306

Gurdon, JB, Lane, CD, Woodland, HR and Marbaix, G (1971) Use of frog eggs and oocytes for the study of messenger RNA and its translation in living cells. Nature 233: 177-182

Hagenbuch, B, Stieger, B, Foguet, M, Lübbert, H and Meier, PJ (1991) Functional expression cloning and characterization of the hepatocyte Na^+/bile acid cotransport system. Proc Natl Acad Sci USA 88: 10629-33

Hediger, MA, Coady, MJ, Ikeda, TS and Wright, EM (1987) Expression cloning and cDNA sequencing of the Na+/glucose cotransporter. Nature 330: 379-381

Kanai, Y and Hediger, MA (1992) Primary structure and functional characterization of a high affinity glutamate transporter. Nature 360: 467-471

Magagnin, S, Werner, A, Markovich, D, Sorribas, V, Biber, J and Murer, H (1993) Expression cloning of human and rat renal cortex Na/Pi-cotransport. Proc Natl Acad Sci USA 90: 5979-5983

Markovich D, Forgo J, Stange G, Biber J and Murer H (1993a) Expression cloning of rat renal Na^+/SO42cotransport. Proc Natl Acad Sci USA 90: 8073-8077

Markovich, D, Stange, G, Bertran, J, Palacin, M, Werner, A, Biber, J and Murer, H (1993b) Two mRNA transcripts (rBAT-1 and rBAT-2) are involved in system bo,+-related amino acid transport. J Biol Chem 268: 1362-1367

Wells, RG and Hediger, MA (1992) Cloning of a rat kidney cDNA that stimulates dibasic and neutral amino acid transport and has sequence similarity to glucosidases. Proc Natl Acad Sci USA 89: 5596-5600

Werner, A, Moore, ML, Mantei, N, Biber, J, Semenza, G and Murer, H (1991). Cloning and expression of a cDNA for a Na/Pi-cotransport system of kidney cortex Proc Natl Acad Sci USA 88: 9608-9612

Chapter 21

Transgenic Animals

ECKART SCHOTT, MARTIN PAUL AND DETLEV GANTEN

Introduction

In recent years, the roles of genes involved in the pathogenesis of numerous diseases have been elucidated. Several of these genes have been sufficiently well characterized to permit experiments in vivo by introducing genomic fragments or DNA-constructs into animals. These transgenic animals can then become valuable models that offer new approaches to understand the genetic basis of disease. Some of the possible strategies in transgenic animal research are the expression of homologous or heterologous genes under their endogenous promoter, the expression of a reporter gene under a specific promoter or the tissue-specific expression using tissue-specific cis-acting regulatory elements. Another possibility is the generation of gene "knock-outs" using homologous recombination in embryonic stem cells.

In general, transgenic techniques can be applied to many species. Until now mice have been used most extensively since the mouse genome has been investigated in detail and microinjection into mouse oocytes does not cause significant technical difficulties. In some cases rats may be the more appropriate model, because they offer several advantages over mice. One advantage is that rats can be easily characterized using the methodological approaches of whole animal physiology and pharmacology, which are more difficult to perform in mice, which are smaller in size.

Experimental approaches in transgenic research

Transgenic animals can be used to study different experimental questions. First, transgenic lines can be established with the purpose of adding an ad-

Eckart Schott, Martin Paul, *Correspondence to* Detlev Ganten, Max-Delbrück-Center for Molecular Medicine, Robert Rössle Straße 10, Berlin, 13122, Germany (*phone* +01-30-9406-3278; *fax* +01-30-949-7008; *e-mail* dganten@mdc-berlin.de)

ditional gene to develop models for the in vivo study of its expression and regulation, as well as possible phenotypic changes induced by the transgene. Several models have been established that use a construct containing a gene under its endogenous promoter, leading to overproduction of a gene product in its physiological site of synthesis. However, a limiting factor in the overproduction of a gene product under its endogenous promoter could be the down regulation of transcriptional elements, that interact with regulatory elements in the promoter region.

A possibile approach to prevent down-regulation in a transgenic model is to place the coding region of a gene under the control of a heterologous promoter. Since extreme overexpression of a gene will prove lethal during fetal life in many cases, inducible promoters are often useful. While these promoters show little activity in the non-activated form and, therefore, do not interfere with normal ontogenesis, expression can easily be induced at the time, overexpression and its phenotypic consequences are to be studied. One of the most widely used promoters for this purpose is the metallothionein promoter, which can be induced by heavy metals such as zinc sulfate supplied in the drinking water. This promoter has been used to overexpress rat renin and angiotensinogen genes in transgenic mice (1) as well as the human growth hormone in transgenic mice, rabbits and sheep (2). Another example for a tissue-selective promoter is the transthyretin promoter, which has been used to overexpress ANP specifically in the liver of transgenic mice (3, 4).

Opposite effects can be obtained by we of antisense constructs, targeted by a tissue-specific promoter sequence, resulting in tissue-specific down-regulation of gene expression in transgenic animals (5). This approach provides opportunity to lower expression rate, since hybridization of the sense mRNA with the antisense construct will occur. Therefore, production and release of the encoded protein is inhibited. However, in practice, these protocols have proven to be extremely difficult.

Another application for transgenic animals is their use for the study of gene regulation in vivo by linkage of a reporter gene to the regulatory sequence of a gene under investigation. Reporter genes encode innocuous gene products that can easily be detected by histochemical staining or other assays. The most frequently used reporter genes are the β-galactosidase (lac-Z) gene derived from *E. coli*, the firefly luciferase gene and the bacterial CAT (cloramphenicol-acetyl-transferase) gene. Since these reporter genes can be used both, in cell culture and in transgenic animals, they open a wide range of possibilities to investigate the tissue-specific responses to regulatory stimuli on the level of transcription.

Lastly, transgenic techniques can be used to establish permanent cell lines derived from transgenic animals by fusing a tissue specific promoter to a gene that has oncogenic potential. The transgenic animals then develop tumors in the tissue where expression occurs. The tumors can be used to isolate cell lines which express the desired gene product in perpetuity (6, 7).

Methods for generating transgenic animals

Several methods are used to generate transgenic animals. The most common are:

- Direct microinjection of foreign DNA into the pronucleus of fertilized oocytes (8-9)
- Use of embryonic stem cells previously transfected with the DNA of interest to produce chimeras via transfer into blastocysts (10-12)
- Retroviral infection of pre-implantation blastocysts (13-15). Infection can be attained either by direct exposure of blastocysts to viral particles or by cocultivation with virus-producing cells.

A common disadvantage of direct microinjection and retroviral infection is that these methods do not permit the delivery of the gene to a specific location of the chromosome, so that integration occurs at random. Therefore, transfer of DNA by direct microinjection into the pronucleus can cause insertional mutations which may interfere with the phenotype. In this context, a wide range of mutations have been decribed previously (16-18). Furthermore, random positioning in the vicinity of other transciptional regulatory sequences may lead to a position effect, resulting in higher or lower expression levels (18).

The microinjection of foreign DNA is widely used and is considered the most efficient method for construction of transgenic animals. Whereas this method has become a routine procedure in mice during the last ten years (9, 13), its application for the generation of transgenic rats has been developed only recently. An approach to obtain a reliable positioning and copy number of the transgene, to regulate the level of expression, and to make the transgene independent of the neighboring chromosome may be the use of recombinant yeast artificial chromosomes (YAC), which allow the introduction of large gene constructs into the germ line (19).

Choice of strain

The development of transgenic mice is well established and the related techniques are well documented elsewhere (13). Since the development of transgenic rats is a relatively new procedure we will focus predominantly on this species in the context of transgenic technology. Superovulation can be induced in mature and immature rats (20-23). The optimum age, for superovulation varies between strains. The period immediately preceeding puberty is the best time to perform superovulation. In Sprague-Dawley rats this is between day 29 and 31, and in Wistar Kyoto rats around day 33. This is in contrast to mice where the optimum age to perform superovulation is 4-6 weeks of age irrespective of the strain chosen.

Procedure

Superovulation

Superovulation is essential to obtain a sufficient number of oocytes with normal developmental potential. Difficulties in inducing efficient superovulation have long prevented the application of transgenic technology to rats. In mice, a single dose of pregnant mare serum gonadotropin (PMSG) is usually sufficient to induce superovulation. The use of PMSG in rats, in contrast, is associated with a high variability in the ovulatory response (24, 25), a low fertilization rate, and a high percentage of degenerated oocytes. Furthermore, oocytes do not develop normally (26). Since PMSG exhibits both, follicle stimulating hormone (FSH) and luteinizing hormone (LH) activity (20-23), the relatively high percentage of LH activity in PMSG is thought to be the reason for the low efficiency of superovulation in rats after treatment with PMSG (10). High LH activity can induce an excessive synthesis of androgens in the ovaries 36 hours after PMSG application, which may cause degeneration of the oocytes (27).

A method to obtain large amounts of normally developed rat oocytes has been developed by Armstrong and co-workers (28). Using highly purified FSH preparations from the porcine pituitary, they showed that a specified ratio of LH/FSH for a given FSH amount is necessary to achieve appropriate superovulation in the rat. Use of a minipump-system, which is implanted subcutaneously, is necessary (28) because of the short half-life of purified FSH (29, 30). According to the protocol of Armstrong et al. (28), a combination of 1U FSH and 0.2 IU HCG per 24 hours is administered to immature rats over a 72 h period. Fifty hours after implantation of the minipump, 30

Table 1. Schedule for the production of transgenic rats

Time	Day	Procedure
9 A.M.	– 3	Implant minipumps into immature rats. Pump rate is 1 U FSH and 0.2 IU human choriogonadotropin (HCG)
11 A.M.	– 1	Administer 30 IU HCG by intraperitoneal injection
4 P.M.	– 1	Place hormone-treated female with fertile male formating; place mature females with vasectomized males for mating
2 A.M.	0	Ovulation and mating completed
7 – 9 A.M.	0	Examine immature and mature females for the presence of a copulatory (vaginal) plug
9 – 11 A.M.	0	Sacrifize immature females and harvest oocytes
1 – 4 P.M.	0	Microinjection
6 P.M.	0	Reimplant surviving oocytes into pseudopregnant females or
7-9 A.M.	1	Reimplant surviving two-cell embryos into day 1 or day 0 pseudopregnant females

IU HCG are injected intraperitoneally for induction and timing of ovulation. The subcutaneous implantation of the minipump is performed under light ether anaesthesia by cutting the skin in the neck-region and preparing a subcutaneous tunnel to the area somewhat lateral to the midline between the posterior extremities.

Mating

The females carrying the minipumps are placed with fertile males for mating the day before microinjection. At the same time, mature females are placed with vasectomized males for induction of pseudopregnancy. Vasectomy is carried out under anesthesia using a mixture of Rompun (Xylasin, 0.1 mg/100 g body-weight) and Ketavet (Ketamin, 11.5 mg/100 g body-weight) in rats and Avertin in mice. Vasectomy is performed according to the method previously decribed by Hogan et al. (13).

Note: Up to 15 mature females are required to obtain 2 -3 pseudopregnant foster mothers. Using females in estrus before mating will increase the yield of recipients.

On the next morning between 7 and 8 A.M., egg donors and mature females are checked for the presence of a copulatory plug. In rats, careful examination is carried out with a speculum; mice are examined using small forceps. Signs that copulation has taken place are:

- wide vaginal introitus
- introitus plastered by secretions
- copulatory plug in the vagina, which is of white color and adheres tightly to the wall.

Collection of oocytes

After the animals are killed, the oviduct is removed surgically, the swollen ampulla is opened, oocytes are collected and immediately placed into tissue-culture medium (M2 medium, (13)) containing Hyaluronidase (4000 IU/ ml) to loosen the adhering cumulus cells. Oocytes are incubated in this medium for 30 seconds to 3 minutes. After the conglomerate has dissolved, the oocytes are washed 2 to 3 times in M2 medium and transferred to pre-warmed M16 medium (13) in a humidified tissue culture incubator with 5% CO_2 at 37°C.

Table 2. Schedule for the generation of transgenic mice

Time	Day	Procedure
11 A.M.	− 3	Inject 5 IU pregnant mare serum gonadotropin (PMSG) into peritoneum of donor mice
11 A.M.	− 1	Inject 5 IU HCG intraperitoneally
2 – 3 P.M.	− 1	Place hormone treated female with fertile male for mating; place mature females with vasectomized males for mating
8 A.M.	0	Examine females for the presence of a copulatory plug
9 – 10 A.M.	0	Kill donor females and collect oocytes
1 – 4 P.M.	0	Microinjection
10 A.M.	1	Reimplant surviving oocytes into pseudopregnant females

Microinjection

Since direct microinjection of DNA into the pronucleus is considered the best established and most extensively used protocol to generate transgenic animals, this protocol refers only to this method (Figure 1).

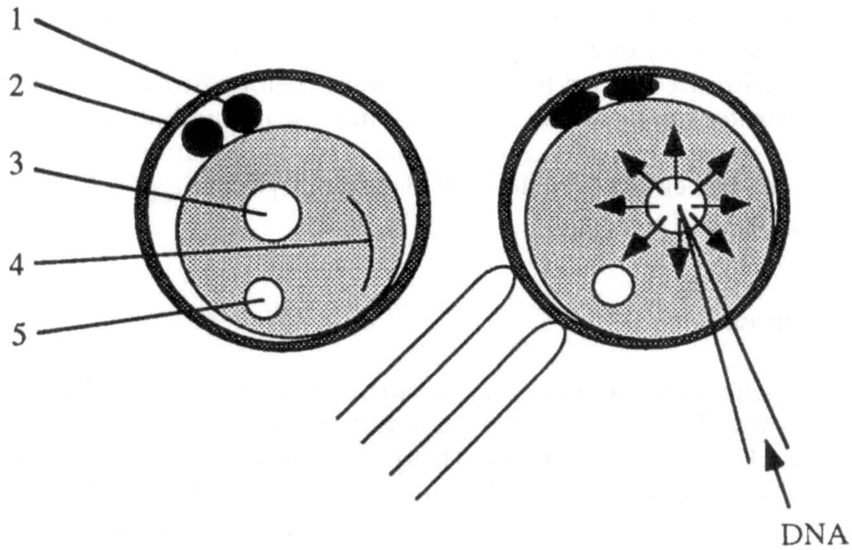

Fig. 1. Microinjection of DNA into the pronucleus. *1,* polar body; *2,* zona pellucida; *3,* male pronucleus ; *4,* rest of sperm; *5,* female pronucleus.

One drop of culture medium is placed in the middle of a glass slide, which is then overlaid with a drop of paraffin-oil. Oocytes are transferred into the drop of medium on the slide using a paraffin-filled glass pipette. Microinjection can only be performed if two pronuclei can be distinguished clearly. If the two pronuclei are not visible, oocytes are returned into medium and are examined again later. If both pronuclei can be identified, microinjection can be performed under 400-fold magnification. Using a micromanipulator, a needle is inserted into the male pronucleus which is slightly larger than the female pronucleus and 1 pl/oocyte (1 fg/oocyte) of DNA is injected. If injection is successful, the pronucleus will swell visibly. After injection, oocytes are incubated in M16 medium until oviduct transfer is performed.

Oviduct transfer

Successful transfer of embryos is dependent on synchronization between embryonic development and timing of pseudopregnancy. The pseudopregnant rat is anesthetized with the Ketavet/Rompun mixture, the mouse with Avertin, as described above.

After opening the body wall by a midline incision in the lower back, the ovaries and the oviducts are surgically exposed. Hemorrhage from ruptured vessels of the ovarian capsule is a frequent problem in rats, whereas this problem normally does not occur in mice. In rats, a few drops of epinephrine therefore are applied to the bursa ovaria, before the bursa is opened. A glass transfer capillary is then inserted into the infundibulum and fixed in position using a microforceps. To obtain a sufficient number of progeny, 30-40 one-cell embryos or 15-30 two-cell embryos are typically transfered bilaterally.

DNA analysis

If transgene integration occurs prior to the first cell cleavage, all cells of the organism will contain the transgene and any tissue can be used for DNA analysis. Using small tail biopsies has proven the most convenient protocol, biopsy is easy to perform and does not cause extensive harm to the animal. 1 cm or less of the tail is cut off and placed in a mixture of buffer and proteinase.

Table 3. Preparation of tail biopsies

Time	Procedure
8 - 18 h	Digest tail at 55 °C in a mixture of 0.7 ml buffer (50 mM Tris-Cl (pH 8.0), 100 mM EDTA, 100 mM NaCl, 1% SDS) and 35 μl proteinase K (10 mg/ml)
10 min	incubate on ice, after 10 min add 0.3 ml saturated NaCl solution, mix
5 min	incubate on ice
10 min	centrifuge at 4 °C at 14000 rpm
15 min	incubate supernatant with 5 μl RNAse A (4 mg/ml) at 37 °C, then precipitate DNA with isopropanol, wash with 70% ethanol, dry pellet, resuspend in H_2O

After isolation of DNA, three different methods are used to assay for the presence of the transgene

- Digestion of DNA with restriction enzymes followed by Southern-blot analysis
- Dot or slot blotting of undigested DNA
- Polymerase chain reaction (PCR).

Southern Blot Analysis: Although Southern blotting is the most labor-intensive of the three methods, it is propably the most reliable way to prove the integration of a transgene, especially if differentiation between a transgene and a closely related endogenous gene is necessary.

Dot or Slot Blotting: Denatured undigested DNA is transfered to a solid membrane and hybridization is performed. Although this method is propably the most convenient, it is only appropriate when the investigated gene exists in a single copy in the animal's genome. Therefore, dot or slot blotting is best suitable when exogenous genes such as reporter genes are used as a transgene or for assessment of the copy number.

Polymerase Chain Reaction: PCR also is a convenient and rapid method to prove the presence of an inserted gene; DNA preparations that are of too poor quality to be used in Southern analysis will still produce results in PCR. However, as in dot and slot blotting, problems can occur if endogenous sequences are present which are highly similar to the transgene. In PCR, therefore, success depends on the ability of the primers to bind specifically to the sequence of the transgene.

Applications

Use of other animal species in transgenic research

Although the mouse has been used extensively in transgenic research, its biggest disadvantage is its small size. Therefore for certain approaches it is necessary to engineer other models for transgenic research using larger animals. However, larger species in general have the disadvantage of longer generation time and smaller litter size (31).

Table 4. Gestation time and litter size in selected animals (from (31))

Species	Gestation time	Litter Size
Mouse	19 - 21 days	5 - 12
Rat	20 - 23 days	6 - 15
Rabbit	28 - 34 days	5 - 10
Goat	5 months	1 - 2
Sheep	5 months	1 - 2
Pig	4 months	9 - 10
Cow	9 months	1 - 2

At the time of this writing only a few protocols for generating transgenics in larger animal species exist:

- Generation of transgenic **rabbits** was first reported in 1985 (2), when Hammer et al. inserted the gene encoding human growth hormone under control of the metallothionein promoter into the rabbit genome. They obtained an integration-rate of 12.8% (ratio of animals whose genome contained the transgene / all animals that resulted from transfered oocytes), which is appoximately half of the rate obtained in mice. The same construct was used by Hammer et al. to generate transgenic **sheep**, resulting in the very low integration rate of 1.3%.
- One approach of transgenic technology is to generate animals that secrete large amounts of protein into their milk. Böhler et al. (32) generated a transgenic **rabbit**, secreting human interleukin-2 into its milk. Clark et al. reported on a transgenic **sheep**, that secreted human factor IX into its milk (33). The possibility of producing protein in the mammary gland of transgenic animals has been recently reviewed by Henninghausen et al. (34).
- Ebert et al. (35) obtained an integration rate of 7% by generating transgenic **goats** that harbored the gene for the human tissue plasminogen activator (tPA) and secreted this protein into their milk.
- In **pigs**, integration rates of DNA are rather low. Wall et al. (36), for example, obtained an integration rate of only 1% in transgenic pigs. Swanson et al. (37) generated a pig line which was transgenic for human hemoglobin. The construct was constructed using the control sequence taken from the human β-globin gene and two copies of the human α_1-globin gene plus a single copy of the β^A-globin gene. Using this protocol, functional human hemoglobin was produced in the pig.

Transgenic animals in hypertension research: example of an application

A large amount of experimental data is now available for transgenic models in hypertension research. As of this writing several transgenic rat lines expressing candidate genes for hypertension have been produced. For a number of reasons that are discussed elsewhere (38), these animals mostly express components of the renin-angiotensin-system (RAS).

Genetic linkage studies have recently demonstrated that the components of the RAS are associated with hypertension (39, 40). Molecular variants of angiotensinogen, for example, which result in significant differences in plasma concentration among hypertensive subjects, have been interpreted as an inherited predisposition to the development of essential hypertension (41). Interestingly, angiotensin-converting enzyme does not appear to be associated with hypertension in humans, but with myocardial infarction in certain patients (42). This suggests that the RAS may not only be involved in the elevation of blood pressure, but also in the sequelae of hypertension such as cardiovascular end-organ disease. For renin, a restriction fragment length polymorphism (RFLP) has been linked to hypertension in the Dahl salt-sensitive rat (43), whereas no linkage could be demonstrated in humans (44). To test the functional relevance of these genes in vivo, transgenic animals have been established. Initally a number of transgenic mouse lines were established to express genes of the RAS, but transgenic rats have been the species of choice more recently.

The first transgenic rat line in hypertension research was generated by microinjection of the mouse Ren-2 gene (45) into oocytes of normotensive rats, resulting in rats suffering from fulminant hypertension. In heterozygous rats of this line, blood pressure increased shortly after weaning at 5 weeks of age, and maximum values of about 240 mmHg were reached at 10 weeks of age (46). Homozygous animals develop even higher blood pressure and exhibited a high mortality rate if not treated with angiotensin converting enzyme (ACE) inhibitors (46) or other anti-hypertensive medications.

Both, heterozygous and homozygous TGR(mREN2)27 rats exhibited a marked gender related polymorphism with respect to blood-pressure, with higher blood pressure in males, probably caused by the higher level of androgens and their stimulatory effect on tissue RAS. The highest expression of the Ren-2 gene was found in the adrenal gland, whereas the kidney (which expresses highest levels of endogenous renin in rats) showed only low renin levels, which were probably influenced by negative feedback mechanisms (47).

Hypertension in TGR(mREN2)27 rats is dependent on angiotensin (ANG) II since treatment with low-dose ACE inhibitors or ANG II antagonists normalized blood pressure (46). Preliminary experiments (48) showed that the animals were extremely sensitive to ANG II receptor (AT_1) antagonists. The plasma RAS is not stimulated in this line and plasma inactive renin (prorenin) is the only substance showing higher plasma concentration. The source of this prorenin is probably the adrenal gland, since bilateral adrenalectomy reduced plasma levels by 80%. Another observation emphasizes the role of the adrenal gland in hypertension: The urinary excretion of deoxycorticosterone, corticosterone and aldosterone was elevated in young TGR(REN2)27 animals during the phase of developing hypertension but not in adult rats. ACTH treatment caused a dramatic increase in steroid excretion indicating that the adrenal glands are more sensitive to ACTH-stimulation in this transgenic line. Since spironolactone treatment did not reduce blood pressure in these animals, hypertension was probably not due to excess of mineralocorticoids. Hilgers et al. (49) demonstrated that the vasculature of the transgenic animals produced increased amounts of ANG II, which was most likely the result of overexpression of the transgene in the vascular wall. This observation was proved by a sensitive RNase protection assay (49).

The hearts of these transgenic animals showed pathological changes such as hypertrophy and fibrosis (50), which only in part were explained by altered blood pressure levels. Current studies, therefore, focus on the investigation of additional effects on cardiac function in these animals.

Other models in transgenic hypertension research include mice and rats expressing human renin (51, 52) or human angiotensinogen (1, 53, 54). These approaches suggest that products of endogenous genes and transgenes do not interact at the measured concentrations. The results from co-expression of the human transgenes in crossbred transgenic animals carrying both genes in one organ supports the possibility of local angiotensin I production from the interaction of the human proteins. The functional role of such human-dependent, tissue-specific RAS remains to be established.

Other recent approaches include development of transgenic mice expressing atrial natriuretic peptide (ANP), resulting in a decrease in mean arterial pressure of 20-30 mmHg (3), and the development of mice that are transgenic for rat arginine-vasopressin (AVP), resulting in mice suffering from SiADH (55, 56). An interesting model was established by Hofmann et al. (57) who generated a transgenic mouse, expressing the gene encoding the human LDL receptor driven by the mouse metallothionein promoter. After administration of cadmium, these animals cleared intravenously-injected ^{125}I-labeled LDL from blood eight to ten times more rapidly than did normal control mice. This protocol may lead to interesting clinical applications in patients who suffer from congenital LDL receptor deficiency that predisposes to the development of artherosclerosis.

Other candidate genes for the development of transgenic animals in hypertension research are the genes encoding human heart chymase and genes of the endothelin family (58).

Future perspectives

One of the major problems in transgenic research is the relative limitation in the size of the constructs. The recently used cloning vectors limit the size of a construct to approximately 40 kb. Therefore in some experiments neither the full genomic sequence nor putative regulatory elements can be included into the construct. Besides the fact that appropriate regulation and tissue specificity cannot be obtained with these constructs, the short "buffering zone" between transgene and endogenous DNA leads to positioning effects as mentioned above.

A new approach in transgenic research that seems to be able to solve this problems is the use of yeast artificial chromosomes (YACs). Recently, a number of experiments have been reported. Jakobovits et al. introduced a 670-kb YAC, containing the human hypoxanthine phosphoribosyl transferase gene into ES cells by spheroblast fusion (59) resulting in appropriate development of transgenic mice and correct gene regulation. Even integration of large amounts of additional yeast DNA was not detrimental to ES cell growth or development of mice.

Schedl et al. generated a transgenic mouse by direct pronuclear injection of a construct containing the 80 kb mouse tyrosinase gene plus 155 kb of flanking sequences. They reported copy-number-dependent and position-independent expression of the transgene. Furthermore, the transgene showed the normal tissue-specific pattern (60).

Choi and co-workers reported a transgenic mouse containing an 85 kb fragment of the human heavy chain immunoglobulin gene (61) engineered by transfection of ES cells followed by blastocyst injection. Serum levels of human μ-chains were very low compared with endogenous mouse IgM, probably due to competition from the endogenous heavy chain gene. The authors suggest that larger "buffering sequences" flanking the gene would solve this problem.

Peterson et al. (62) reported a mouse, containing an intact 82 kb human ß-globin locus flanked by a 5' sequence of 148 kb. The construct was inserted into the mouse genome using direct microinjection. Since large molecules such as YACs are sensitive to mechanical shearing, they used a solution of high ionic strength (100 mM NaCl, according to Gnirke et al. (63), which served as a protective agent. They reported correct developmental regulation of the gene-locus and expression in the appropriate tissue-specific pattern.

Another approach to obtain tissue-specificity in transgenes is to link the structural gene to a promoter of known tissue-specificity. One example is the specific expression in heart-muscle of genes that are linked to the regulatory elements of the gene encoding the ventricular myosin light chain 2 (MLC_{2V}) (64, 65). Other regulatory sequences leading to specific expression in cardiac myocytes are the sequences derived from the genes encoding myosin heavy chain (MHC) α and β or ANP (66, 67). For a more detailed review see (68).

In summary the availability of techniques allowing integration of larger DNA fragments will offer new avenues for the application of transgenic research to medicine and biology.

References

1. Ohkubo H, Kawakami H, Kakehi Y, Takumi T, Arai H, Yokota Y, Iwai M, Tanabe Y, Masu M, Hata J, Iwao H, Okamoto H, Yokoyama M, Nomura T, Katsuki M, Nakanishi S (1990) Generation of transgenic mice with elevated blood pressure by introduction of the rat renin and angiotensinogen genes. Proc Natl Acad Sci USA 87:5153-5157
2. Hammer RE, Pursel VG, Reynolds RC, Rexroad CEJr, Wall RJ, Bolt DJ, Ebert KM, Palmiter RD, Brinster RL (1985) Production of transgenic rabbits, sheep and pigs by microinjection. Nature 315:680-683
3. Steinhelper ME, Cochrane KL, Field LJ (1990) Hypotension in transgenic mice expressing atrial natriuretic factor fusion genes. Hypertension 16:301-307
4. Field LJ, Veress AT, Steinhelper ME, Cochrane K, Sonnenberg H (1991) Kidney function in ANF-transgenic mice: effect of blood volume expansion. Am J Physiol 260:R1-R5
5. Katsuki M, Sato M, Kimura M, Yokoyama M, Kobayashi K, Nomura T (1988) Conversion of normal behavior to shiverer by myelin basic protein antisense cDNA in transgenic mice. Science 241:593-595
6. Jat PS, Noble MD, Ataliotis P, Tanaka Y, Yannoutsos N, Larsen L, Kioussis D (1991) Direct derivation of conditionally immortal cell lines from a H-2Kb-tsA58 transgenic mouse. Proc Natl Acad Sci USA; 88 (12):5096 - 100
7. Hayes C, Kelly D, Murayama S, Komiyama A, Suzuki K, Popko B (1992) Expression of the neu oncogene under the transcriptional control of the myelin basic protein gene in transgenic mice: generation of transformed glial cells. Neurosci Res 31(1):175 - 87
8. Babinet C, Morello D, Renard J (1989) Transgenic mice. Genome 31:938-949
9. Gordon JW, Ruddle FH (1983) Gene transfer into mouse embryos: production of transgenic mice by pronuclear injection. Methods Enzymol 101:411-433
10. Evans G, Armstrong D (1984) Reduction in fertilization rate in-vitro of oocytes from immature rats induced to superovulate. J Reprod Fertil 70:131-135
11. Robertson E (1991) Using embryonic stem cells to introduce mutations into mouse germ line. Biol Reprod 44:238-245
12. Williams DA (1990) Embryonic stem cells targets for gene transfer: a new approach to molecular manipulation of the murine hematopoietic system. Bone marrow transplantation 5:141-144
13. Hogan B, Costantini F, Lacy E (1986) Manipulating the mouse embryo. Cold Spring Harbor Laboratory, Cold Spring Harbor, New York 11724

14. Jaenisch R, Jahner D, Nobis P, Simon T, Lohler J, Harbers K, Grotkopp D (1981) Chromosomal position and advance of retroviral genomes inserted into germ line of mice. Cell 24:519
15. Soriano P, Cone R, Mulligan R, Jaenisch R (1986) Tissue-specific and ectopic expression of genes introduced into transgenic mice by retroviruses. Science 234:1409
16. Keller SA, Liptay S, Hajra A, Meisler MH (1990) Transgene-induced mutation of the murine steel locus. Proc Natl Acad Sci USA 87:10019-10022
17. Gordon JW, Uehlinger J, Dayani N, Talansky BE, Gordon M, Rudomen GS, Neumann PE (1990) Analysis of the hotfood (ho) locus by creation of an insertional mutation in a transgenic mouse. Dev Biol 36:159-169
18. Costantini F, Radice G, Lee JL, Chada K, Perry W, Son HJ (1989) Insertional mutagenesis in transgenic mice. Prog Nucleic Acids Res Mol Biol 36:159-169
19. Schedl A, Beerman F, Thies E, Montoliu L, Kelsey G, Schütz G (1992) Transgenic mice generated by pronuclear injection of a yeast artificial chromosome. Nucleic Acids Res 20:3073-3077
20. Armstrong DT, Opavsky MA (1988) Superovulation of immature rats by continuous infusion of FSH. Biol Reprod 39:511-518
21. Jong-Ho L, Kang DJ, Park CS (1988) Ovulation in timing and PMSG dose response in superovulation of immature rats. Korean J Anim Sci 30:714-719
22. Leveille MC, Armstrong DT (1989) Preimplantation embryo development and serum steroid levels in immature rats induced to ovulate or superovulate with pregnant mares serum gonadotropin injection or FSH infusion. Gamete research 23:127-138
23. Seong H, Yun C (1987) Study on the transfer of embryos in mature rats. Induction of superovulation by PMSG and HCG treatment in mature rats. Korean J Anim Sci 29:112-117
24. Walton EA, Evans G, Armstrong DT (1983) Ovulation response and fertilization failure in immature rats induced to superovulate. J Reprod Fertil 67:91-96
25. Young WY, Yuen BH, Moon YS (1987) Effects of superovulatory doses of pregnant mare serum gonadotropin on oocyte quality and ovulatory and steroid responses in rats. Gamete Research 16:109-120
26. Papkoff H (1974) Chemical and biological properties of the subunits of pregnant mare serum gonadotropin. Biochem Biophys Res Commun 58:397-404
27. Yun YW, Yuen BH, Moon YS (1988) Effects of antiandrogen flutamide on oocyte quality and embryo development in rats superovulated with pregnant mare serum gonadotropin. Biol Reprod 39:279-286
28. Armstrong D, Siuda A, Opavsky M, Chandrasekhar Y (1989) Biomodal effects of luteinizing hormone and role of androgens in modifying superovulatory responses of rats to infusion with purified porcine follicle stimulating hormone. Biol Reprod 41:54-62
29. Bogdanove E, Gay V (1969) Studies of the disappearance of LH and FSH in the rat, a quantitative approach to adenohypophyseal secretory kinetics. Endocrinology 84:1118-1138
30. Laster DB (1972) Disappearence and uptake of 125 J-FSH in the rat, rabbit, ewe, and cow. J Reprod Fertil 30:407-415
31. Mullins JJ, Mullins LJ (1993) Transgenesis in nonmurine species. Hypertension 22:630-633
32. Bühler T, Bruyere T, Went D, Stranzinger G, Bürki K (1990) Rabbit beta-casein promoter directs secretion of human interleukin-2 into the milk of transgenic rabbits. Bio/Technology 8:140-143

33. Clark A, Bessos H, Bishop J, Brown P, Harris S, Lathe R, McClenaghan M, Prowse C, Simons J, Whitehaw C, Wilmut I (1989) Expression of human anti-hemophilic factor IX in the milk of transgenic sheep. Bio/Technology 7:487-492
34. Henninghausen L, Ruiz L, Wall R (1990) Transgenic animals:production of foreign proteins in milk. Curr Opin Biotech 1:74-78
35. Ebert KM, Selgrath JP, DiTullio P, Denman J, Smith TE, Memon MA, Schindler JE„ Monastersky GM, Vitale JA, Gordon K (1991) Transgenic production of a variant of human tissue-type plasminogen activator in goat milk: Generation of transgenic goats and analysis of expression. Bio/Technology 9:835-838
36. Wall R, Pursel V, Shamay A, McKnight R, Pittius C, Henninghausen L (1991) High-level synthesis of a heterologous milk protein in the mammary gland of transgenic swine. Proc Natl Acad Sci USA 88:1696-1700
37. Swanson M, Martin M, O'Donnel J, Hoover K, Lago W, Huntress V, Parsons C, Pinkert C, Pilder S, Logan J (1992) Production of functional human hemoglobin in transgenic swine. Bio/Technology 10:557-559
38. Ganten D, Takahashi S, Lindpaintner K, Mullins J (1991) Genetic Basis of Hypertension: The Renin-Angiotensin Paradigm. Hypertension 18:III 109-114
39. Hilbert P, Lindpaintner K, Beckmann JS, Serikawa T, Soubrier F, Dubay C, Cartwright P, DeGouyon B, Julier C, Takahasi S, Vincent M, Ganten D, Georges M, Lathrop GM (1991) Chromosomal mapping of two genetic loci associated with blood-pressure regulation in hereditary hypertensive rats. Nature 353:521-529
40. Jacob HJ, Lindpaintner K, Lincoln SE, Kusumi K, Bunker RK, Mao Y-P, Ganten D, Dzau VJ, Lander ES (1991) Genetic mapping of a gene causing hypertension in the stroke-prone spontaneously hypertensive rat. Cell 67:213-224
41. Jeunemaitre X, Soubrier F, Kotelevtsev YV, Lifton RP, Williams CS, Charru A, Hunt SC, Hopkins PN, Williams RR, Lalouel J-M, Corvol P (1992) Molecular basis of human hypertension: role of angiotensinogen. Cell 71:169-180
42. Cambien F, Poirier O, Lecerf L, Evans A, Cambou JP, Arveiler D, Luc G, Bard JM, Bara L, Ricard S, Tiret L, Amouyel P, Alhenc-Gelas F, Soubrier F (1992) Deletion polymorphism in the gene for angiotensin-converting enzyme is a potent risk factor for myocardial infarction. Nature 359:641-644
43. Beermann F, Ruppert S, Hummler E, Schütz G (1991) Tyrosinase as a marker for transgenic mice. Nucleic Acids Res 19:958
44. Naftilan AJ, Williams R, Burt D, Paul M, Pratt RE, Hobart P, Chirgwin J, Dzau VJ (1989) A lack of genetic linkage of renin gene restriction fragment length polymorphisms with human hypertension. Hypertension 14:614-618
45. Mullins JJ, Peters J, Ganten D (1990) Fulminant hypertension in transgenic rats harbouring the mouse Ren-2 gene. Nature 344:541-544
46. Bader M, Zhao Y, Sander M, Lee M, Bachmann J, Böhm M, Djavidani B, Peters J, Mullins JJ, Ganten D (1992) Role of tissue renin in the pathophysiology of hypertension in TGR(mREN2)27 rats. Hypertension 19:681-686
47. Peters J, Münter K, Bader M, Hackenthal E, Mullins JJ, Ganten D (1993) Increased adrenal renin in transgenic hypertensive rats, TGR(mREN2)27, and its regulation by cAMP, angiotensin II, and calcium. J Clin Invest 91:742-747
48. Böhm, M, Lippoldt A, Wienen W, Ganten D, Bader M (1996) Reduction of cardiac hypertrophy in TGR (mREN2) 27 by angiotensin II receptor blockade. Mol Cell Biochem; 163 - 164:217 - 21

49. Hilgers KF, Peters J, Veelken R, Sommer M, Rupprecht G, Ganten D, Luft FC, Mann JFE (1992) Increased vascular angiotensin formation in female rats harboring the mouse Ren-2 gene. Hypertension 19:687-691
50. Bachmann S, Peters J, Engler E, Ganten D, Mullins J (1992) Transgenic rats carrying the mouse renin gene - Morphological characterization of a low renin hypertension model. Kidney Int 41:24-36
51. Fukamizu A, Seo MS, Hatae T, Yokoyama M, Nomura T, Katsuki M, Murakami K (1989) Tissue-specific expression of the human renin gene in transgenic mice. Biochem Biophys Res Commun 165:826-832
52. Sigmund CD, Jones CA, Kane CM, Wu C, Lang JA, Gross KW (1992) Regulated tissue- and cell-specific expression of the human renin gene in transgenic mice. Circ Res 70:1070-1079
53. Ganten D, Wagner J, Zeh K, Bader M, Michel J-B, Paul M, Zimmermann F, Ruf P, Hilgenfeldt U, Ganten U, Kaling M, Bachmann S, Fukamizu A, Mullins JJ, Murakami K (1992) Species specificity of renin kinetics in transgenic rats harboring the human renin and angiotensinogen genes. Proc Natl Acad Sci USA 89:7806-7810
54. Takahashi S, Fukamizu A, Hasegawa T, Yokoyama M, Nomura T, Katsuki M, Murakami K (1991) Expression of the human angiotensinogen gene in transgenic mice and transfected cells. Biochem Biophys Res Commun 180:1103-1109
55. Grant F, Reventos J, Gordon J, Kawabata S, Miller M, Majzoub J (1993) Expression of the rat arginine vasopressin gene in transgenic mice. Mol Endocrinol 7:659-667
56. Habener JF, Cwikel BJ, Hermann H, Hammer RE, Palmiter RD, Brinster RL (1989) Metallothionein-vasopressin fusion gene expression in transgenic mice. J Biol Chem 264:18844-18852
57. Hofmann SL, Russell DW, Brown MS, Goldstein JL, Hammer RE (1988) Overexpression of low density lipoprotein (LDL) receptor eliminates LDL from plasma in transgenic mice. Science 239:1277-1281
58. Paul M, Zimmermann F, Ruf P, Zintz M, Maior D, Yanagisawa M (1993) Transgenic rats expressing the human endothelin-2 gene. Naunyn-Schmiedeberg's Arch Pharmacol 347(Suppl.):R93(Abstract)
59. Jakobovits A, Moore AL, Green LL, Vergara GJ, Maynard-Currie CE, Austin HA, Klapholz S (1993) Germ-line transmission and expression of a human-derived yeast artificial chromosome. Nature 362:255-258
60. Schedl A, Montoliu L, Kelsey G, Schütz G (1993) A yeast artificial chromosome covering the tyrosinase gene confers copy number-dependent expression in transgenic mice. Nature 362:258-261
61. Choi TK, Hollenbach PW, Pearson BE, Ueda RM, Weddell GN, Kurahara CG, Woodhouse CS, Kay RM, Loring JF (1994) Transgenic mice containing a human heavy chain immunoglobin in a yeast artificial chromosome. Nature Genet 4:117-123
62. Peterson KR, Clegg CH, Huxley C, Josephson BM, Haugen HS, Furukawa T, Stamatoyanopoulos G (1993) Transgenic mice containing a 248-kb yeast artifical chromosome carrying the human beta-globin locus display proper developmental control of human globin genes. Proc Natl Acad Sci USA; 90(16):7593 – 7
63. Gnirke A, Huxley C, Peterson K, Olson MV (1993) Microinjection of intact 200- to 500-kb fragments of YAC DNA into mammalian cells. Genomics 15:659-667
64. Henderson S, Spencer M, Sen A, Kumar C, Siddiqui M, Chien K (1989) Structure, organization, and expression of the rat cardiac myosin light chain-2 gene: identification of a 250 bp fragment which confers cardiac specific expression. J Biol Chem 264:18142-18148

65. Zhu H, Garcia A, Ross R, Evans S, Chien K (1991) A conserved 28-base-pair element (HF-1) in the rat cardiac myosin light-chain-2 gene confers cardiac-specific and alpha-adrenergic-inducible expression in cultured neonatal rat myocardial cells. Mol Cell Biol 11:2273-2281
66. Subramaniam A, Jones WK, Gulick J, Wert S, Neumann J, Robbins J (1991) Tissue-specific regulation of the α-myosin heavy chain gene promoter in transgenic mice. J Biol Chem 266:24613-24620
67. Katz EB, Steinhelper ME, Delcarpio JB, Daud AI, Claycomb WC, Field LJ (1992) Cardiomyocyte proliferation in mice expressing α-cardiac myosin heavy chain-SV40 T-antigen transgenes. Am J Physiol 262:H1867-H1876
68. Maniatis T, Goodbourn S, Fischer JA (1987) Regulation of inducible and tissue-specific gene expression. Science 236:1237-1245

Chapter 22

Expression in Eukaryotic Cells: Gene Transfer Using Retroviral Vectors

MARTIN MARX

Introduction

Retroviral vectors are the most efficient gene delivery systems available for in vitro and in vivo studies. Retroviral vectors are particularly useful for achieving stable and highly efficient integration of a gene in the genome of target cells. They have a wide host range, and infection does not lead to cell death. Transfected or infected packaging cell lines may produce replication-incompetent infectious virus over an indefinite time period. The DNA introduced into the target cells is driven by a strong promoter, leading to efficient expression of the foreign gene. In addition, inducible promoters are available. For a complete description of the retroviral life cycle and construction of vectors, see Weiss et al. 1985 (1). Briefly, replication-incompetent retroviral vectors are constructed by deletion of the genes encoding the structural proteins *gag*, *env* and *pol*. These deletions make the retrovirus replication-incompetent and provide room for the insertion of foreign genes that can be expressed in the target cell or used as a selectable marker. In the vector described in this chapter, expression of neomycin phosphotransferase encoded by the bacterial transposon Tn5 (2) provides resistance to the aminoglycoside antibiotics kanamycin, neomycin and geneticin. Despite the extensive deletions, the vectors retain the *cis*-acting viral sequences that are necessary for transmission and the ψ packaging sequence essential for encapsidation into the viral particle. The deleted *gag*, *pol* and *env* genes are provided by a helper or packaging cell line that contains mutant provirus with non-functional packaging signals due to a deletion of the ψ packaging sequence. When the plasmid form of the retroviral vector is transfected into the packaging cell line, genetic complementation between the primary provirus (ψ-, *gag*+, *env*+, *pol*+) and the transfected provirus

Martin Marx, Universität Köln, Medizinische Klinik I, Ostmerheimerstrasse 200, Köln, 51109, Germany (*phone* +49-221-8907-2200; *fax* +49-221-8907-2335; *e-mail* martin.marx@uni-koeln.de)

(ψ+, *gag-*, *pol-*, *env-*) leads to production of replication-incompetent retroviruses which contain an RNA copy of the plasmid-encoded provirus transfected into the packaging cell line.

This chapter outlines the components necessary to perform in vitro gene transfer using animal or human cells in tissue culture as targets. A standard and well established protocol for the use of a Moloney Murine Leukemia Virus (MoMuLV)-derived vector is described. Finally, potential biohazards and related biological safety issues are discussed.

Outline

Standard transfection protocol

No single retroviral vector will be appropriate for all experimental purposes. However, similar methods are used when employing different vectors to generate replication-incompetent viruses able to integrate a foreign DNA in to the genome of target cells. In the following, a standard protocol is described that was used frequently to insert a specific DNA into the retroviral vector pLXSN (Figure 1) which was developed by A.D. Miller (3) and to generate ecotropic and amphotropic viral stocks suitable for the infection of a broad range of mammalian cells. The vector contains long terminal repeats derived from the Moloney Murine Leukemia Virus (MoMLV) and from the Moloney Murine Sarcoma Virus (MoMSV). The EcoRI, HpaI, XhoI or BamHi restriction sites in the cloning region of pLXSN can be used for the insertion of foreign DNA. Further downstream the neomycin resistance gene is driven by the SV40 promoter enhancer. The use of this and related vectors in human and animal cells has previously been extensively reported (4-12).

Materials

Components of a retroviral gene delivery system

The following components are required to set up a retroviral gene delivery system:

- The plasmid containing specific MoMuLV sequences
- The DNA to be ligated into the plasmid and inserted into the genome of the target cell

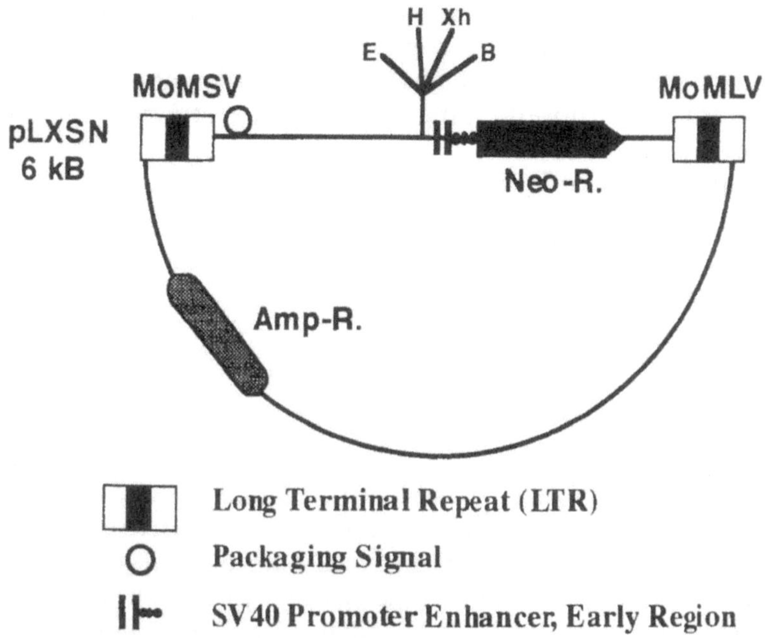

Fig. 1. Retroviral Vector pLXSN

- Competent bacteria used to replicate the plasmid
- A packaging cell line for the production of replication-incompetent ecotropic retroviral particles. These can be used to infect rodent cells. For the production of amphotropic replication-incompetent viruses that have a broad host specificity a second packaging cell line must be infected with the ecotropic viruses.
- The target cell

Subprotocol 1
Synthesis of the Insert

Procedure

The DNA that is to be inserted into the genome of the target cells is usually synthesized by PCR (see Chapter 14). Suitable restriction sites to facilitate the ligation of the DNA into the vector need to be considered when designing the appropriate PCR primers. If possible, we preferentially include an

EcoRI site at the 5' end of the coding strand and a BamHI site at the 5' end of the complementary strand. This allows the insertion of the cDNA into the multiple cloning site of pLXSN in the "sense" orientation. If BamHI or EcoRI cut within the insert, some of the other restriction sites may be used. Before further use in transfection studies, the quality of the PCR product should be assessed by DNA sequencing (see Chapter 13).

Subprotocol 2
Purification

Procedure

DNA purification methods are described in Chapter 3. The vector DNA and the PCR product should be cut with the appropriate restriction enzymes and gel-purified. The DNA may be transferred onto a DEAE-cellulose membrane and eluted, or excised from the agarose gel and electroeluted. The amount of DNA recovered from the final phenol/chloroform extraction and ethanol precipitation must be determined (see Chapter 3 for DNA quantitation).

Subprotocol 3
Ligation of the Insert into the Vector

Materials

- Gel-purified vector and insert, both digested with appropriate restriction enzymes
- Ultra pure H_20
- Ligase
- 10× ligase buffer
- Sterile microcentrifuge tubes

Procedure

1. Use 0.025 pmol vector DNA-ends per ligation (1 µg of a 2.5 kb DNA corresponds approximately to 0.025 pmol DNA ends), dilute the vector DNA to 0.025 pmol ends in 2 µl.
2. Dilute the insert to 0.025 pmol ends in 1 µl.
3. Set up five ligations (A - E) with the volumes (in µl) given in the table:

	A	B	C	D	E
Vector 0.0125 pmol ends/ml	2	2	2	2	-
10x ligase buffer	2	2	2	2	2
Ligase	1	1	1	1	1
Insert 0.025 pmol ends/ml	2	6	12	-	12
H2O	13	9	3	15	5
Total Volume	20	20	20	20	20
Insert/Vector ratio	2	6	12	-	-

4. Incubate 2 hours at room temperature.

Subprotocol 4
Transfection of *E. coli* HB101 and Plasmid DNA Isolation

Materials

- Competent *E. coli* HB101 (e.g. Gibco/BRL)
- Ligation mixtures A - E
- LB media
- LB agar
- Ampicillin
- TE

Procedure

1. Prepare five 15 ml Falcon tubes with 50 µl competent HB101 (or any other suitable bacteria). See Chapter 15 for the preparation of competent bacteria.
2. Add 20 µl of the ligation mixture A, B, C, D or E.
3. Incubate 30 minutes on ice.
4. Heat shock at 42°C for two minutes.
5. Add 500 µl LB media (see Subprotocol 1) into each tube.
6. Shake the tubes at 250 rpm at 37°C for 30 minutes.
7. Plate the bacteria on 100 mm petri dishes with LB agar supplemented with 100 µg/ml ampicillin (see Subprotocol 1, Materials for preparation of LB agar). Set up four dishes with the ligation mixtures A, B and C each, one dish with the mixture D ("positive control") and E ("negative control")
8. Incubate the petri dishes overnight at 37°C.
9. The number of colonies on the plate should not decrease with increasing insert/vector ratio. The positive control D (vector alone) will probably render the highest number of colonies, whereas the negative control E (insert alone) should allow only the growth of a few, or ideally, no colonies.
10. Pick a suitable number (usually 3-5) colonies from each of the plates A, B and C, pick two or three colonies from plate D (vector alone). Transfer each colony to a master plate with LB agar and 100 µg/ml ampicillin (touch the plate first) and then, with a "second touch", to a 15 ml Falcon tube with 5 ml LB media supplemented with 50 µg/ml ampicillin.
11. Incubate the master plates at 37°C overnight, then store at 4°C for possible future use.
12. Grow up the bacteria in the Falcon tubes in an orbital shaker (250 rpm) at 37°C overnight.
13. Isolate the plasmid DNA by the "miniprep" method (see Subprotocol 4) followed by phenol/chloroform extraction and ethanol precipitation.
14. Resuspend the precipitated DNA in 60 µl of TE and store it at -70°C.

Subprotocol 5
Assay for the Presence of the DNA Insert

The presence of the insert can easily be confirmed by digestion of the plasmid DNA recovered from the miniprep. Plasmids are digested with the same restriction enzymes used prior to the ligation of the insert into the vector. In the following, a standard protocol is given which assumes that the EcoRI and BamHI sites in the multiple cloning region of pLXSN have been used.

Materials

- Ethanol precipitated plasmid DNA
- EcoRI and BamHI (10 U/μl)
- Digestion buffer
- Ultra pure water
- TE
- RNase stock solution (1 mg/ml)

Procedure

1. Resuspend the ethanol precipitated plasmid DNA in 60 μl TE.
2. Transfer 14 μl to a new microcentrifuge tube.
3. Add 2 μl EcoRI, 2 μl BamHI, 3 μl digestion buffer and 9 μl ultra pure water.
4. Incubate 2 hours at 37°C.
5. Add 3 μl RNase and incubate for an additional 15 minutes at 37°C.
6. Analyze the samples on a horizontal agarose gel and detect the plasmid DNA and the insert DNA with ethidium bromide.

Subprotocol 6
Transfection of ψ2 Packaging Cells

ψ2 (13) and ψAM (14) cells are available from ATCC (American Type Culture Collection, Rockville, MD) or from laboratories that routinely work with retroviral vectors. These cells originate from 3T3 fibroblasts and are already transfected with the coding sequences for the viral envelope proteins. ψ2 cells produce ecotropic viral particles that are only capable to infect rodent cells. ψAM cells produce amphotropic viral particles that are able to infect a broad range of mammalian cells, including those of human origin. The potential biohazard due to the possible generation of replication competent viruses has stimulated the design of new packaging cell lines that eliminate the recombination events leading to helper virus production. With the ecotropic packaging cell lines ψCRE (15) and GP+E86 (16) helper virus rescue has never been reported. These packaging cell lines should preferentially be used, when the target cells are of rodent origin. For the generattion of amphotropic viruses, the packaging cell line ψCRIP (15) is probably a safe choice. The protocols described here for the first generation packaging cells ψ2 andψ AM are also applicable, when the new second generation packaging cell lines are used. It is preferable to transfect the ψ2 cells first and then use the ecotropic viral stocks to infect the ψAM cells for the production of amphotropic viral stocks. The reasons are:

- Infected cells, but not transfected cells, all produce infectious virus. Thus, the highest titer virus stocks are produced from infected cells.
- Transfection of proviral DNAs apparently enhances the rate of recombination which leads to helper virus rescue, so transfection of the packaging cells which produce amphotropic virus (ψAM, ψCRIP and PA317) should be avoided.
- ψ2 cells have a lower rate of helper-virus rescue than the ψAM cells. Therefore transfection of ψ2 carries only a small risk of generation of replication-competent (helper) virus.

Note: For the transfection of ψ2 cells we usually use the DMSO shock method, but other methods such as lipofection, electroporation or calcium phosphate coprecipitation may be used.

Materials

- ψ2 cells
- cDMEM
- Calf serum
- PBS
- 75 cm^2 tissue culture flasks
- Penicillin/Streptomycin
- Trypsin/EDTA
- LB medium
- Ampicillin
- Geneticin (G418)
- Polybrene (Abbott Labs., Sigma))
- DMSO
- Sterile 0.2 and 0.45 μm filters
- Ultrapure water

Procedure

Transfection

Pick a positive clone which has been analyzed for the presence of the insert from the master plate (see above). Grow the bacteria in 5 ml LB media supplemented with 20 μg/ml ampicillin overnight at 37°C in an orbital shaker at 250 rpm.

1. Prepare the plasmid DNA using the "miniprep" protocol (see Subprotocol 4). Phenol/chloroform extract the DNA at least four times and ethanol precipitate the DNA. Wash the DNA overnight with 70% ethanol.
2. Grow ψ2 cells in 75 cm^2 tissue culture flasks in cDMEM (complete Dulbecco's modified Eagle's medium) supplemented with penicillin, streptomycin and 10% bovine calf serum (available from Gibco or Sigma). Be

sure not to use fetal calf serum because this will detach the cells immediately! You will need two flasks of cells: one for the transfection, the other as a control.

3. Grow the cells until about 75-90% confluent.
4. Prepare a stock solution of Polybrene (30 mg/ml) in PBS, filter through a 0.2 μm filter.
5. Resuspend the ethanol-precipitated DNA in 50 μl of sterile water or TE. Use all DNA recovered from the miniprep.
6. Add the DNA aseptically to 4 ml complete medium. Add polybrene to the medium to a final concentration of 30 μg/ml and mix.
7. Remove the old medium from the cells in the 75 cm^2 flask, and replace with the DNA-polybrene containing medium. 4 ml should be sufficient to just cover the cell layer in a 75 cm^2 flask. Allow contact with the cells for 6-8 hours at 37°C in a humidified atmosphere. Rock the flask occasionally.
8. Remove the old medium by aspiration.
9. Add 4 ml of complete medium containing 20% DMSO. This medium should be used at room temperature, not at 37°C! Gently rock the flask two or three times and allow the DMSO to act on the cells for exactly 4 minutes.
10. Wash the cells three times with 10 ml complete medium. Feed with 10 ml complete medium and return flasks to the incubator.

Note: There is a small chance that the cells may detach during the DMSO shock procedure. For this reason, the flask should be handled with extreme care. For the DMSO shock and subsequent washes, avoid touching or rinsing the monolayer directly. Add the DMSO medium to the ceiling of the inverted flask, then invert the flask slowly and carefully. Wash the cells the same way.

Selection of successfully transfected ψ2 cells with G418

Since pLXSN carries the neomycin resistance gene, succesfully transfected cells can be easily selected for in medium supplemented with the toxic aminoglycoside geneticin (G418).

1. Prepare a stock solution of geneticin (16 mg/ml in PBS), filter through a sterile 0.2 μm filter, store at -20°C. Be aware that the geneticin powder consists only of about 50-65% active substance, the exact proportion is given on the label. Your stock solution should be 16 mg/ml of the *active* substance.

2. You may begin selection 24-48 hours after transfection. The cells should be confluent at this time point.

3. Add G418 to complete medium to a final concentration of 400 μg/ml.

4. Remove old medium by aspiration and replace with G418 medium. Do not forget to treat the control cells not exposed to the virus with G418. After 1 or 2 days the non-resistant cells start to round up and detach. Feed the cells twice or three times per week with fresh medium supplemented with G418. After 5-10 days, all cells in the control flask should be detached and dead. At this time point outgrowth of succesfully transfected ψ2 cells should be easily visible.

5. Grow the cells until confluency, then split in a 1:2-1:4 ratio, depending on their growth capacity. About two weeks after transfection the selection medium is replaced with normal medium.

Harvesting of ecotropic replication incompetent viruses

Wait until the cells are subconfluent, remove old medium by aspiration and cover the cell layer with 5 ml of fresh medium. Be sure not to use medium supplemented with geneticin! Do not use larger volumes of medium to achieve higher concentrations of viral particles. After 24 hours, remove the medium, filter through a 0.45 μm filter to remove cellular debris (do not use 0.2 μm membranes to avoid loss of viruses!) and store in a cryo tube at -70°C. Subsequently, the viruses may be harvested a second time in the same way. Afterwards, the monolayer will be confluent and the cells should be passaged at a 1:2-1:4 ratio. When they reach subconfluency, a new harvesting cycle may start. The viral stocks may be stored at -70°C indefinitely. The transfected ψ2 cells can be stored in liquid nitrogen for future use.

Subprotocol 7
Infection of ψAM Packaging Cells

Procedure

1. Grow up the ψAM cells in 75 cm^2 tissue culture flasks containing cDMEM supplemented with penicillin, streptomycin and 10% bovine calf serum. Again, be sure not to use fetal calf serum because this will detach the ψAM cells immediately! You will need two flasks of cells: one for the infection, the other for the monitoring of the selection process.
2. Grow until the cells are about 75-90% confluent.
3. Thaw a cryo tube containing 5 ml of ecotropic viral particles (supernatants from transfected ψ2 cells). Add polybrene to a final concentration of 8 μg/ml; this will make the viruses sticky. Filter through a 0.45 μm filter. Remove old medium by aspiration and replace with the viral stock-polybrene mixture. Return the cells to the incubator and let the viruses act overnight.
4. Remove the viral stock by aspiration and replace with regular growth medium.
5. After 24 hours, start selection with G418 in the same way as described above.
6. When the control cells are completely destroyed and detached, start harvesting the amphotropic viral particles as described above for the ecotropic viruses.

Subprotocol 8
Infection of Target Cells

Procedure

The procedure is exactly the same as described for the generation of amphotropic virus stocks. We usually incubate subconfluent target cells with the virus overnight. However, infections may be achieved in one hour. Short incubation times may be preferable if the medium used during the virus harvest (cDMEM with 10% bovine calf serum) does not meet the growth requirements of the target cells. The G418 concentration necessary to select the target cells should be determined in advance and may be different from the 400 μg/ml used for selection of ecotropic and amphotropic packaging cells. A G418 concentration that kills the non-resistant target cells slowly over a period of 5-8 days should be used.

Results

Assessment of transfection efficacy

The presence of the protein encoded by the cDNA inserted into the genome of the target cell can be easily detected by Western blotting or immunoprecipitation (see Chapter 24), if a suitable antibody has been prepared or if one is commercially available. Expression levels in different target cells or using different transfection strategies may be compared in cell lysates after adjusting for total protein content. The cellular localization of the protein can be investigated by immunocytochemistry. If the protein expressed by the target cells is secreted into the medium, its presence in the supernatants can be investigated by ELISA, Western blotting or immunoprecipitation. If the protein of interest exhibits enzymatic activity, a substrate assay can be used instead or in combination with the immunological methods.

Comments

Biohazard issues

The specific biosafety issues raised when working with retroviral vectors should be discussed in advance for any given protocol with the local biosafety comittee. In several countries, strict governmental standards or laws have been set up which need to be considered carefully. The review of experimental protocols by governmental institutions may require several months. These time requirements need to be considered when planning the experimental work.

The ecotropic viruses cannot replicate in non-rodent cells. Release of replication-competent helper viruses has never been reported with the second generation ecotropic packaging cell lines ψCRIP CRE (15) and GP+E86 (16), whereas the use of ψ2 cells (13) carries a small but well known risk. The amphotropic viruses can bind to human cells and can integrate the provirus into the human genome. However, due to other levels of host range restriction the helper virus probably cannot replicate in human cells. To date, replication in humans has not been reported. Both ecotropic and amphotropic viruses must be treated as potential biohazards. The minimal requirements when working with all the viruses include the exclusive use of laminar flow hoods, protective gloves and clothing, disinfection of all utensils under the laminar flow hood, disposal in a biohazard bag and sterilization of all waste in an autoclave.

The transfected target cells must be assayed for the presence of replication-competent helper virus. Several radioactive and non-radioactive assays for the detection of retroviral reverse transcriptase activity are commercially available. Alternatively, helper virus can be detected through horizontal spread of drug resistance. In this sensitive assay the supernatants of the transfected target cells are assayed for the presence of virus carrying the selectable marker neomycin resistance. The most suitable assay for a given experimental protocol should be discussed with the local biosafety comittee or with an experienced retrovirologist.

References

1. Weiss K, Teich N, Varmus H, Coffin J (Eds.), (1985) Retroviruses as Vectors of Foreign Genes, in Molecular Biology of Tumor Viruses, 2nd Edition, pp. 36-73, Cold Spring Harbor Laboratory 1985, New York
2. Jorgensen RA, Rothstein SJ, Reznikoff WS (1979) A Restriction Enzyme Cleavage Map of Tn5 and Location of a Region Encoding Neomycin Resistance. Mol Gen Genet 177:65-72
3. Miller AD, Rosman GJ (1989) Improved Retroviral Vectors for Gene Transfer and Expression. BioTechniques 7:980-990
4. Rosenberg SA, Aebersold P, Cornetta K, Kasid A, Morgan RA, Moen R, Karson EM, Lotze MT, Yang JC, Topalian SL, Merino MJ, Culver K, Miller AD, Blaese RM, Anderson WF (1990) Gene Transfer into Humans - Immunotherapy of Patients with Advanced Melanoma, Using Tumor-Infiltrating Lymphocytes Modified by Retroviral Gene Transduction. N Engl J Med 323:570-578
5. Veillette A, Fournel M (1990) The CD4 Associated Tyrosine Protein Kinase p56lck Is Positively Regulated through its Site of Autophosphorylation. Oncogene 5:1455-1462
6. Lynch CM, Clowes MM, Osborne WRA, Clowes AW, Miller AD (1992) Long-Term Expression of Human Adenosine Deaminase in Vascular Smooth Muscle Cells of Rats: A Model for Gene Therapy. Proc Natl Acad Sci USA 89:1138-1142
7. Nolta JA, Xiao JY, Bahner I, Kohn DB (1992) Retroviral-Mediated Transfer of the Human Glucocerebrosidase Gene into Cultured Gaucher Bone Marrow. J Clin Invest 90:342-348
8. Haughn L, Gratton S, Caron L, Sekaly RP, Veillette A, Julius M (1992) Association of Tyrosine Kinase p56lck with CD4 Inhibits the Induction of Growth through the $\alpha\beta$ T-Cell Receptor. Nature 358:328-331

9. Watkins DC, Johnson GL, Malbon CC (1992) Regulation of Differentiation of Teratocarcinoma Cells into Primitive Endoderm by $G\alpha_{i2}$. Science 258:1373-1375
10. Moreb JS, Schweder M, Gray B, Zucali J, Zori R (1998) In Vitro Selection for K562 Cells with Higher Retrovirally Mediated Copy Number of Aldehyde Dehydrogenase Class-I and Higher Resistance to 4-Hydroperoxycyclophosphamide. Hum Gene Ther 9:611-619
11. Juang SH, Xie K, Xu L, Shi Q, Wang Y, Yoneda J, Fidler IJ (1998) Suppression of Tumorigenicity and Metastasis of Human Renal Cell Carcinoma Cells by Infection with Retroviral Vectors Harboring the Murine Inducible Nitrit Oxide Synthase Gene. Hum Gene Ther 9:845-854
12. Abboud SL, Woodruff KA, Choudhury GG (1998) Retroviral-Mediated Gene Transfer of CSF-1 into op/op Stromal Cells to Correct Defective in Vitro Osteocoastgenesis J Cell Physiol 176:323-331
13. Mann R, Mulligan RC, Baltimore D (1983) Construction of a Retrovirus Packaging Mutant and its Use to Produce Helper-Free Defective Retrovirus. Cell 33:153-159
14. Cone RD, Mulligan RC (1984) High-Efficiency Gene Transfer into Mammalian Cells: Generation of Helper-Free Recombinant Retrovirus with Broad Mammalian Host Range. Proc Natl Acad Sci USA 81:6349-6353
15. Danos O, Mulligan RC (1988) Safe and Efficient Generation of Recombinant Retroviruses with Amphotropic and Ecotropic Host Ranges. Proc Natl Acad Sci USA 85:6460-6464
16. Markowitz D, Goff S, Bank A (1988) A Safe Packaging Cell Line for Gene Transfer: Separating Viral Genes on Two Different Plasmids. J Virol 62: 1120-1124

Chapter 23

Positional Cloning and Linkage Analysis

FRIEDHELM HILDEBRANDT AND HEYMUT OMRAN

Positional Cloning

Positional cloning is employed for identification of a disease gene if there is no protein defect or functional defect known, which may help in expression cloning or cloning by sequence similarity (Chap. 16). In positional cloning a disease gene may be mapped by information from cytogenetic abnormalities. If no such information is available, **linkage analysis** is employed (see Figure 1).

Theory of Linkage Analysis

Recombination in meiosis

Meiosis occurs in the egg or sperm and leads to the formation of the gamete. In the zygotene of the first phase of meiosis, the 23 chromosomes that are inherited from the individual's father will pair with the 23 homologous chromosomes that are inherited from the individual's mother. In this phase each chromosome consists of two chromatids that are identical copies of each other. The so-called bivalents that are formed by pairing of the homologues, each contain 4 chromatids. In the pachytene that follows the zygotene, crossing over (or chiasmata) occur between a chromatid of one homologous chromosome and a chromatid of another homologous chromosome. Crossing over is the process that leads to genetic recombination. It can be observed directly in the pachytene (see Figure 2).

During meiosis (formation of gametes) there is random assortment of chromosomes. In other words, whether one gamete ends up with the chromosome from the individual's father or with the homologous chromosome from the individual's mother is random and independent between different

Correspondence to Friedhelm Hildebrandt, Universitäts-Kinderklinik, Mathildenstrasse 1, Freiburg, 79106, Germany (*phone* +49-761-270-4301; *fax* +49-761-270-4533; *e-mail* hildebra@kkl200.ukl.uni-freiburg.de), Heymut Omran

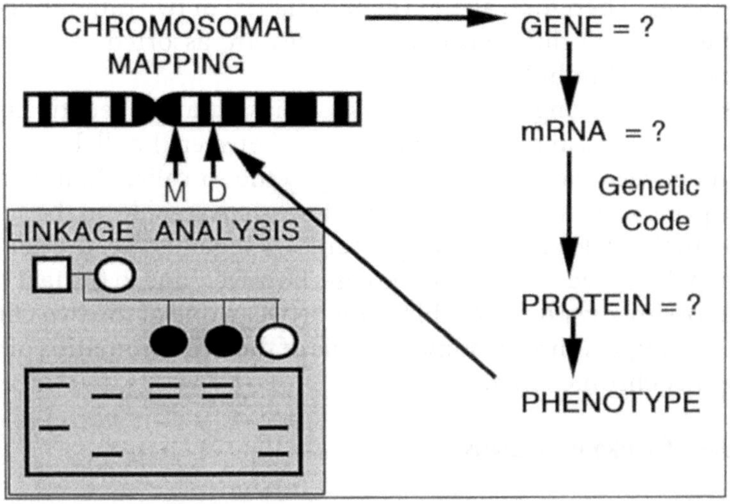

Fig. 1. The approach of positional cloning of a disease gene. The column on the right represents the flow of genetic information from the structural gene to its transcription into mRNA, and its translation into protein, which, as the gene product, will determine the disease phenotype. In linkage analysis (see stippled insert) the "family method" is used (see below), in which the segregation of the affected status (filled symbol) is compared to the segregation of alleles of polymorphic DNA markers (e.g. bands on electrophoresis), thereby mapping the disease locus to the known chromosomal region of the marker (*M*, DNA marker locus; *D*, disease locus). The disease gene is then isolated by physical mapping techniques. cDNAs are isolated and sequenced and the amino acid sequence is deduced using the genetic code. Finally, the axis of pathogenesis from the gene and its expression as a defective protein and the phenotype can be studied.

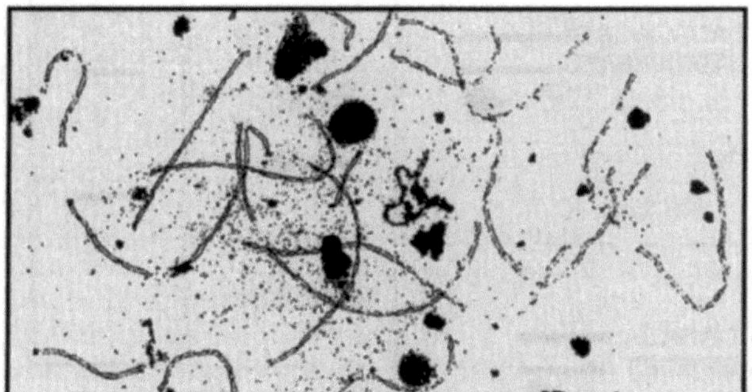

Fig. 2. Chiasmata (crossovers) can be observed cytogenetically in the pachytene of meiosis.

chromosomes. Therefore, alleles (gene variants) of two different genes that are located on two different chromosomes are assorted randomly. Such genes are called "unlinked".

Instead of observing crossing overs as chiasmata cytogenetically in a microscope, in linkage analysis recombination between the alleles of two loci is observed between parents and offspring by the so-called "family method". Each offspring represents one meiosis. If two genes reside on the same chromosome, their alleles would always segregate together, i.e. the genes would be completely linked. This is not the case, however, due to the fact that there may be crossing over or recombination between one of the two chromatids of one homologous chromosome and one of the two chromatids of the other homologous chromosome.

Techniques of Linkage Analysis

In linkage analysis, we compare the segregation (i.e. passing on to the offspring) of the alleles of a polymorphic DNA marker with the segregation of disease states (affected versus healthy), which represent the different alleles at a disease gene locus.

The two most important types of polymorphic DNA markers are restriction fragment length polymorphisms (RFLPs) and microsatellite markers (Figure 3).

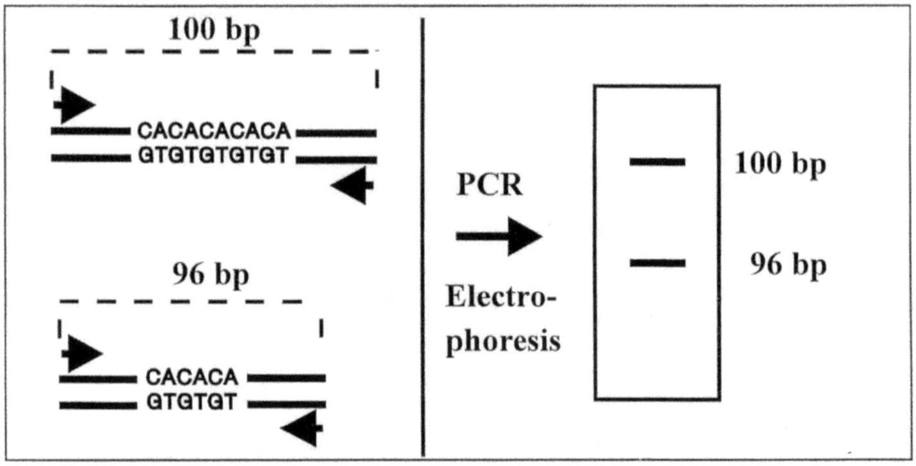

Fig. 3. For linkage analysis studies, two oligonucleotide primers flanking a CA repeat (a microsatellite marker) are used, which in a polymerase chain reaction (PCR) amplify specifically one region in the human genome. The products amplified from the two homologous chromosomes of an individual differ in size, due to microdeletions that have occurred in evolution. This leads to two distinct bands when the PCR products are size-separated on a polyacrylamide gel, the two bands representing the two alleles of this marker.

Strategy of mapping a gene by linkage analysis

The strategy of mapping a disease gene by linkage analysis is described in Figure 4. The first requirement is genomic DNA from families affected with the disease. The second requirement is a set of polymorphic DNA markers, which are relatively evenly distributed over the human chromosomes and for which the localization is well known.

In the example given a microsatellite marker (see Figure 3) that localizes to chromosome 1p13.3 has been typed in a family with an autosomal recessive disease, the localization of which is indicated by an "X" on chromosome 1. The father has the alleles 100 bp and 98 bp, while the the mother has the alleles 96 bp and 94 bp.

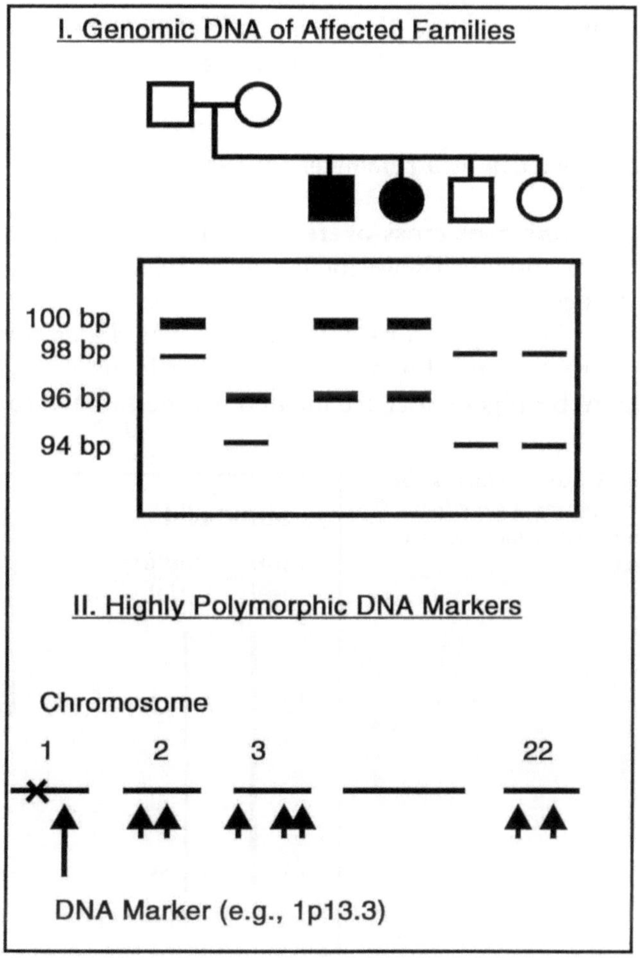

Fig. 4. Strategy of mapping a disease gene by linkage analysis (see text).

In this example it happens that the two affected children inherit the 100 bp allele from the father and the 96 bp allele from the mother. In contrast, the two healthy children, inherit the alternative alleles from father and mother. Thus, for this marker there seems to be a correlation between the disease status (affected or healthy) and the type of allele inherited. The explanation for this phenomenon is that by chance a marker has been chosen which happens to be so close to the formerly unknown disease locus (shown as an "X" on chromosome 1) that the chance is very low for a crossing-over to occur between the marker locus and the disease locus. In other words, linkage has been demonstrated between the marker and the disease, or the disease locus has been mapped to the vicinity of this DNA marker.

A statistical criterion that can be tested for such findings is the LOD score Z (see below). If it is higher than 3, linkage has been proven to exist between the marker and the disease locus. If the LOD score is below -2, the disease locus has been excluded from a region around this marker.

Recombination as a parameter for genetic distance

The number of cross-overs (recombinations) between two loci (e.g. one marker and one disease locus) can be regarded as a measurement of genetic distance.

In Figure 5 we consider a disease gene locus *D* with the alleles (= gene variants) D and d and a DNA marker locus *A* (with the alleles A and a), where both gene loci are located on the same chromosome. The paternal

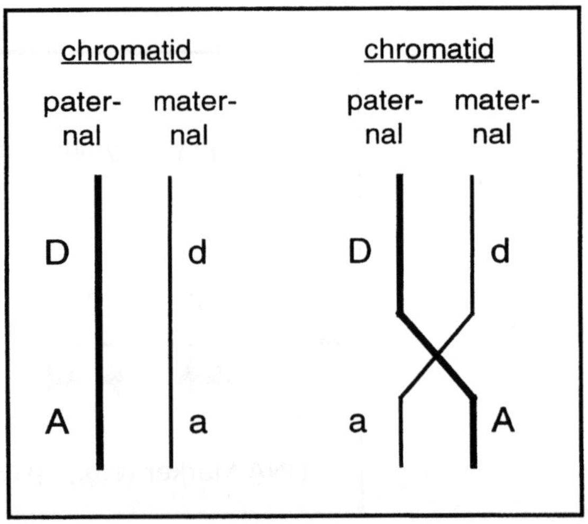

Fig. 5. Recombination between disease gene locus *D* and marker locus *A* (see text).

chromosome has (at locus *D*) the allele D and (at locus *A*) the allele A. The maternal chromosome has (at locus *D*) the allele d and (at locus A) the allele a. Due to a crossing over between the two loci of the homologous chromosomes, there could be recombination such that D and a now lie on the same chromosome and d and A lie on the same chromosome. If the two gene loci are positioned far enough apart on the same chromosome, there may be multiple crossing overs occurring between the two loci. If there were an uneven number of crossovers we would observe recombination in the gametes or in their offspring, if there were an even number of crossovers we would not observe a recombination. The occurrence of an uneven or even number of crossovers is equally likely (50%). Therefore, the likelihood of observing a recombination between two distant loci is 50%, i. e. the recombination fraction $\theta = 0.5$. The recombination fraction is θ defined as the number of recombinations observed between two gene loci divided by the number of meioses examined.

The closer two gene loci are to each other, the lower is the likelihood of a crossover (recombination) between the two, i. e. the two loci are linked. Therefore, genetic linkage is present if the observed recombination fraction is <0.5. If the two loci were directly next to each other we would never observe a recombination (θ=0). Between the two extremes (θ=0.5 and θ=0)

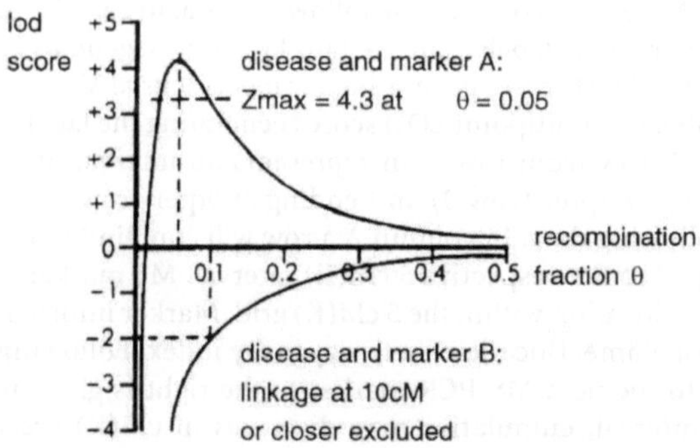

Fig. 6. LOD score curve between two loci (disease gene locus and marker loci A). Note that the peak LOD score of 4.3 is found at a θ of 0.05. The confidence interval is given by the recombination fractions at which the LOD score is one unit below the peak (0.025 and 0.10). A disease gene has been excluded from marker B at a distance of 10 cM on both sides of the marker. Note that the LOD score Z(θ=0.5) is always zero, since Z = log [L(θ=0.5)/L(θ=0.5)] = log [1] = 0. Note also that, if there is at least one crossover observed in the data, Z(θ=0) is -∞, since Z(θ=0) = log [likelihood of linkage at θ=0)/L(no linkage)] = log [zero/L(no linkage)] = -∞. (From A. Read: Medical Genetics, Gower Publ., 1989)

values between 0 and 0.5 will be observed, depending on the distance between the two loci. Therefore, the recombination fraction is related to the genetic distance x between two gene loci. This relation is given by a so-called map function, the simplest of which is the function by Morgan: x=.The genetic distance x is measured in cM (centiMorgans), where 1cM 1% recombination.

The likelihood method: the LOD score

Since recombination is a stochastic event, a statistical test has to be performed when interpreting observed recombinations. The likelihood method is used to calculate such a statistical parameter which is called the LOD score:

The likelihood $L(\theta)$ that there is linkage between two loci of unknown distance is calculated for recombination fractions θ. Theta is varied between $\theta=0$ (complete linkage) and $\theta=0.5$ (no linkage). The likelihood of linkage at a certain recombination fraction (equivalent to

Fig. 7. "LODVIEW", an evaluation programm for total genome linkage analysis. LODIN5, the LODVIEW input file for human Chromosome 5. LOD score data were entered into the preformatted file LODIN5. Row 1 contains the following data: in cells B1 to L1 the names of families 1 – 11, which generate the labels "fam 1 – fam 11" in the legend to Figure 8; in cells M1 to O1 the labels for 2-point LOD scores generating the labels "curve M, N, O" in the legend to Figure 8; in cell P1 the label for multipoint LOD scores generating the label "multipoint" in the legend of Figure 8. Each row from row 2 on represents an interval of 5 cM(K) on the abscissa in Figure 8, starting at 5pter (row 2) and ending at 5qter (row 37).
Rows 2 – 37 contain the following data: In column A a row will contain MS marker information if this marker is mapped to the respective 5 cM(K) interval. MS markers are positioned relative to their approximate location within the 5 cM(K) grid. Marker information is given in the following order: symbol, name, (location) heterozygosity index. Following this information, the genetic distance to the next MS-PCR-marker to the right is given in cM(K). If no marker is known for an interval, cumulative map distances in cM(K) are given at every 10 cM(K) intervals.
Thus, column A generates the labels appearing above the abscissa in Figure 8. Columns B to L contain LOD scores at a recombination fraction of $\theta = 0.05$ which are represented in Figure 8 as histograms segmented by family. Columns M to O contain LOD scores calculated at θ = 4.98, 9.87, 14.57, 19.00, etc., yielding LOD score curves surrounding each marker in Figure 8 at genetic distances of xK = 5, 10, 15, 20 cM(K) etc., respectively. Column P contains LOD scores calculated for 3-point linkage between each of two adjacent markers and a disease locus calculated with the program LINKMAP as taken from the output program LRP of the LINKAGE package. The results are displayed as multipoint curves in Figure 8. All cells in the table are preformatted in order to automatically produce the graphical representation of data shown in Figure 8.

	A	B	C	D	E	F	G	H	I	J	K	L	M	N	O	P
1	Marker	fam 1	fam 2	fam 3	fam 4	fam 5	fam 6	fam 7	fam 8	fam 9	fam 10	fam 11	curve M	curve N	curve O	multip oint
2	D5S117 Mfd48 (5p15.3-15.1)0.55* 0.0cM															
3	D5S208 Mfd88 (5p15.3-15.1)0.69*29.7cM	-1,90	-0,24	-1,44	-0,76	-0,67	-0,50	0,62	-0,46	0,62	0,62	0,31	-12,00			
4	10												-7,19			
5													-4,01			
6	20												-2,42	-1,79		
7													-1,47	-2,63		
8	30													-4,21		
9	D5S111 Mfd40 (5p14.1-13.1)0.53*3.9cM	-0,77	-0,68	0,00	-0,76	-1,44	0,00	0,02	-0,72	0,45	0,00	0,22		-12,00		-12,00
10	D5S108 Mfd34 (5p14.1-13.1)0.51* 18cM													-4,21		-5,79
11														-2,63	-1,96	-4,80
12	50													-1,79	-2,93	-5,06
13															-4,77	-6,47
14	D5S118 Mfd63 (5cen-q11.2)0.56*17.9cM	-1,12	0,00	-0,62	0,22	0,00	-0,24	-0,64	-0,46	0,24	0,00	-2,21			-12,00	-12,00
15													-1,65		-4,77	-6,99
16	70												-2,30		-2,93	-6,20
17													-3,33		-1,96	-6,30
18	D5S357 Mfd151(5q11.2_13.3)0.64*13.9cM												-4,97			-7,40
19													-8,10			-10,77
20	D5S107 Mfd27 (5q11.2-13.3)0.82*35.7cM	-0,89	-1,48	0,00	-0,24	-1,44	-0,04	-0,67	-0,46	-0,53	-0,35	-2,21	-12,00			-12,00
21													-8,10			-8,99
22	100												-4,97			-5,94
23													-3,33			-4,30
24	110												-2,30			-3,37
25													-1,65		-1,13	-3,05
26	120												-1,22		-1,75	-3,15
27	IL9 (5q22.3-31.3)0.62* 9cM														-2,79	-3,69
28	130														-4,89	-5,77
29	D5S210 Mfd122(5q31.3-33.3)0.75* 0.0cM	0,00	-0,50	0,10	-1,35	-0,35	-0,76	-0,35	-0,46	-0,46	0,62	-0,76	-1,39		-12,00	-12,00
30	D5S207 Mfd43 (5q31.3-33.3)0.69* 4cM												-2,46		-4,89	
31	CSF1R(5q31.3-33.3)0.86* 6cM	-0,01	-0,50	-0,35	-0,55	0,62	-0,24	-0,67	-0,72	-0,47	0,12	-0,76	-4,69		-2,79	
32	D5S209 Mfd116(5q31.3_33.3)0.71*22.6cM	-0,94	-0,51	0,10	-0,54	0,62	0,74	-0,35	-0,46	-1,44	-0,60	-0,80	-12,00	0,80	-1,75	
33													-4,69	0,74	-1,13	
34	160												-2,46	0,55		
35													-1,39	0,10		
36	170													-0,98		
37	D5S211 Mfd154(5q33.3-qter)0.72*	0,56	0,49	0,37	0,72	0,62	-0,50	-0,86	-0,46	-0,50	-0,04	-0,10		-12,00		

genetic distance) $L(\theta)$ is divided by the likelihood of no linkage $\mathbf{L(\theta=0.5)}$. This is called the

odds ratio $= L(\theta)/L(\theta = 0.5)$ or

odds ratio $=$

$$\frac{\text{odds that the 2 loci linked at a recombination fraction } \theta}{\text{odds that the 2 loci are unlinked } (\theta=0.5)}$$

where L is a likelihood or odds.

In words: "The odds ratio is the likelihood that there is linkage at an assumed θ divided by the likelihood that there is no linkage."

The logarithm of the odds ratio is called the

LOD score $Z = \log\ [L(\theta)/L(\theta=0.5)]$

The LOD score is evaluated at the recombination fraction θ where it is maximal. This is called the maximum LOD score Z. A maximum LOD score >3 is considered significant linkage between two loci. This is equivalent to odds of 1:1000 for linkage. If the LOD score is <-2, this is equivalent to significant exclusion of linkage between two loci.

When calculating LOD scores the data can be expressed in the LOD score curve (see Figure 6).

Exclusion mapping

In exclusion mapping one tries by consecutively examining many markers to exclude large areas of the human genome from linkage with a disease

Fig. 8. LODOUT5, the LODVIEW graphical display for human Chromosome 5. The abscissa represents a map of Chromosome 5 extending from the p terminus (left) to the q terminus (right) divided into intervals of 5 cM(K) of sex-averaged genetic distance (xK) according to Kosambi's map function. MS markers are positioned relative to their approximate location within the 5 cM(K) grid. Information on MS markers is shown above their respective positions in the following order: symbol, name, (location) heterozygosity index. The sex-averaged genetic distance to the next MS marker towards the q-terminus is given in cM(K).
At the position of each marker, LOD scores $Z(\theta=0.05)$ are shown as stacked histograms, each segment representing the LOD score of one family (see legend), where positive LOD scores are stacked on top of each other above the zero line and negative LOD scores are stacked on top of each other below the zero line. Thus, the histograms allow direct ascertainment of which families give positive or negative LOD scores at $\theta=0.05$. Early-onset families are shown in hatched segments (fam 1 - 6). Late-onset families are shown in shades of gray (fam 7 - 11). In the position of each marker that generated a histogram, funnel-shaped curves surrounding each histogram in a symmetrical fashion show tracings of LOD scores at x(K) = 5, 10, 15, 20 cM(K) etc., on both sides of the marker position. A thick dashed line is drawn at $Z(\theta) = -2$. Its intersections with the two sides of a LOD score curve delimits the regions of exclusion assuming the absence of non-allelic heterogeneity in the families examined. Thus, graphical assessment of the width of an exclusion area is facilitated. The LOD score curve for multipoint linkage is denoted by (●).

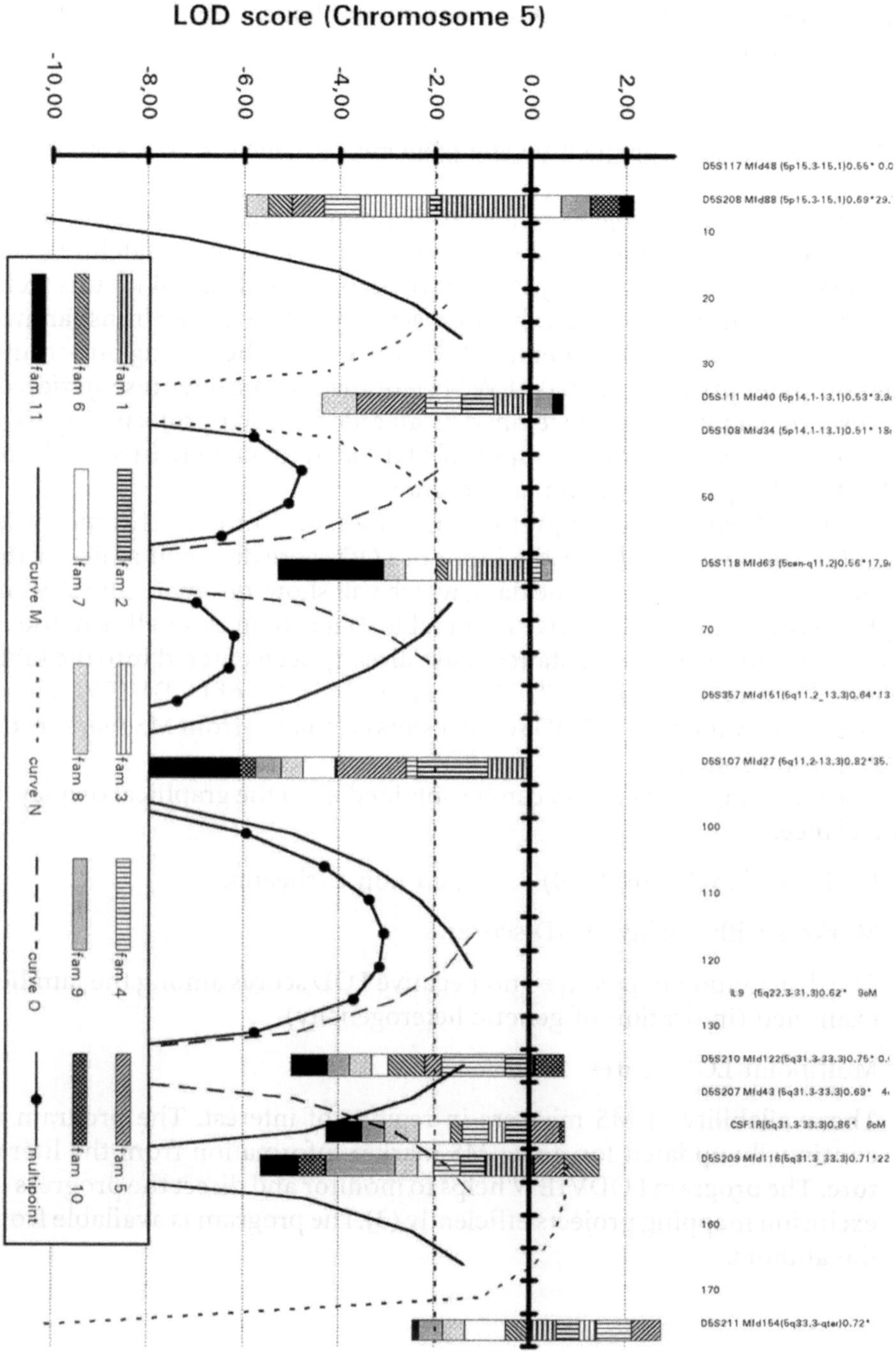
LOD score (Chromosome 5)
-10,00
-8,00
-6,00
-4,00
-2,00
0,00
2,00
fam 1
fam 2
fam 3
fam 4
fam 5
fam 6
fam 7
fam 8
fam 9
fam 10
fam 11
curve M
curve N
curve O
multipoint

gene (by the criterion of a LOD score of -2), until one finds a marker that is linked to the disease.

LODVIEW: a computer program for the graphical evaluation of LOD score results

For linkage analysis projects aimed at mapping hereditary diseases in humans, more than twothousand of highly polymorphic microsatellite (MS) markers which can be typed by PCR, have become available. With this technical improvement, the availability of a technique allowing for transparency in the handling of rapidly generated LOD score data is becoming important. The computer program "LODVIEW" (1) for the graphical representation of LOD score data has been developed as an aide for such projects (Figure 7 and 8). It is designed for the input of LOD score data generated with the LINKAGE (2) package or similar programs.

LODVIEW consists of 24 preformatted files, one for each chromosome. Each file contains a table for the input of LOD score data and a file for the graphical representation of the data, which will show automatically any entry that is made in the respective input table. Data on microsatellite markers with known inter-marker distances have already been entered into the table (Figure 7) and graph (Figure 8). The graphical display of LODVIEW allows for the rapid evaluation of LOD score results calculated from MS markers on each chromosome.

The following information can be obtained from the graphical display at one glance:

- Regions of exclusion ($Z(\theta) < -2$) and non-exclusion.
- Markers with positive LOD scores.
- The distribution of positive and negative LOD scores among the families examined (indication of genetic heterogeneity).
- Multipoint LOD scores.
- The availability of MS markers in regions of interest. The program is continually updated for novel MS marker information from the literature. The program LODVIEW helps to monitor and direct the progress of exclusion mapping projects efficiently (3).The program is available from the authors.

References

1. Hildebrandt F, Pohlmann A, Omran H (1993) LODVIEW: A computer program for the graphical evaluation of lod score results in exclusion mapping of human disease genes. Comput Biomed Res 26:529-599
2. Lathrop GM, Lalouel J-M, Julier C, Ott J (1984) Strategies for multilocus linkage analysis in humans. P.N.A.S. USA 81:3445-3446
3. Zerres K, Mücher G, Bachner L, Deschennes et al. (1994) Mapping of the gene for autosomal recessive polycystic kidney disease (ARPKD) to chromosome 6p21-cen. Nature Genet 7:429-432

Part III

Protein Methods

Chapter 24

Protein Techniques

ROBERT F. REILLY

Subprotocol 1
In Vitro Translation and Immunoprecipitation

Cell-free translation systems contain all the components necessary for protein synthesis: tRNAs; ribosomes; amino acids; initiation, elongation and termination factors; GTP; ATP; and magnesium. They require only the addition of a transcript to be translated. Single-stranded RNA for translation can be transcribed from a wide variety of plasmid vectors, such as pBluescript®, which contain a multiple cloning site downstream from a powerful promoter for a bacteriophage DNA-dependent RNA polymerase. The two most common cell-free translation systems in use are the rabbit reticulocyte lysate system, and the wheat germ extract system. Of these the rabbit reticulocyte lysate system is more commonly used. The major drawback of the wheat germ system is its tendency to produce incomplete translation products due to premature termination (Shih and Kaesberg 1976). As a result, it is generally unsuitable for producing polypeptides greater than 60 kDa. One advantage of the wheat germ system is that it does not require the removal of endogenous mRNAs. Therefore, it may translate weakly-initiating mRNAs more efficiently than the reticulocyte lysate system. Recent modifications of the wheat germ system have made it more suitable for the translation of higher molecular weight proteins (Morch et al. 1986).

Robert F. Reilly, University of Colorado Health Science Center, Division of Nephrology, 4200 East Ninth Avenue, DenverCO, 80262, USA (*phone* +01-303-399-8020 ext. 3101; *fax* +01-303-393-4611; *e-mail* Robert.Reilly@uchsc.edu)

Procedure

The reticulocyte lysate system

Reticulocyte lysate is prepared from rabbits made anemic by repeated injection of phenyl-hydrazine (3). Blood is collected, washed, and lysed in water. After a high speed spin the supernatant is collected. Endogenous globin mRNA is removed by a brief treatment with calcium-dependent micrococcal nuclease, and the nuclease is subsequently inactivated by chelation of divalent cations with EDTA (3). This procedure does not alter the protein synthetic activity of the extract. The requirement for the removal of endogenous mRNA is the major disadvantage of this system. If significant levels of mRNA remain after the digestion, including sequences that contain initiation sites, exogenous mRNAs will have to compete with these fragments for ribosomal binding sites. Weakly initiating mRNA species may do so inefficiently. Unfortunately, the activity of commercially available translation systems is variable and is never specified for a given lot. The lysate is optimized by the addition of an energy generating system (phosphocreatine kinase and phosphocreatine), a mixture of tRNAs to expand the range of mRNAs that can be translated, and hemin to prevent inhibition of initiation (4, 5) (hemin is a suppressor of an inhibitor of the initiation factor eIF2a). The lysate should be stored in liquid nitrogen since it slowly loses activity even at -70°C. It should also be stored in aliquots, since it dramatically loses activity on repetitive freeze-thawing.

Reticulocyte lysate

Template RNA is first heated at 65°C for 10 min and cooled on ice (to relieve secondary structure). In a 500 µl microcentrifuge tube, assemble the following components:

Nuclease treated lysate	35µl
DEPC-treated water	7µl
RNAsin	1µl
1 mM amino acid mixture minus the labeled amino acid	1µl
Translated RNA	2µl
35S-methionine (1200 Ci/mmole) at 10 mCi/ml (40 mCi)	4µl
total volume	50µl

The reaction is incubated at 30°C for 60 min. One should always include a negative control containing no added mRNA. Based on the synthesis of globin the reticulocyte lysate system translates about 80 amino acids per min. Therefore, a 100-kDa protein would require about 10-15 min to synthesize. Protease inhibitors can be added to the reaction although the choice is not broad. Soybean trypsin inhibitor (1 mg/ml) and ovomucoid (0.5 mg/ml) can be used safely. EGTA (2.7 mM) and phenylmethylsulfonylfluoride (PMSF; 17 μg/ml) can also probably be used. It should be noted that different lots of methionine may contain an inhibitor of translation. Therefore, it is best to use either translation-grade methionine from New England Nuclear or Translabel from ICN. The New England Nuclear product has a higher specific activity (800-1200 Ci/mmol) than Translabel (800 Ci/mmol), but Translabel also contains radiolabeled cysteine. Methionine is very sensitive to oxidation and should be stored as single-use aliquots at -70°C.

Reaction products are then analyzed by TCA precipitation to determine the number of counts incorporated. Under ideal conditions one should incorporate $3\text{-}5 \times 10^6$ counts (0.5-1.5 μCi) of ^{35}S-methionine per reaction mixture. A small aliquot of the reaction (5μl) is added to 20 μl of sample buffer, heated, and the products of the reaction are separated on an SDS-polyacrylamide gel. Potential artifacts include labeled peptidyl-tRNA resulting from the translation of small fragments of globin mRNA (6). This migrates as a group of bands at 25-30 kDa. Their labeling is suppressed by inhibitors of protein synthesis and by the addition of efficiently translated mRNAs to the assay. As a result these fragments are most troublesome when one is trying to translate poorly initiating or scarce messages. Fortunately, these fragments can easily be eliminated if the sample is treated with 50-100 μg/ml of RNase A for 10 min at 30°C. Another artifactual polypeptide migrating at 42-45 kDa may also be seen (6). It is thought to result from a tRNA-dependent but ribosome-independent addition of methionine to a lysate protein. If this polpeptide is a major problem its mobility can be altered by omitting the customary heating step before loading the gel. It will then migrate at 220 kDa. Omitting the heating step has little effect on the mobility of the translation products. Using ^{35}S-methionine, it should be possible to detect 100-200 ng/ml of protein in 1-2 hrs employing fluorography. The limit of detection of this method is determined by the small amount of lysate that can be loaded in each lane (2-5 μl), since the lysate itself has a very high protein concentration (50-60 mg/ml). In comparison, immunoprecipitation can detect the translation of as little as 5-20 ng/ml of protein.

Other important points include:

- One can use a variety of templates for cell-free translation. Cellular RNA requires 5-10 μg; for poly $(A)^+$ or transcribed RNA one needs only 0.25-1.0 μg.
- Known inhibitors of translation include oxidized thiols, low concentrations of double stranded RNA (inhibits initiation by stimulating phosphorylation of eIF2), and acidic polysaccharides.
- The concentrations of magnesium (range 0.5-3.5 mM; concentration in lysate 0.5 mM) and potassium (range 60-135 mM; concentration in lysate 79 mM) may need to be optimized for certain transcripts. One of the more extreme examples of this is poliovirus mRNA which requires a final concentration of potassium of 60 mM and is translated very poorly at 125 mM. Different initiation sites may have different magnesium optima, and termination may be suppressed by high concentrations of magnesium.
- Spermidine (0.4 mM) has been noted to increase translation efficiency in some cases (7).
- Processing events such as signal peptide cleavage, membrane insertion, translocation, and core glycosylation can also be examined by adding canine microsomal membranes to the standard reaction (8). The microsomes are stripped of endogenous membrane-bound ribosomes and mRNA. Depending on the mRNA assayed the translation efficiency may drop by 10-50% in the presence of microsomal membranes. The following is a sample protocol for the use of microsomal membranes with the reticulocyte lysate system. In a 500 μl microcentrifuge tube, assemble the following components:

Nuclease treated lysate	17.5 μl
DEPC-treated water	2.2 μl
1mM amino acid mixture minus the labeled amino acid	0.5 μl
Translated RNA	1.0 μl
35S-methionine (1200 Ci/mmole) at 10 mCi/ml (40 mCi)	2.0 μl
Microsomal membranes	1.8 μl
total volume	25 μl

Immunoprecipitation

Immunoprecipitation involves the use of antibodies in solution to bind and precipitate specific antigens. The advantages of this technique compared to Western blotting include: relatively mild denaturing conditions are used, more native conformations are detected, the use of radiolabeled material is permitted, high specificity and efficiency, and further biochemical analysis of isolated proteins is possible. Its disadvantages include: dependence on the ability to radiolabel the antigen of interest, relatively high avidity antibodies are required, and the method is relatively time consuming.

Immunoprecipitation involves four basic processes: binding of the antigen to the antibody; precipitation of the antigen-antibody complex, washing; and elution.

Binding of the antigen to the antibody: Binding occurs equally well at any temperature but is generally carried out at 4°C to avoid proteolytic degradation. The precipitation should be performed on the same day as the labeling, since freeze-thawing of the antigen-containing mixture may cause protein denaturation and aggregation. If the antigen must be frozen, it should be snap frozen in an ethanol-dry ice bath and stored at -70°C. Antibody should be present in large excess; this usually requires 5-50 µl of polyclonal sera or 20-200 µl of hybridoma supernatant. The duration of incubation with the primary antibody is often not critical (30-60 min). Incubation is carried out at physiological salt concentration and pH. Occasionally, individual monoclonal antibodies may precipitate in the cold or at low ionic strength.

Precipitation of the antigen antibody complex: Although there are two methods generally available for the precipitation of antigen-antibody complexes, equivalence complexes and solid-phase immunoadsorption, only the latter is currently in widespread use. Solid-phase immunoadsorption involves the selective absorption of the antigen-antibody complex to an affinity matrix. Complexes are often adsorbed onto protein A-Sepharose (9). Protein A is a bacterial protein that binds the F_c domain of many IgGs, and some IgMs and IgAs. It has high affinity to most but not all IgG isotypes. Protein A binds immunoglobulins from certain species better than others: human, pig › mouse › rabbit, goat, sheep ›› rat. A similar protein, protein G, has a broader species specificity and binds certain rat IgGs well. Solid-phase methods have the advantage that there is less IgG in the final pellet which results in less overloading of the SDS-PAGE gel and less nonspecific precipitation. Shorter incubation times are also possible since binding is virtually instantaneous. 100 μl of packed protein A-Sepharose will bind about 2 mg of IgG. The IgG concentration in immune sera is approximately 10 mg/ml. Beads with a significant number of charged groups or hydrophobic spacer arms should be avoided since they are more likely to exhibit high levels of nonspecific binding.

Washing: Usually 2-3 washes are sufficient. Each wash should dilute the soluble phase by about 100-fold. Protein A-Sepharose requires very little centrifugation. Many monoclonal antibodies are of low affinity, and if antigen is not precipitated the number of washes should be decreased. Alkaline pH stabilizes the binding of immunoglobulin to protein A, and the strength of the interaction is independent of salt concentration over a wide range.

Elution: Resuspend in SDS-PAGE sample buffer, boil, and load.

Radiolabeling the antigen

One must first decide how to radiolabel the antigen. There are three options:

- **Iodination**
  This technique is used for surface antigens. It has the advantage that only a small percentage of total cell proteins are membrane proteins. Therefore, a considerable enrichment is achieved from the labeling process itself. Cytoplasmic labeling can occur if dead cells are present in the sample.

- **Carbohydrate Labeling**
  This can be carried out using either metabolic or enzymatic methods with tritiated mannose, fucose, galactose or glucosamine. Sialic acid cannot be used because it is not taken up by cells. Enrichment is achieved since glycoproteins make up less than 20% of total cell protein. The technique suffers from low specific activity. Analogously, one can also use sulfation or phosphorylation.

- **Metabolic Labeling With Amino Acids**
  ^{35}S-methionine is the label of choice. The sulfur- containing amino acids have a high specific activity (500-1000 Ci/mmol), a relatively strong decay, a half-life of 87 days, and a fairly low cost. Methionine and cysteine are supplied in aqueous solution containing 0.1% β-mercaptoethanol (12 mM). Cells can only tolerate 0.8 mM β-mercaptoethanol or 500 μCi/ml. One should avoid repetitive freeze-thawing of the isotope. The tritiated amino acids are the second choice (20-200 Ci/mmol). The ^{14}C-amino

acids should never be used because of their low specific activity (300 mCi/mmol) and high cost.

Cell lysis

Cells can be lysed with a variety of detergents. SDS (0.5-1.0%) will disrupt the nuclear membrane and should be followed by either enzyme treatment or shearing (sonication) to disrupt DNA prior to the addition of the primary antibody.

Sample immunoprecipitation protocol

The following is a sample protocol for a membrane protein. No single protocol will be appropriate for all antigens.

Metabolic labeling with amino acids

1. Prepare Selectamine media (Gibco-BRL), that is deficient in the amino acid to be labeled, according to the instructions provided by the manufacturer.
2. Add dialyzed fetal calf serum to a final concentration of 10%.
3. Wash the cells in this medium.
4. Add the isotope to the smallest possible volume of media (10 mls for a T-175 flask), filter, and place in the flask overnight.

Preparation of microsomal membranes

1. Rinse the cells in 10 ml of K-Hepes solution (200 mM mannitol, 41 mM K^+, 80 mM *N*-2-hydroxyethylpiperazine-*N'*-2-ethenesulfonic acid, (pH 7.5)) containing the protease inhibitors leupeptin (0.5 μg/ml), pepstatin (0.7 μg/ml), phenylmethylsulfonylfluoride (PMSF) (40 μg/ml), and EDTA (1mM) (solution A) and harvest by scraping with a disposable plastic policeman.
2. Rinse the flask with an additional 10 ml of solution A. Centrifuge at 2,300*xg* for 10 min. Discard the supernatant and freeze the pellet in an ethanol-dry ice bath.
3. Resuspend the pellet in 10 ml of solution A and homogenize 40 strokes in a Potter-Elvehjem homogenizer.
4. Centrifuge the homogenate at 2,300 x g for 10 min.
5. Collect the supernatant and centrifuge at 23,600 x g for 1 hr. Resuspend the pellet in 100 μl of solution A (protein concentration approximately 15-25 mg/ml).

Immuno-precipitation

1. Solubilize the membrane preparation in Tris-saline solubilization buffer or RIPA solubilization buffer for 1 hr at 4°C (final protein concentration 2 mg/ml). Tris-saline solubilization buffer is 50 mM Tris, 150 mM NaCl, 1% Triton X-100, pH to 7.2 with HCl. RIPA solubilization buffer is 50 mM Tris, 150 mM NaCl, 1 mM EDTA (disodium salt), 1% Triton X-100, 0.1% SDS, 1% sodium deoxycholate, pH to 7.2 with HCl.
2. Spin for 10 min in a microfuge (14,000 rpm) and discard the pellet.
3. Preclear the lysate by incubating with a protein A-Sepharose slurry (3 parts protein A-Sepharose and 1 part phosphate buffered saline) at 4°C. One should incubate three parts solubilized protein with one part protein A-Sepharose slurry. Place 500 μl of precleared lysate in a 1.5 ml tube.
4. Add polyclonal antisera to the supernatant and incubate with constant agitation on a Nutator (Adams) or similar device overnight at 4°C. Antibody should be present in large excess, this usually requires 5-50 μl of polyclonal sera or 20-200 μl of hybridoma supernatant. For affinity-purified monclonal antibodies we have used a final concentration of 200 μg/ml. Add 200 μl of a 50:50 slurry of protein A-Sepharose in phosphate buffered saline (PBS) and incubate for 2 hr in the cold with constant agitation.
5. Wash twice with 1 ml ice cold PBS. Centrifuge at 1500 rpm for 3 min after each wash.
6. Centrifuge the beads as above, decant the supernatant, add 120 μl of SDS-PAGE sample buffer, boil for 3 min, and load. SDS-PAGE sample buffer is 10 mM Tris HCl, pH 6.8, 0.6 M β-mercaptoethanol, 2% SDS, and 20% glycerol. Boiling should be eliminated for hydrophobic membrane proteins since they often aggregate when boiled. After electrophoresis is complete stain the gel with Coomassie Blue and destain. Rinse the gel for 15 min in water. Add Autofluor((National Diagnostics) to completely immerse the gel. Agitate gently for 1-2 hrs depending on the gel thickness. Pour off the Autofluor(. Place the gel directly on filter paper and dry under heat and vacuum.

Subprotocol 2
Western Blotting

Western blotting employs antibodies to detect antigens that have been separated by SDS-polyacrylamide gel electrophoresis (10) following transfer to a suitable matrix (11). The matrix is quenched and probed using antibodies to visualize antigens of interest.

Electrophoretic transfer is the most commonly used method to transfer proteins (12). Potential problems include: proteins can be "over transferred" leading to electrophoresis through the matrix (this is minimized by including 20% methanol in the transfer buffer), and high molecular weight proteins often transfer much less efficiently than low molecular weight proteins. Nitrocellulose or Immobilon-P((Millipore) can be stained with Ponceau S (13), or amido black (12) to verify transfer. We use Immobilon-P((Millipore) because it can be stripped and reprobed.

Western blotting assumes that the transferred proteins were either not completely denatured during SDS-PAGE or were able to partially renature during transfer in order to permit detection by the antibody. Potential problems include:

- Many epitopes are conformation-dependent and sensitive to denaturation. As a result an antibody may not detect antigens after transfer.
- Inclusion of methanol in transfer buffers may cause further denaturation of antigens.
- Renaturation is assisted by running SDS-PAGE gels under nonreducing conditions. The optimal dilutions for each primary and secondary antibody must be determined empirically.

Procedure

1. Strips are cut from the nitrocellulose or Immobilon-P((Millipore) transfer, three per lane, and incubated with 1 ml of BLOTTO (0.87% NaCl, 7.2 mM dibasic phosphate, 2.2 mM monobasic phosphate, 5% nonfat dried milk) for at least 2 hr with constant agitation in an Accutrans Incubtray (14).
2. The BLOTTO is aspirated and the primary antibody is added at a 1:500 dilution and incubated overnight. BLOTTO used for overnight incubations should contain 0.02% sodium azide.

3. Wash each lane with 1 ml of BLOTTO five times at 10 min intervals.
4. Add a 1:1000 dilution of secondary antibody in 1 ml of BLOTTO to each lane and incubate with constant agitation for 1 hr. The secondary antibody employed will depend on both the species of animal in which the primary antibody was generated, and the detection system used. We have used two enzymatic detection systems (alkaline phosphatase (15) and horseradish peroxidase). If using a horseradish peroxidase detection system BLOTTO without azide should be used from this point onward since azide inhibits this enzyme.
5. Wash again with BLOTTO as in step 3.
6. For alkaline phosphatase-conjugated secondary antibodies, wash with 1 ml of substrate buffer (0.1 M NaCl, 0.1 M Tris, 5 mM $MgCl_2$, (pH 9.5)). To develop add 500 μl of substrate (ratio is 4 mls of water, 0.5 ml of 5-bromo-4-chloro-3-indoyl phosphate concentrate, and 0.5 ml nitro blue tetrazolium concentrate) to each lane. Allow the reaction to proceed. Stop the reaction by rinsing with distilled water when desired. The substrate can be obtained from Kirkegaard and Perry Laboratories, 2 Cessna Court, Gaithersburg, MD, USA.
7. We have developed horseradish peroxidase-conjugated secondary antibodies with the ECL Western blotting detection system as recommended by the manufacturer (Amersham). We currently use this system exclusively. Its advantages compared to the alkaline phosphatase detection system include: greater sensitivity; transfers carried out using Immobilon-P(can be stripped and reprobed (16); exposure time can be varied to obtain a better image.

References

1. Shih D, and Kaesberg P (1976) Translation of the RNAs of brome mosiac virus: the monocystronic nature of RNA1 and RNA2. J Mol Biol 103: 77-88
2. Morch MD, Drugeon G, Zagorski W, and Haenni AL (1986) The synthesis of high-molecular-weight proteins in the wheat germ translation system. Meth in Enzymol 118: 154-164
3. Pelham HRB, and Jackson RJ (1976) An efficient mRNA-dependent translation system from reticulocyte lysates. Eur J Biochem 67: 247-256
4. Adamson SD, Herbert E, and Godschaux W (1968) Factors affecting the rate of protein synthesis in lysate systems from reticulocytes. Arch Biochem Biophys 125: 671-683

5. Zucker WV, and Schulman HM (1968) Stimulation of globin chain initiation by hemin in the reticulocyte cell-free system. Proc Natl Acad Sci USA 59: 582-589
6. Jackson RJ, and Hunt T (1983) Preparation and use of nuclease-treated rabbit reticulocyte lysates for the translation of eukaryotic messenger RNA. Meth in Enzymol 96: 50-74
7. Roberts BE, and Paterson BM (1973) Efficient translation of tobacco mosaic virus RNA and rabbit globin 9S RNA in a cell-free system from commercial wheat germ. Proc Natl Acad Sci USA 70: 2330-2334
8. Walter P, and Blobel G (1983) Preparation of microsomal membranes for cotranslational protein translocation. Meth In Enzymol 96: 84-93
9. Kessler SW (1975) Rapid isolation of antigens from cells with a Staphylococcal protein A-antibody adsorbent: parameters of the interaction of antibody-antigen complexes with protein A. J Immunol 115: 1617
10. Laemmli UK (1970) Cleavage of structural proteins during the assembly of the head of bacteriophage T4. Nature Lond 277: 680-685
11. Gershoni GM (1988) Protein blotting: a manual. Meth Biochem Anal 33: 1-58
12. Towbin HT, Staehelin T, and Gordon J (1979) Electrophoretic transfer of proteins from polyacrylamide gels to nitrocellulose sheets: procedure and some applications. Proc Natl Acad Sci USA 76: 4350-4354
13. Muilerman HG, Ter Hart HGJ, and Van Dijk W (1982) Specific detection of inactive enzyme protein after polyacrylamide gel electrophoresis by a new enzyme-immunoassay method using unspecific antiserum and partially purified active enzyme and application to rat liver phosphodiesterase 1. Anal Biochem 120: 46-51
14. Johnson DA, Gautsch JW, Sportsman JR, and Elder JH (1984) Improved technique utilizing nonfat dry milk for analysis of proteins and nucleic acids transferred to nitrocellulose. Gene Anal Tech 1: 3-8
15. Blake MS, Johnston KH, Russell-Jones GJ, and Gotschlich EC (1984) A rapid, sensitive method for detection of alkaline phosphatase-conjugated anti-antibody on Western blots. Anal Biochem 136: 175-179
16. Kaufmann SH, Ewing CM, and Shaper JH The erasable Western blot. (1987) Anal Biochem 161: 89-95

Subject Index

A
acid citrate dextrose (ACD) 58
affinity chromatography, oligo (dT) 274
agarose 73
- low melting point 76, 120
agarose/formaldehyde gel 78
allele-specific oligonucleotides (ASOs) 221
alphoid probes 169
AMCA 170
amino acid sequence 257, 259
ammonium
- acetate 55
- ions 56
angiotensinogen 322
annealing 209, 212
anode 71
antibody
- choice 296
- dilutions 296
- preadsorbtion 296
- secondary 296
antisense strand 17
ASOs (allele-specific oligonucleotides) 221
autofluorescence 376

B
BACs (bacterial artificial chromosomes) 99
BAP (bacterial alkaline phosphatase) 36
Barth's solution, modified 313
bases, complementary 207
β-galactosidase gene 190
biohazard issues 351
biopanning 297
biotinylation of DNA 172
bivalents 354
blastocysts 323
blood
- samples 59
- smears 155
BLOTTO 377, 378
blue plaque 292
BNN97 302
bone marrow smears 156
break point 219
bromphenol blue 74
BTA282 302

C
CA repeat 356
Ca^{++}
- precipitation method 234
- transformation 234
canine microsomal membranes 372
5'-cap 17
carcinogen 61
CAT (chloramphenicol-acetyl-transferase) gene 322
CCD (charge-coupled device) camera 170
cdc2 kinase 15
cDNA 275 pp.
- expression library 315
- insert 279
- library 257, 275
- - vector selection 283
- single stranded 277
cell
- cycle 13, 15

- division 13
- - cycle genes 15
- line
- - transfected 259
- - permanent 323
central dogma 17
CG
- buffer 80
- clamps 220
CHCL3
- remove 307
- tests 303
CHEF (contour clamped homogenous electric fields) 99, 103
chiasmata (*see also crossing over*) 354-356
chimera/chimerism 226, 273, 323
chimeric clones 279
chloroform 51
chloroform:isoamyl alcohol 50
chorionic villus biopsy 218
chromatids 354
chromatin 11
- structure of 12
chromosomal in situ suppression (CISS) hybridization 179
chromosome 3, 11, 13
- bivalent 13
- homologous 354
- metaphase 5, 11
- walking 262, 267, 269
CIP (calf intestine alkaline phosphatase) 36
cloning 226
- strategies 257 pp.
- vectors 226 pp.
codon 259
- usage 259
cohesive ends 279
colcemid solution 151
co-linearity 17, 19
co-migration 100
complexity 127
composite probe 169
confocal laser scanning microscopy 183
contamination 213
cos sites 233, 279
cosmid 233, 261, 262
cRNA synthesis, in-vitro 316
crossing over (*see also chiasmata*) 354, 358
cyclin 15
cytogenetic abnormalities 260, 354

D
DAPI 170
ddNTPs 187
7-deaza-dGTP:dGTP 213
degeneracy 259
denaturating 72
- gradient gel electrophoresis (DGGE) 221
denaturation 209, 212
deoxynucleotide triphosphates (dNTPs) 209
DEPC (diethyl pyrocarbonate) 61
dephosphorylation 36
DGGE (denaturating gradient gel electrophoresis) 221
- parallel 87
- perpendicular 87
diagnostic techniques 14
diethyl pyrocarbonate (DEPC) 61
differential
- cloning 260
- display 260, 295
digestion, partial 261, 271
digoxigenin 294
diploid genome 11
disease gene 354, 356
- genetic 257
DMSO 215
- shock 346
DNA (deoxynucleotide acid)
- biotinylation of 172
- circular 72
- complementary 261
- conformation 72
- degraded 218
- genomic 54, 259
- high-molecular-weight 50, 98, 103
- - genomic 54
- markers, known size 98
- microinjection of 326
- - foreign 323
- migration 71

- primary structure of 6, 7
- purification 50
- recovery of 75
- repair 15
- secondary structure of 6, 9
- superhelical 72
- tertiary structure of 10
DNA polymerase 14, 56, 190
- avian myeloblastosis virus reverse transcriptase 191
- heat stable 209
- *Klenow fragment* 190
- T7 DNA polymerase 191
- Taq polymerase 191
DNAse I 33
dNTPS (deoxynucleotide triphosphates) 209, 211, 215
- removal of 56
dot blot 129
- assay 174
- hybridization 221
- reverse 221
double
- digestion 32
- helix 6
dyad symmetry 8

E
E. coli HB101 234
EDTA 56
electrophoresis 71
- buffer 74
electroporation 238
EMBL 3 232
EMBL 4 232
end-labeling 38, 114
enhancers 19
epifluoroscence microscopy 170
ethanol 55
ethidium bromide 75
eukaryotic cells 50
- expression in 339
evolution 220
exchange kinase reaction 115, 116
exclusion mapping 361
exon-intron boundaries 17
exons 17
exonuclease activity 222
expression
- cloning 259, 312
- - examples of 318
- library 284, 291
- vectors 295
extension 212
- time 216
extraction buffer 58, 63

F
FIGE (field inversion gel electrophoresis) 98, 103
filling of 3' ends 40
first strand cDNA, synthesis 286
FISH (fluoroescence in situ hybridization) 148, 166
FITC 170
5'-flanking region 19
fluorochromes 170
footprinting 219
formaldehyde 78
formamide loading buffer 84
formamide-formaldehyde buffer 78
freeze-squeeze 76
full length cDNA 219
fusion proteins 295

G
GAL4 284
βgalactosidase (lac-Z) gene 232, 322
gametogenesis 13
GC
- clamp 87
- content 216
gene 13
- expression 17
- - tissue-specific 19
- families 219
- knock-outs 262
- locus 358
- mapping 99
- product 17, 257
- transfer 339 pp.
genetic
- code 17, 22
- - degeneracy of 22, 257
- distance 358
- information 15

- recombination 354
genome 3
- mitochondrial 220
- DNA 14, 56, 259
- - isolation of 56
- library 259
germ line 323
glass milk 77
glucocorticoid receptor 19
glycogen 66
glycoxal 78
λgt10 230, 291
λgt11 230, 291, 295
guanidinium thiocyanate 64

H
hairpin
- formation 8
- loops 204
haploid genome 11
heteroduplex molecules 87
heterologous genes 321
heteronuclear RNA (hnRNA) 17
hexamers, random 120
histones 11
Hoechst 3327 170
homologous genes 321
host
- cells 233
- - JM103 233
- - HB101 234, 235
- - XL1-blue 233
- strains 295
hot
- PCR 221
- start 214
human genome projects 219
hybrid 226
- primer-template 209
hybridization 301
- DNA blots 135
- parameters 293
- RNA blots 143
- subtractive 260, 295
- successive 100
Hydrolink 73
hypotonic solution 151

I
immobilon-P 377
inosine 259
in-situ hybridization 219
interphase
- analysis 148
- cytogenetics 166
intron 17, 259
intron-exon boundaries 262
IPTG 295
isoamyl alcohol 51
isopropanol 56
isoschizomers 27

J
JM103 233

K
Klenow
- enzyme 40
- fragment 118, 120, 190
knock-outs 321

L
labeling 294
lactamase gene 190
lacZ 292
- α-peptide gene 229
lagging strand 15
λ bacteriophage 261, 262, 279
- cloning cDNA into 288
leading strand 15
library 219
- amplified 279
- chimerism 261
- commercial 274
- complexity 261, 272
- construction 261
- expression 283, 284, 291
- genomic 261
- normalized 282
- oligod(T) primed 280
- partially digested 268, 269
- phage 291
- plasmid 290
- redundancy 261
- screening 261
LiCl 56
likelihood 359

- method 360
linkage 356, 358
- analysis 260, 354, 356
- - strategy of mapping a disease gene 357
- significant 360
linker 279
- primers 220
- adapters 280
LOD score 358, 360
- curve 359
LODVIEW 362
lysis 230
- buffer 63
lytic cycle 230

M
M13 188, 232
M13-derived cloning vectors 232
magnesium 215
magnetic beads, streptavidin-coated 277
mammalian cells 57
manganese 222
marker locus 358
MCR (multiple cloning region) 190, 228
MDE 85
meiosis 13, 354
melting
- point 86
- temperature 216
membrane stripping 147
β-mercaptoethanol 64
messenger RNA (*see mRNA*) 15
metaphase chromosome 5, 11
methyl mercuric hydroxide 78
methylase 280
methylation patterns 100
5-methyl-dCTP 280
microdissection 219
microsatellite marker 356, 357, 362
mineral oil 212
misincooperation 213
mismatch 127, 217
mispriming 217
mitochondrial genome 220
mitosis 13
mobility 71
- zone of limiting 102
modification, posttranslational 17
molecular weight 71
- markers 75
Moloney Murine
- Leukemia Virus 340
- Sarcoma Virus 340
MoMulV 340
MOPS 79
3-morpholinopropanesulfonic acid 79
mRNA (messanger RNA) 15, 276
- copies 281
multicolor fluorescence in situ hybridization 182
multiple cloning region (MCR) 190, 228
mutations 260
- disease-causing 82
MVR-PCR 218

N
neomycin phospho-transferase 339
nick translation 118, 170, 171
nitrogenous bases 4
Northern
- blot 77, 140 pp., 281
- hybridization 67
nuclease S1 33
- assays 67
nucleic acids 3, 4, 50
- extraction 51
nucleosides 4
nucleosome 11
nucleotide 4
- analogs 191
- dITP 191
- 7-deaza-dGTP 191

O
OD 54
OFAGE (orthogonal field agarose gel electrophoresis) 98, 103
Okazaki fragments 15
OLA (oligomer ligation assay) 221
oligo(dT)
- cellulose 67, 69
- column 69

oligonucleotide 209, 257
- synthetiser 217
oocytes
- collection of 326
- Xenopus laevis (*see also there*) 312
orientation 295
oviduct transfer 327

P
P1 clones 99, 262
PACE (programmable autonomously-controlled electrodes) 97, 99
pachytene 354, 355
packaging 279
- cell
- - infection of 350
- - line 341
- - transfection of 346
- extract 233
PAGE (polyacrylamide gel electrophoresis) 73
partial digestion 100, 102
pBR322 229
PCR (polymerase chain reaction) 207 pp.
- arbitrary primed 219
- competitive 222
- components of 207
- cycle 209
- in situ 219
- inverse 218
- jumping 218
- modifications 218
- multiplex 221
- one-sided 220
- protocol 211
- quantitative 222
- radioactive 221
- repeat 218
- single cells 219
- specificity 214
- transposon-related 218
- troubleshooting 212
- yield 214
pepsin digestion 153
PFGE (pulsed field gel electrophoresis) 74, 96 pp., 103 pp.
- application of 99
- mapping 100
- pulse times 97
- resolution 97
- switch interval 97
λphage 230
- λgt10 230, 291
- λgt11 230, 291
phage library 291
phagemid 297
phenol 50, 52
- salt-saturated 52
phenol/chloroform 50
- extraction 51
phosphodiester bond 6, 15, 209
physical mapping 103, 260
plaque density 296
plasmid 227
- library 290
- vectors 227
point mutations 221
poly(A) tail 17, 276
poly(A)$^+$RNA, preparation 67
polyacrylamide 73
- gel electrophoresis (PAGE) 73
polyacrylamide gels
- isolation of DNA 77
- nondenaturating 82
poly-L-lysine 157
polymerase 15
- activity 215
- chain reaction (*see PCR*) 207 pp.
- fidelity 215
- processivity 215
- thermostable 222
polymorphic DNA markers 356, 357
polymorphism 82
positional cloning 260, 354, 355
prehybridization 301
preparative gel 75
primer
- biotinylated 221
- concentration 217
- degenerate 217', 219, 220
- dimers 213
- extension 209
- fluorescence-labeled 221
- mismatch 217

- nested 220
- pair 209
prometaphase 11
promoter 17, 220, 321
- alternative 22
- metallothionein 322
- transthyretin 322
- unducible 322
pronucleus 323
- microinjection into 327
proof-reading 40
propidium iodide 170
protein
- biosynthesis 22
- purification 257
protein A-sepharose 373, 376
proteinase K 55, 56, 156
protein-protein interaction 284
pseudopregnant foster mothers 325
pSP6/T7 230
pUC 229, 291
- plamids 190
pulsed field gel electrophoresis (*see PFGE*) 74
purification 50

R
RACE 219
random primers 280
- labeling 118
rare cutter 27
reading frame 295
rearrangements 219
recognition sequence 100
recombination 358
- fraction 359, 360
region of interest 207, 209
regulatory elements, cis-acting 321
Ren-2 gene 331
renin 322
repeat PCR, inter-alu 218
replication 16
- error rate 15
- fork 15
- origin of 228
reporter genes 322
retriction enzyme, recognition sites 217
retroviral vectors 339
reverse transcriptase 44, 55, 219, 277, 279
RFLPs 356
rhodamine 170
ribonuclease 60
riboprobes 121
ribosomal RNA (*see rRNA)* 77
ribosomes 17, 22
R-loop 17
RNA (ribonucleotide acid)
- in vitro transcription 121
- isolation of 60
- migration rate 77
- polymerase 17, 121
- rapid preparation 62, 64
- secondary structure of 8
- storage 70
- template 219
- total cellular 276
RNA blot hybridization 139
RNA dot blots 139
RNAse 78, 51
- block II 122
RNAse A 34
RNAse H 287
ROFE (rotating field inversion gel electrophoresis) 103
rRNA (ribosomal RNA) 67, 77, 276

S
samples, paraffin embedded 218
satellite probe 169
second strand
- synthesis 261
- - cDNA 287
secondary structure 72, 204
sense strand 17
sequenase 42, 194
sequence
- difficult to clone 264
- information 209
- similarity 219, 354
- - cloning 257
- tagged sites (STSs) 219
sequencing
- automated fluoroscent 193
- chain termination 186, 192
- gel 200

- multiplexing 193
shearing 98
shuffle
- clones 214
- vectors 297
signal peptide 17
silencers 19
single
- cell suspensions from solid tumor, preparation of 158
- strand
- - binding protein 204
- - conformation polymorphism (*see SSCP*) 82
site-directed mutagenesis 220
size fractionation 271
sodium
- acetate 55
- lauryl sarkosinate 66
solenoid 11
somatic cell hybrids 262
Southern blot 131 pp.
spermine 55
S-phase 13
spin columns 76
splicing, alternative 17, 22
splicosomes 17
SSCP (single strand conformation polymorphism) 82, 83, 221
- analysis 83
start codon 22
stem cells, embryonic 323
Stoffel fragment 222
stop codon 22
stringency 128
STSs (sequence tagged sites) 219
sucrose density gradient 271, 271
supercoiling 8
superhelix 8
superovulation 324

T
T4
- ligase 34
- polymerase 42
- polynucleotide kinase 114, 115
T7
- DNA polymerase 42, 194
- SP6 RNA polymerase 47
TAE 74
Taq DNA polymerase 43, 197, 209, 215
TBE 74
TE-buffer 57, 65
- RNAse-free 62
temperature
- annealing 216
- gradient gel electrophoresis (TGGE) 221
template 207
terminal transferase 56, 114, 117
texas red 170
TGGE (temperature gradient gel electrophoresis) 221
thimerosal 304
tissue samples 59
total RNA 67
tracking dye 74
transactivation 284
transcription 15, 17 - 18
- factors 17, 18
- regulation of 20
transfection 230
transfer RNA (*see tRNA*) 8
transformation
- efficiencies 264, 283
- electroporation 238
- recombinant bacteriophage DNA 241 pp.
transgenic
- animals 321
- - hypertension research 330
- mice 324, 326
- rats 324, 325
- research, species in 329
translabel 371
translation 22
- mechanisms of 21
transport assay 315
transposon 218
- Tn5 339
tRNA 8, 23, 55, 67, 276
- structure of 10
- synthetase 23
trypsin-EDTA solution 154
two-hybrid 297
- system 258
type II restriction enzyme 26

U
unclonable segments 264
uracil N-glycosylase 214
UV light 75

V
vectors 261
- blunt-ended 279
- selection 283
voltage 72

W
walking efficiencies 263
wheat germ extract system 369

X
Xenopus
- laevis
- - oocytes 260, 312, 319
- - system 312
- toad 314
X-Gal 304
XL1-blue 233
xylene cyanole 74

Y
Y1088 298
Y1090 298
YACs (yeast artificial chromosomes) 99, 103, 261, 262, 333
- recombinant 323

Z
λZAP 291, 297
zygotene 354

MIX
Papier aus verantwortungsvollen Quellen
Paper from responsible sources
FSC® C105338

If you have any concerns about our products,
you can contact us on
ProductSafety@springernature.com

In case Publisher is established outside the EU,
the EU authorized representative is:
Springer Nature Customer Service Center GmbH
Europaplatz 3, 69115 Heidelberg, Germany

Printed by Libri Plureos GmbH
in Hamburg, Germany